T0214079

Communications
in Computer and Information Science   1413

More information about this series at http://www.springer.com/series/7899

Dmitry Balandin · Konstantin Barkalov ·
Victor Gergel · Iosif Meyerov (Eds.)

Mathematical Modeling and Supercomputer Technologies

20th International Conference, MMST 2020
Nizhny Novgorod, Russia, November 23–27, 2020
Revised Selected Papers

Editors
Dmitry Balandin ⓘ
Lobachevsky State University of Nizhni
Novgorod
Nizhny Novgorod, Russia

Konstantin Barkalov ⓘ
Lobachevsky State University of Nizhni
Novgorod
Nizhny Novgorod, Russia

Victor Gergel ⓘ
Lobachevsky State University of Nizhni
Novgorod
Nizhny Novgorod, Russia

Iosif Meyerov ⓘ
Lobachevsky State University of Nizhni
Novgorod
Nizhny Novgorod, Russia

ISSN 1865-0929   ISSN 1865-0937 (electronic)
Communications in Computer and Information Science
ISBN 978-3-030-78758-5   ISBN 978-3-030-78759-2 (eBook)
https://doi.org/10.1007/978-3-030-78759-2

This Springer imprint is published by the registered company Springer Nature Switzerland AG
The registered company address is: Gewerbestrasse 11, 6330 Cham, Switzerland

Preface

The 20th International Conference and School for Young Scientists "Mathematical Modeling and Supercomputer Technologies" (MMST 2020) was held during November 23–27, 2020, in Nizhni Novgorod, Russia. The conference and school were organized by the Mathematical Center "Mathematics of Future Technologies" and the Research and Educational Center for Supercomputer Technologies of the Lobachevsky State University of Nizhni Novgorod. The conference was supported by reputable sponsors (Intel, Huawei, and RSC). It was organized in partnership with the International Congress "Russian Supercomputing Days".

The topics of the conference and school covered a wide range of problems related to mathematical modeling of complex processes and numerical methods of research, as well as new methods of supercomputing aimed at solving state-of-the-art problems in various fields of science, industry, business, and education.

This edition of the MMST conference was dedicated to the 100th birthday of Professor Yuri Neimark. Professor Neimark's contribution to science and higher education is truly immense. He has left considerable heritage in the qualitative theory of differential equations, the theory of stability, the theory of adaptive and robust control, pattern recognition, nonholonomic and gyroscopic systems mechanics, optimization, and mathematical modeling. Yuri Neimark founded the USSR's first Faculty of Computational Mathematics and Cybernetics and the Research Institute of Applied Mathematics and Cybernetics. He is the author of about 600 published works, including at least 12 for inventions and 10 monographs.

The scientific program of the conference featured the following plenary lectures given by outstanding scientists:

- Alexander Boukhanovsky (Russia): Generative design of value-based systems.
- Vladimir Voevodin (Russia): AlgoWiki, the structure of algorithms and the impossible becomes possible.
- Alexander Gorban (UK): The new centaur: man and artificial intelligence.
- Andrei Gritsun (Russia): Predicting Earth climate change using the INM RAS Earth system model.
- Aleksey Eliseev (Russia): The IAP RAS Earth system model: state of the art and review of key findings.
- Aleksey Koronovskii (Russia): Multistability in an intermittent generalized chaotic synchronization regime.
- Evgeniy Mareev (Russia): Current problems in geophysical electrodynamics.
- Alexander Moskovsky (Russia): Unique and effective RSC solutions for complex problems.
- Sergey Pavlov (Russia): ARM Ecosystem Development for AI, Cloud and High Performance Computing.
- Arkady Pikovsky (Germany): Low-dimensional reduction for ensembles of noisy oscillators.

- Yaroslav Sergeyev (Italy): Recent advances in Lipschitz global optimization.
- Victor Stepanenko (Russia): Mathematical modelling of the land surface.
- Dmitry Turaev (UK): On triple instability.
- Igor Belykh (USA): Synchronization in multilayer networks: when good links go bad.
- Alexander Hramov (Russia): Functional networks of the brain: methods for connectivity restoration and their analysis.

These proceedings contain 25 full papers and 8 short papers carefully selected to be included in this volume from the main track and special sessions of MMST 2020. The papers accepted for publication were reviewed by three referees who were either members of the MMST 2020 Program Committee or selected independent reviewers.

The proceedings editors would like to thank everyone involved in the conference, especially the other members of the Organizing Committee and the Program Committee, as well as the external reviewers for their contributions. We also thank Springer for producing these high-quality proceedings of MMST 2020.

May 2021

Dmitry Balandin
Konstantin Barkalov
Iosif Meyerov
Victor Gergel

Organization

Program Committee Chairs

Voevodin, V. V.   Lomonosov Moscow State University, Russia
Gorban, A. N.   University of Leicester, UK

Program Committee

Balandin, D. V.   Lobachevsky State University of Nizhni Novgorod, Russia
Barkalov, K. A.   Lobachevsky State University of Nizhni Novgorod, Russia
Belykh, I. V.   Georgia State University, USA
Boukhanovsky, A. V.   ITMO University, Russia
Cherepennikov, V. V.   Huawei Nizhni Novgorod Research Center, Russia
Denisov, S. V.   Oslo Metropolitan University, Norway
Eliseev, A. V.   Lomonosov Moscow State University, Russia
Feygin, A. M.   Institute of Applied Physics RAS, Russia
Gergel, V. P.   Lobachevsky State University of Nizhni Novgorod, Russia
Gonoskov. A. A.   University of Gothenburg, Sweden
Gonchenko, S. V.   Lobachevsky State University of Nizhni Novgorod, Russia
Gritsun, A. S.   Institute of Numerical Mathematics RAS, Russia
Hramov, A. E.   Innopolis University, Russia
Ivanchenko, M. V.   Lobachevsky State University of Nizhni Novgorod, Russia
Kazantsev, V. B.   Lobachevsky State University of Nizhni Novgorod, Russia
Korniakov, K. V.   Intel Corporation, Russia
Koronovskii, A. A.   Saratov State University, Russia
Kuzmin, I. Yu.   Intel Corporation, Russia
Malyshkin V. E.   Institute of Computational Mathematics and Mathematical Geophysics SB RAS, Russia
Mareev, E. A.   Institute of Applied Physics RAS, Russia
Moshkov, M. Y.   King Abdullah University of Science and Technology, Saudi Arabia
Meyerov, I. B.   Lobachevsky State University of Nizhni Novgorod, Russia
Nekorkin, V. I.   Institute of Applied Physics RAS, Russia
Osipov, G. V.   Lobachevsky State University of Nizhni Novgorod, Russia
Pikovsky, A. S.   University of Potsdam, Germany
Sergeyev, Ya. D.   University of Calabria, Italy; Lobachevsky State University of Nizhni Novgorod, Russia
Slavutin, I. G.   Huawei Nizhni Novgorod Research Center, Russia
Tsaregorodtsev, A. Y.   Particle Physics Center of Marseille, France
Turaev, D. V.   Imperial College London, UK
Wyrzykowski R.   Czestochowa University of Technology, Poland
Yakobovskii, M. V.   Institute of Applied Mathematics RAS, Russia
Zaikin, A. A.   University College London, UK

Organizing Committee

Gergel, V. P. (Chair)	Lobachevsky State University of Nizhni Novgorod, Russia
Balandin, D. V.	Lobachevsky State University of Nizhni Novgorod, Russia
Barkalov, K. A.	Lobachevsky State University of Nizhni Novgorod, Russia
Kozinov, E. A.	Lobachevsky State University of Nizhni Novgorod, Russia
Lebedev, I. G.	Lobachevsky State University of Nizhni Novgorod, Russia
Levanova, T. A.	Lobachevsky State University of Nizhni Novgorod, Russia
Meyerov, I. B.	Lobachevsky State University of Nizhni Novgorod, Russia
Oleneva, I. V.	Lobachevsky State University of Nizhni Novgorod, Russia
Osipov, G. V.	Lobachevsky State University of Nizhni Novgorod, Russia

Contents

Computation in Optimization and Optimal Control

Supercomputer Simulation

Computational Methods for
Mathematical Models Analysis

Mathematical Rigor in Applied Mathematics Based on the Nonstandard Analysis

Evgeny Gordon$^{(\boxtimes)}$ (iD)

Eastern Illinois University, Charleston, IL 61920, USA

Abstract. This article contains an extended exposition of my talk "Yu. I. Neimark and mathematical rigor" at the special session of the conference MMCT-2020. It discusses the role of mathematical rigor in applied mathematics. The discussion focuses on the question of relationship between continuous processes and their computer modeling. The question of mathematically rigorous formalization of physical theories, which goes back to Hilbert's 6th problem and the widespread the point of view that continuous mathematics is an approximation of the discrete one, and not vice versa are also discussed. A new axiomatic of set theory is introduced that includes vague definitions and concepts at the same level of rigor as in modern classical mathematics, which operates only with well-defined objects. This allows to consider some non-rigorous arguments of applied mathematics as rigorous ones and, thus, to be sure that they are consistent.

Keywords: Nonstandard analysis · Infinitely large numbers · Hyperfinite set

1 Introduction

The special session dedicated to the centennial of the outstanding Russian mathematician Yu. I. Neimark was held within the framework of the MMCT-2020 conference on November 24, 2020 at Nizhny Novgorod State University, Russia. This article contains an extended exposition of my talk "Yu. I. Neimark and mathematical rigor" at this session. I dedicate it to the blessed memory of my teacher and friend Yuri Isaakovich Neimark.

Skepticism of the importance of mathematical rigor in applied mathematics began to spread in the second half of the last century in connection with the development of the use of computers in applied and theoretical research and it is quite widespread at the present time. I first heard about it from Yu. I. Neimark, A. D. Myshkis and A. S. Kronrod, when I was a student. At the same time, I learned from Kronrod the point of view that continuous mathematics is an approximation of the discrete one, and not vice versa. Disputes about the role of mathematical rigor continue to this day. See, for example, D Zeilberger's Letter

© Springer Nature Switzerland AG 2021
D. Balandin et al. (Eds.): MMST 2020, CCIS 1413, pp. 3–18, 2021.
https://doi.org/10.1007/978-3-030-78759-2_1

to the Editor against the importance of mathematical rigor in the December 2013 (11) issue of the Notices of the AMS, p. 1431, a criticism of this letter and a response to it in the 2014 (2) of the Notices, pp. 128 – 129. He also explicitly formulated and justified the idea about interrelation of the continuous and the discrete world [1]: "Continuous analysis and geometry are just degenerate approximations to the discrete world... While discrete analysis is conceptually simpler ... than continuous analysis, technically it is usually much more difficult. Granted, real geometry and analysis were necessary simplifications to enable humans to make progress in science and mathematics....". Continuous analysis contains some good formulas that do not exist in the discrete one. These formulas make it possible to calculate *approximately* the values of quantities whose direct calculation is impossible without powerful computers. Such quantities include, for example, ratios of very small numbers or very large sums of very small numbers. So, with the development of computer technology, the role of the continuous analysis decreases, and of the discrete analysis increases.

D. Hilbert formulated at number 6 the problem of constructing a rigorous mathematical axiomatization of physical theories, in which mathematics plays an important role, in his famous talk "Mathematical Problems" at the II International Mathematical Congress in (Paris, 1900). Among physical theories, he first of all singled out the theory of probability and statistical mechanics and posed the problem of mathematically rigorous substantiation of limiting processes, which lead from the atomistic view to the laws of the rigid body motion in the Boltzmann's book on principles of mechanics.

The problem of axiomatization of probability theory was first solved by Kolmogorov in 1933 [2]. His axiomatics based on countably additive measures on probability spaces is conventional now.This axiomatics is generally accepted. Most mathematically rigorous books on probability theory are based on it. However, applied mathematicians and those who taught probability theory to students who were not pure mathematicians were not satisfied by Kolmogorov's axiomatics. The best explanation of this dissatisfaction was formulated by E. Nelson to his book [3]:

"The foundations of probability theory were laid just over fifty years ago, by Kolmogorov. I am sure that many other probabilists teaching a beginning graduate course have also had the feeling that these measure-theoretic foundations serve more to salve our mathematical consciences than to provide an incisive tool for the scientist who wishes to apply probability theory."

Rigorous mathematical foundations in the spirit of Hilbert's 6th problem have been proposed for quantum mechanics as well. The formation of quantum mechanics was almost complete to the end of 20s of the last century. Its main principles were formulated by P. Dirac at the physical level of rigor [4], and by von Neumann at the mathematical level of rigor [6]. This book may be considered as another impact in the 6th Hilbert's problem. Despite the fact that almost any course of quantum mechanics is based on these axioms, for a long time many physicists and mathematicians were dissatisfied with them due to the lack of physical meaning in their formulations, since this axioms of Dirac and von Neumann themselves do not express any experimental laws. Also the rigorous theory of self-adjoint unbounded operators in Hilbert spaces used in von Neumann axiomatic is not intuitively clear and technically complicated for physicists. See Note to Chapter VIII.11 [7] for discussion of various approaches to rigorous axiomatization of the quantum mechanics. We can say that at present there are few mathematically rigorous proofs of physical results. The existing proofs are often technically complex, lacking physical intuition, and useless for physicists. The famous mathematical physicist F. A. Berezin wrote in the preface to the first edition of the book [5] (1972):

"I believe that at the present time statistical physics has not yet found its own adequate mathematical language .... As for the rigorous results available now, they are valuable from my point of view if they can compete in their simplicity and naturalness with those heuristic considerations that they are intended to replace. Such results are available in classical statistical physics ... In quantum statistical physics there are no such results yet."

This situation did not change significantly over the next 35 years.

D. A. Leites – the editor of the second edition of the book [5], (2006) wrote in the preface: "Berezin ... always insisted on proving theorems that could have direct applications to physics. It was unrealistic then, and experts believe that it is unrealistic even now."

A similar situation can be observed in computer simulations of continuous processes. As a rule, the convergence of the most effective numerical methods that give results in excellent agreement with the experiments, is either unproven or non-exists.

The reason for this state of affairs is that by strict proofs in modern mathematics we mean proofs formalized in Cantor's Set Theory where a set is understood by Cantor "as a combining of objects that satisfy some well-defined property in a single whole". Here a property is **well-defined** (a wd-property) if it is possible to say about any object unambiguously whether it has this property or not. Otherwise, the property is called *vague*. However, the objects and collection of objects in natural science as a rule, are determined by vague properties, and sometimes do not have any definitions at all. Recall that the field $\mathbb{R}$ is defined axiomatically as a complete linearly ordered field that satisfies the Axiom of the Least Upper Bound (LUB). Real numbers (reals) in Cantor's Set Theory are defined as Dedekind cuts – the specific subsets of the field $\mathbb{Q}$ of all rationals. The Axiom LUB is a theorem there. Compare this theorem with the following statement taken from the Chap. 26.1 of the Feynman Lectures on Physics, vol. I:

"There are no actual boundaries between one range of wavelengths and another, because nature did not present us with sharp edges. The numbers associated with a given name for the waves are only approximate and, of course, so are the names we give to the different ranges."

Similar difficulties are encountered when studying the correspondence between computer and continuous solutions of both applied and theoretical mathematical problems. In this case the difficulties arise since we have to deal with numbers that are not very close to the computer's memory boundaries. This property is vague. The computer operations with numbers that do not satisfy this property do not approximate operations in the field of reals, and even do not satisfy the usual laws of arithmetic. This circumstance makes it very hard, if not to say impossible to study the correspondence between continuous mathematics and its computer simulation within the framework of classical mathematics. Section 2 contains a detailed discussion of modern mathematical rigor including the brief introduction to Axiomatic Set Theory.

In the 60s of the last century A. Robinson constructed a proper extension of the standard model of mathematical analysis [11]. In his construction the proper extension of the field $\mathbb{R}$ is the linearly ordered field $^*\mathbb{R}$. It is an easy exercise to prove that $^*\mathbb{R}$ contains elements that are greater by absolute value, than any element of $\mathbb{R}$. These elements are called **infinitely large numbers**, and their inverse elements are called **infinitesimals**. Obviously the properties of a real to be infinitely large and to be infinitesimal are vague properties. Thus, Robinson's book contained the first consistent mathematically rigorous introduction of vague properties into mathematics. The analysis developed on the base of Robinson's extensions was called by him Nonstandard Analysis (NSA). See e.g. [12] for a brief introduction to Robinson's NSA. In NSA many intuitive mathematical formulations that go back to Leibniz and later to Cauchy, such as, for example, the definition of limit:

"$\lim_{x \to a} f(x) = L$ means that if x is infinitely close to a but $x \neq a$, then $f(x)$ is infinitely close to L", received the status of rigorous mathematical statements. This made it possible to simplify significantly the proofs of many theorems of standard analysis and even obtain new results in the standard mathematics using NSA. Most of these applications were associated with the use of **hyperfinite sets** i.e., sets, whose cardinalities, are infinitely large natural numbers in the sense of NSA. Retaining many properties of usual finite sets, hyperfinite sets make it possible to use intuition of finite mathematics in the continuous one more directly, and to obtain rigorous results. However, Robinson's model-theoretic approach to NSA relies heavily on advanced mathematical logic. This significantly complicates its use by mathematicians, especially those that are focused on applications. In order to facilitate the perception of the NSA for a wide range of mathematicians, E. Nelson developed an axiomatic version of the NSA based on axiomatic **ZFC** of the Set Theory that is acceptable for most mathematicians at least in its non-formal version – the Naïve set theory [8]. He called the introduced theory "Internal set theory" (**IST**) [13]. In 1987 Nelson published the book [3] where he developed a new approach to the foundations of probability theory based on hyperfinite probability spaces. The presentation there is based on **IST**, but only some fragments of it are used, which are formulated informally, but intuitively clear. The fundamental results of the theory of probability, on which its applications are based, are formulated and proved in **IST**. The formulations have a clear physical meaning and are available not only to pure mathematicians, but also to all kinds of applied mathematicians. This favorably distinguished them from formulations within the framework of

Kolmogorov's axiomatic based on measure theory. The appendix to the book contains a derivation of the theorems of the standard theory of probability from the "non-standard" theorems proved in the book. Nelson told me that he included this appendix in the book purely for social reasons, as he considers to be correct just the nonstandard versions of these theorems. Despite the availability of the basic concepts of **IST**, many proofs of its theorems contain very sophisticated logical reasoning to prove. This is partly due to the fact that there are no variables for collections defined by vague properties (classes) in the **IST**. They are defined only by specific formulas of **IST**. A new version of the nonstandard axiomatic set theory – Nonstandard Theory of Classes (**NCT**), which allows such variables, is presented in [14]. It is similar to **IST**, but it is based on the theory **NBG**. The theory **NBG** allows variables that are interpreted as wd-classes, and quantifiers over classes. The theory **NCT** allows to simplify some proofs in **IST**. Also one more version of the NSA – the Theory of Hyperfinite Sets (**THS**) – was developed on the base of **NCT** [15]. This theory deals only with hyperfinite sets and their subclasses, including the proper ones. The proper subclasses of sets are called **quasi-sets** in **THS**. Hyperfinite sets that do not contain proper subclasses are called **small**. Otherwise they are called **large**. Small sets were introduced in P. Vopenka's Alternative Set Theory (**AST**) [16], where they are called semisets. The theory **AST** is a version of the NSA that differs both from Robinson's (model-theoretic) NSA, and from the axiomatic versions of the NSA considered in the fundamental book [17]. An important notion of an **indiscernibility relation** on a hyperfinite set that is defined below allows to formalize standard continuous structures in the **THS** using factorization of hyperfinite sets by indiscernibility relations. Moreover, a significant part of theorems of **ZFC** can be formalized and proved in **THS**. In a sense, **THS** can be considered as a formalization of Plato's famous allegory about the cave, presented in his work "Republic" (360 BC) as a dialogue between Plato's brother Glaucon and his mentor Socrates. In the allegory, Socrates describes a group of people who have lived chained to the wall of a cave all of their lives, facing a blank wall. The people watch shadows projected on the wall from "real world" objects passing in front of a fire behind them and give names to these shadows. The shadows are the prisoners' reality but are not accurate representations of the real world. In **THS** the real world (the world of ideas according to Plato) is thought of as a discrete (atomistic) world – the world of finite sets, while continuous objects are shadows that prisoners see. I hope that this theory may be appropriate for investigation of the questions, formulated in the Hilbert's sixth problem. A short discussion in favor of this hope is given at the end of Sect. 3.

The theory **THS** can be used in studies the relationship between discrete and continuous mathematics. In my opinion it is the most adequate formalization of the dialectic of discrete and continuous in mathematics. A new axiomatic version of the NSA – the Theory of Quasi-sets (**TQS**) presented in Sect. 3 is mainly intended to simplify the proofs of theorems in **THS**, many of which are technically difficult. In addition, this axiomatic system is as informal as possible, which makes it accessible to the wast majority of mathematicians.

2   What is the Modern Mathematical Rigor

Modern rigor in mathematical analysis developed during the 19th century. Its development is primarily associated with the names of Cauchy, Bolzano, Weierstrass and Cantor. When studying some problems about convergence of Fourier series, Cantor introduced the concept of a set mentioned in Introduction and developed Set Theory, which was destined to become the basis for the rigorous formalization of all mathematics. In fact, any mathematical statement can be formalized and proved in terms of Set Theory. Such formalization may be very long and unnatural. In practice no one conducts it. However, the overwhelming majority of mathematicians do not doubt that, in principle, this is possible.

Remark 1. *This does not mean that the way of thinking in mathematics is reducible to one in set theory.*

At the beginning of the 20th century, some well-defined properties were discovered such that the statements about the existence of the sets defined by them led to a contradiction. These were the well known **paradoxes of the set theory**. So, it was necessary to impose axiomatically some restrictions on the properties, that define sets. The following is a brief and incomplete survey of the two main systems of axioms of Set Theory. A clear detailed presentation can be found in the Chap. 2 of the book [9].

The first axiomatic **ZC** was introduced by Zermelo in 1908. Here C stays for the **Axiom of Choice** introduced by E. Zermelo a bit earlier. After about 20 years, this axiomatics was expanded by adding two more axioms: the **Axiom of Replacement** added by A.Fraenkel, and the **Axiom of Regularity** added by J. von Neumann. These axioms are not needed in this article. The extended axiomatics is called the Zermelo-Fraenkel axiomatics. It is generally accepted that all mathematics can be formalized in the Zermelo-Fraenkel's Set Theory and, accordingly, a rigorous proof is such a proof that can be formalized in this theory. Indeed a huge part of mathematics, including Calculus, ODE, PDE, etc. can be formalized within **ZC** .

The primary objects of **ZFC** are sets. The basic relations are equality and inclusion $\in$. As a rule lowercase Latin letters denote sets or set variables. A collection $\mathcal{X}$ of sets x that satisfy a wd-property P is denoted by $\{x \mid P(x)\}$. We mentioned above that not for every wd-property P the collection $\mathcal{X}$ is a set. However, the following **Axiom of Separation** (AS) holds: For a well-defined property $P(x)$ and a set y there exists the set $z = \{x \in y \mid P(x)\}$. The equality of sets is defined by the
Axiom of Extensionality (AE): $x = y \Longleftrightarrow \forall z\, z \in x \longleftrightarrow z \in y$.
The uniqueness of sets, whose existence is stated in the next three axioms follows from (AE).

Axiom of Union Set. (AU) For any set x there exist the set
$\bigcup x = \{y \mid \exists\, z \in x\,(y \in z)\}$. E.g. the union $\bigcup\{x,y\} = x \cup y$.

Axiom of Power Set. (APs) For any set x there exist the set
$\mathcal{P}(x) = \{y \mid y \subseteq x\}$.

Axiom of Pair. (AP) For any two sets a and b, there exists the set $\{a, b\}$.
By (AE) $\{a, b\} = \{b, a\}$

The ordered pair of sets x and y in **ZC** is defined as follows: $\langle x, y \rangle :=$
$\{\{x\}, \{x, y\}\}$. It follows from (AE) that $\langle x, y \rangle = \langle u, v \rangle \iff x = u \land y = v$
It is easy to see that the set $\langle x, y \rangle \in \mathcal{P}(\mathcal{P}(\mathcal{P}(x \cup y)))$. So, the Cartesian product
$a \times b = \{\langle x, y \rangle \mid x \in a, y \in b\}$ is a set by the axiom (AS).

The statement $f : a \to b$ is formalized in **ZC** as follows:
$f \subseteq a \times b \land \forall x \in a \exists! y \in b \, \langle x, y \rangle \in f$.
Here $\exists! y \, P(y) := \exists y \, (P(y) \land \forall z (P(z) \longrightarrow z = y))$.

Axiom of Infinity $\exists x (\emptyset \in x \land \forall y \in x \, (y \cup \{y\} \in x))$. The minimal set that
satisfies the Axiom of infinity is the set $\{\emptyset, \{\emptyset\}, \{\{\emptyset, \{\emptyset\}\} \dots \}$. This set is the
set $\mathbb{N} = \{0, 1, 2, \dots\}$ of natural numbers in **ZC**. The Principle of Mathematical
Induction follows from the minimality.

Properties of mappings (injectivity, surjectivity, etc.), as well as the set b^a of
all mappings from a to b, are formalized in an obvious way. The formulation of
Axiom of Choice (AC) is usual. The listed axioms constitute the axiomatics
ZC.

Strictly speaking the axiomatic **ZFC** (as well as **ZC**) is not perfectly rigorous, since the notion of a property is not well defined. For this axiomatic the
term "Naïve set theory" is used. This is also the title of the book [8] containing
a detailed exposition of the basic set theory for non-logicians.

All properties and statements of the Naïve set theory including the axioms of
ZFC can be written as formulas of the formal language of the first order predicate
logic with equality and a single extra-logical symbol of a binary membership
predicate $\in$ This language is denoted by $\mathcal{L}$ below. If a variable ξ occurs in a
formula φ under the universal or existential quantifier, it is called bounded,
otherwise it is called free. Notation $\varphi(\xi_1, ..., \xi_k)$, means that each free variable
included in φ is a variable from the list $(\xi_1, ..., \xi_k)$. This formula is interpreted
as a statement about variables $(\xi_1, ..., \xi_k)$, i.e. after assigning specific values to
these variables, the corresponding statement becomes either true or false. A
truth value of a statement does not depend on bounded variables. Formula that
does not contain free variables is called a **proposition**. Axioms and theorems
of **ZFC** formulated in $\mathcal{L}$ are propositions.

The set of axioms of **ZFC** is infinite since the Axiom of Separation is not
a one axiom, but a countable set of axioms – its own axiom for each formal
sentence.

According to Gödel's second incompleteness theorem, the consistency of
ZFC
cannot be proved in **ZFC**. Therefore, all statements about the consistency (independence) of sentences are conditional, i.e. begin with the words "If **ZFC** is
consistent, then ..."

The following axiomatic **NBG** (von Neumann-Bernays-Gödel) was introduced by K. Gödel in [10]. The formal language of **NBG** is also $\mathcal{L}$. The primary

objects of **NBG** are interpreted as wd-properties, that are called in **NBG** classes. Sets are defined as classes that can be included in other classes as elements. It is easy to see that this is a formalization of Cantor's intuitive understanding of the concept of a set on the page 3. The classes that are not sets, are called *proper* classes. The following agreement is adopted to simplify the writing of $\mathcal{L}$-formulas. The variables assuming values of sets are denoted by lowercase Latin letters, while variables denoted by uppercase Latin letters can assume values of both sets and proper classes depending on a context. For example, the statement "X is a set" can be formalized in $\mathcal{L}$ as $\exists\, Y\, X \in Y$, or using the agreement above $\exists\, y\, X = y$. The **NBG Axiom of Separation** is formulated here as a single axiom: $\forall\, x, X \,\exists\, y\, x \cap X = y$. Variables denoted by lowercase Latin letters are called set-type variables.

Definition 1. *We say that $\varphi(x_1, ..., x_k, Y_1, ...Y_m)$ is a set-type formula if all its bounded variables are set-type variables.*

We say that a set-type formula $\varphi(\bar{x}, \bar{Y})$ defines the set-type class $A_\varphi(\bar{Y}) = \{\bar{x} \mid \varphi(\bar{x}, \bar{Y})\}$, where
$$\bar{x} = \langle x_1, ..., x_k \rangle, \quad \bar{Y} = \langle Y_1, ...Y_m \rangle.$$

In this article the following Axiom is included in **NBG** .

Axiom of Existence of Set-type Classes (AEC). For every set-type formula $\varphi(\bar{x}, \bar{Y})$ the class A_φ defined by this formula exists.

Proofs of the following theorem and its corollary can be found in [9].

Theorem 1. *Every set-type proposition is a theorem of* **NBG** *, if and only if it is a theorem of* **ZFC** *.*

Corollary 1. *The axiomatic* **NBG** *is consistent if and only if the axiomatic* **ZFC** *is consistent.*

3   Theory of Quasi-sets – A Version of NSA

The theory **THS** can be used in studies of the relationship between discrete and continuous mathematics. In my opinion it is the most adequate formalization of the dialectic of discrete and continuous in mathematics. A new axiomatic version of the NSA – the Theory of Quasi-sets (**TQS**) presented in this section is mainly intended to simplify the proofs of theorems in **THS** , many of which are technically difficult. In addition, the axioms of **TQS** are as informal as possible, which makes it accessible to a wide range of mathematicians.

Similar to **NBG** the primary objects of **TQS** are classes and sets are defined to those classes that are included as elements in other classes. The **Axiom of Extensionality** is the same as in **NBG** . The principle difference between **NBG** and **TQS** is that the **TQS** allows classes for which the **Axiom of Separation** of **NBG** fails. The classes, for which it holds are called wd-classes, i.e. a class X is a wd-class, if $\forall\, x\, \exists\, y\, x \cap X = y$. If X is not a wd-class, then there exists a set x such that $Y = x \cap X$ is not a set. A class Y that is a subset of a set is called a **quasi-set**. Obviously each set is a quasi-set. A quasi-set that is not a set is called a **proper** quasi-set.

1° **Axiom of Weak Transfer** [Weak Transfer Principle] (WTP). *Every theorem of* **NBG** *is a theorem about wd-classes in* **TQS** .

Remark 2. *In the Robinson's version of nonstandard analysis and in axiomatic versions, where the predicate of standardness is involved, the strong transfer principle (STP) is included as an axiom. The axiom STP is formulated approximately as follows:any sentence that does not contain a standard predicate is a theorem of the standard theory if and only if its version relativized to the predicate of standardness is a theorem of non-standard theory. The STP is a key axiom for proving standard theorems using NSA. The* **TQS** *is intended for proving theorems in non-standard mathematics, i.e. theorems containing vague concepts (sometimes using standard mathematics). Therefore,* **TQS** *does not need the predicate of standardness and the STP.*

Question 1. *The absence of a strong Transfer Principle in* **TQS** *gives grounds to assume the existence of* **NBG** *statements, unprovable in* **NBG** *, but provable in* **TQS** *. It would be interesting to find meaningful examples of such statements. Can it be e.g. the statement P=NP?*

2° **Axiom of Class Formation (ACF).** *If $P(x)$ is any statement that involves a free variable x and such that all quantifiers are applied only to set and quasi-set variables, then there exists a class $\{x \mid P(x)\}$.*

The statement $\mathbf{S}(x) := \forall X \subseteq x \, \exists \, y \, X \in y$ satisfies the conditions of ACF. A set a such that $\mathbf{S}(a)$ is true is called a **small** set. The existence of the **S** follows from ACF.

The class of small sets is denoted by **S**. Its existence follows from the ACF. It is easy to see that the class **S** is not a quasi-set. For any class X the notation $\mathbf{S}(X)$ is used for $\mathbf{S} \cap X$.

A natural number $n \in \mathbb{N}$ is small if the set $\bar{n} = \{0, 1, ..., n-1\} \in \mathbf{S}(\mathbb{N})$ is small. We denote $n \cup \{n\}$ by $n + 1$.

Proposition 1 *(Strong Induction Principle). If $X \subseteq \mathbf{S}(\mathbb{N})$, $0 \in X$ and $\forall n \in X \, n + 1 \in X$, then $X = \mathbf{S}(\mathbb{N})$.*

▷ Let $m \in \mathbf{S}(\mathbb{N}) \setminus X$, then $Y = \overline{m+1} \setminus X \subseteq \overline{m+1}$. Thus Y is a set, since $\overline{m+1} \in \mathbf{S}(\mathbb{N})$. Thus, $\exists k = \min Y$ by the WTP. Then $k \neq 0$ and $k - 1 \in X$. So $k \in X$ by the conditions of the proposition. Contradiction. ◁

Remark 3. *By the WTP $\mathbb{N}$ is the set in* **TQS** *that satisfies standard Induction Principle:*

$$\forall \, x \subseteq \mathbb{N} \, (0 \in x \wedge \forall n \in x \ n + 1 \in x) \longrightarrow x = \mathbb{N}.$$

Thus, in the Strong Induction Principle all subclasses of $\mathbf{S}(\mathbb{N})$ are involved, while in the standard one only all subsets of $\mathbb{N}$. This does not imply yet that $\mathbb{N} \neq \mathbf{S}(\mathbb{N})$. This inequality follows from the Axiom of Compactness, that will be introduced later.

We also define the quasi-set of small integers as follows: $\mathbf{S}(\mathbb{Z}) := \mathbf{S}(\mathbb{N}) \cup (-\mathbf{S}(\mathbb{N}))$. Large natural numbers (integers) are called **infinitely large**. Notations: $\mathbb{N}_\infty := \mathbb{N} \setminus \mathbf{S}(\mathbb{N})$ $(\mathbb{Z}_\infty = \mathbb{Z} \setminus \mathbf{S}(\mathbb{Z}))$

Proposition 2. $\mathbb{N}_\infty \neq \emptyset$.

▷ This Proposition will be proved later. ◁

Proposition 3. *A set a is hyperfinite, if and only if the cardinality $|a| \in \mathbf{S}(\mathbb{N})$* ◁.

Definition 2. *1. A real number $\alpha \in \mathbb{R}$ is called bounded ($\alpha \in \mathbb{R}_b$), iff $|\alpha| < n$ for some $n \in \mathbf{S}(\mathbb{N})$.*
2. A real $\alpha \in \mathbb{R}$ is called an infinitesimal ($\alpha \in \mathcal{M}_0$), iff $|\alpha| < n^{-1}$ for all $n \in \mathbf{S}(\mathbb{N})$. The quasi-set $\mathcal{M}_0$ is called the monad of 0. Reals $\alpha, b \in \mathbb{R}$ a said to be infinitesimally close ($\alpha \approx \gamma$), if $\alpha - \gamma \in \mathcal{M}_0$. Sometimes we write $\alpha \approx 0$ for $\alpha \in \mathcal{M}_0$ and $|\alpha| \gg 0$ for $\alpha \notin \mathcal{M}_0$.
3. A real $\Omega \in \mathbb{R} \setminus \mathbb{R}_b$ is called infinitely large ($\Omega \in \mathbb{R}_\infty$). Obviously
$$\mathbb{R}_\infty = (\mathcal{M}_0 \setminus \{0\})^{-1}.$$

Proposition 4. *The monad $\mathcal{M}_0$ is the maximal ideal in the ring $\mathbb{R}_b$.*

▷ The statement "$\mathcal{M}_0$ is an ideal in $\mathbb{R}_b$" is similar to the theorem "The sum of two infinitesimal functions is an infinitesimal function and the product of an infinitesimal function by a bounded one is an infinitesimal function" of Calculus I. The proofs are also similar.

If $\mathcal{I}$ is an ideal in $\mathbb{R}_b$ such that $\mathcal{M}_0 \subsetneqq \mathcal{I}$ and $a \in \mathcal{I} \setminus \mathcal{M}_0$, then obviously $1/a \in \mathbb{R}_b$, so $1 \in \mathbb{R}_b$. Thus, $\mathcal{I} = \mathbb{R}_b$. ◁

In what follows the quotient class $\alpha + \mathcal{M}_0$ for $\alpha \in \mathbb{R}_b$ is denoted $\mathcal{M}_\alpha$.

Remark 4. *Since any monad $\mathcal{M}_\alpha$ is a proper quasi-set one can not use the quotient class $\mathbb{R}_b/\mathcal{M}_0$ that consists of monads. However, it will be shown later that the existence of complete system of representative of monads $\mathcal{M}_\alpha, \alpha \in \mathbb{R}_b$ exists.*

Proposition 5. *The classes $\mathbb{N}_\infty, \mathbb{Z}_\infty, \mathbb{R}_\infty, \mathbb{R}_b, \mathcal{M}_0$ are proper quasi-sets.*

▷ Suppose that $\mathbb{N}_\infty$ is a set. Then, since $\mathbb{N}_\infty \subseteq \mathbb{N}$ there exist a number $N = \min \mathbb{N}_\infty$ by WTS, so $N - 1 \in \mathbf{S}(\mathbb{N})$. Thus $N \in \mathbf{S}(\mathbb{N})$ by Proposition 1, while $N \in \mathbb{N}_\infty$. The contradiction. So, $\mathbb{N}_\infty$ is a proper quasi-set as well as $\mathbb{Z}_\infty$.

If $\mathbb{R}_\infty$ is a set, then $\mathbb{Z}_\infty = \{\lfloor \alpha \rfloor | \alpha \in \mathbb{R}_\infty\}$ is a set by WST. Here $\lfloor \alpha \rfloor$ denote the integer part of α. So $\mathbb{R}_\infty$ is a proper quasi-set. The proof of this Proposition for $\mathbb{R}_b$ and $\mathcal{M}_0$ follows immediately from Definition 2 (3). ◁

Proposition 6. *If X, Y are quasi-sets, then $dom(X)$, $range(X)$, $X \cap Y$, $X \cup Y$, $X \setminus Y$, $X \times Y$ Y^X are also quasi-sets.*

▷ Let $X \subseteq x$, $Y \subseteq y$. Then first five operations applied to X and Y are subclasses of the same operations applied to x and y. The latter are sets by WTP. So, the first ones are quasi-sets. For the Cartesian product the Proposition follows the Kuratowski's definition of an ordered pair. By the definition the operation X^Y denotes the quasi-set of all functions f, whose graph is a subset from $X \times Y$. Since f is a set, then $dom(f)$ and $range(f)$ are sets, thus $X^Y \subseteq x^y$. ◁

Definition 3. *We say that a quasi-set X is sharp, if any its subset is small.*

Proposition 7. *The quasi-set $\mathbf{S}(\mathbb{N})$ is sharp.*

▷ Let $x \subseteq \mathbf{S}(\mathbb{N})$. By Proposition 2 there exists $N \in \mathbb{N}_\infty$. If x is an infinite set, then $\forall n \in \mathbb{N} \exists m \in x \; m > n$. This is an **NBG** theorem and by WTP there exists $m \in x$ such that $m > n$. This contradicts the assumption above. So, x is a finite set. Since $x \subseteq [\min x, \max x] \subseteq \mathbf{S}(\mathbb{N})$, x is small. ◁

3 °**Axiom of Compactness** (AC). Let X be a sharp quasi-set that has finite intersection property (FIP), i.e. every subset of X has a nonempty intersection, then $\bigcap X \neq \emptyset$.

AC implies Proposition 2. Indeed: consider the sharp quasi-set $X = \{\mathbb{N} \setminus \{0, 1, ..., n - 1\} \mid n \in \mathbf{S}(\mathbb{N})\}$ that exists by ACF. By AC $\bigcap X = \mathbb{N} \setminus \mathbf{SN} \neq \emptyset$.

Proposition 8. *Let F be a function such that $dom(F)$ is a sharp quasi-set. Then there exists a function f such that $F \subseteq f$.* □

Though we cannot include classes in sets we can consider classes that are indexed families of some classes.

Definition 4. *We say that a class X is a family of classes $\{X_i \mid i \in I\}$, indexed by elements of the set I (I-family), if $dom(X) = I$ and $\forall i \in I \; X_i = \{x \mid \langle i, x \rangle \in X\}$*

4 ° **Axiom of Choice for Quasi-sets** (ACh). If F is a family of quasi-sets, indexed by a sharp quasi-set I (i.e. $dom(F) = I$) and $\forall i \in I \; F_i \neq \emptyset$, then there exists a function $G \subseteq F$, such that $dom(G) = I$ and $\forall i \in I \; G(i) \in F_i$. This function is called a choice function as usual.

5 ° **Axiom of Exponentiation** (AExp). *If X is a sharp quasi-set, then there exists the sharp family $\mathbf{P}(X)$ of all subquasi-sets of X. This means that*

$$\forall i \in I \; \mathbf{P}(X)_i \subseteq X \text{ and } \forall Y \subseteq X \; \exists i \in I \; \mathbf{P}(X)_i = Y, \tag{1}$$

where $I = dom(\mathbf{P}(X))$. This axiom was suggested by P. Andreev for **THS**. It is included in **TQS** without any changes. The theory **TQS** can be interpreted in **NCT**, whose consistency with respect to **ZFC** is proved in [7]. So, the theory **TQS** is consistent with **ZFC** as well. AExp implies the following

Proposition 9. *1. Any subquasi-set of a sharp quasi-set is sharp.*
2. The family of sharp quasi-sets is closed under operations listed in Proposition 6
3. If $F : X \to Y$ is a surjective map and X is a sharp quasi-set, then Y is a sharp quasi-set as well.
4. If F is a function and $dom(F)$ is sharp, then F is a sharp quasi-set.

Sharp quasi-sets simulate standard mathematics within **TQS** as follows. For any sharp quasi-sets X and Y set:

$$X \varepsilon Y := \exists i \in dom(Y) \; X = Y_i, \quad X \equiv Y := \forall Z \; Z \varepsilon X \longleftrightarrow Z \varepsilon Y \tag{2}$$

Theorem 2. *Every **ZC** -theorem written in $(\varepsilon, \equiv)$-language is a theorem about sharp quasi-sets in **TQS** .*

This theorem was proved for **THS** in [15]. The proof can be modified for the case of **TQS** .

Now we able to formulate rigorously and prove the existence of complete system of representatives of monads $\in \mathbb{R}_b$ (c.s.r.m.), mentioned in Remark 4.

Theorem 3. *There exists a sharp class $R \subseteq \mathbb{R}_b$, whose intersection with every monad $\mathcal{M}_\alpha$, $\alpha \in \mathbb{R}_b$ has only one element.*

▷ For every $\alpha \in \mathbb{R}_b$ put

$$C_\alpha = \{\langle q, q' \rangle \in \mathbf{S}(\mathbb{Q})^2 \mid \alpha \in (q, q') \subseteq \mathbb{R}_b\}$$

Obviously, if $\alpha \approx \gamma$, then $C_\alpha = C_\gamma$ and $\mathcal{M}_\alpha = \bigcap\{(q, q') \mid \langle q, q' \rangle \in C_\alpha\}$. The family $\mathcal{C} = \{C_\alpha \mid \alpha \in \mathbb{R}_b\}$, is a subfamily of the class $\mathbf{P}(\mathbf{S}(\mathbb{Q}))$. By the Andreev's Axiom of Exponentiation the latter is a sharp class, thus $\mathcal{C}$ is a sharp class as well. So, the family of monads in $\mathbb{R}_b$ indexed by the sharp class $\mathcal{C}$ has a choice function by the Axiom of Choice for quasi-sets. The range of this choice function is a c.s.r.m. in $\mathbb{R}_b$. ◁

There are infinitely many c.s.r.m.'s. since if $f : \mathcal{C} \to \mathbb{R}_b$ is a choice function and $g : \mathcal{C} \to \mathbb{R}_b$ is a function such that $\forall C \in \mathcal{C}$ $f(C) \approx g(C)$, then g is also a choice function and the range of g is a c.s.r.m. as well. Moreover, there does not exists any definable in **TQS** c.r.s.m. This statement was proved in [15] (Proposition 6.3.) in the framework of **THS** . The proof presented there can be easily modified for **TQS** .

Operations in any c.s.r.m R are uniquely determined based on the requirement that R must be a field isomorphic to $\mathbb{R}_b/\mathcal{M}_0$. Thus, for $a, b \in R$

$$a +_{sh} b = c \in R, \ c \approx a + b; \quad a \cdot_{sh} b = d \in R, \ d \approx a \cdot b \tag{3}$$

Here $+$ and $\cdot$ are operation in $\mathbb{R}$. The linear order in R is inherited from $\mathbb{R}_b$. The proof of the following Proposition is easy.

Proposition 10. *For every c.s.r.m. R the algebraic system $\langle R; +_{sh}, \cdot_{sh}, < \rangle$ is a complete linearly ordered field.*

The next proposition follows immediately from the construction of c.s.r.m.

Proposition 11. *For every $a \in \mathbb{R}_b$ there exists a number $sh(\alpha) \in R$ such that $sh(a) \approx \alpha$.*

We see that, generally speaking, operations in R, are not equal to those in $\mathbb{R}$ but only infinitesimally close to them. They depend on a choice of a set R. In other versions of NSA, in which there is an external (i.e. not definable in standard terms) predicate of standardness, the external set (quasi-set in our version) of standard elements is a c.s.r.m. that is a subfield of $\mathbb{R}_b$. In the **TQS** we can always choose a c.s.r.m. R so that the field $\mathbf{St}(\mathbb{Q})$ is a subfield of R. It follows from the non-definability of c.s.r.m. in **TQS** , that no any definable c.s.r.m. R can be a subfield of $\mathbb{R}_b$. In what follows some c.s.r.m. R that has this property is fixed. It is denoted by $\mathbb{R}^{sh}$.

As it was mentioned in the Remark 4 the field $\mathbb{R}^{sh}$ is obtained by identification of infinitesimally close elements. Recalling the discussion of Plato's allegory on page 5, we can say that the field $\mathbb{R}$ of the theory of quasi-sets is an element of the "real world", and field $\mathbb{R}^{sh}$ is its shadow, which the prisoners see. In fact, prisoners only see the shadows of the numbers they can access (the elements of R_b. Infinitely large numbers (the elements of $\mathbb{R}_\infty$) are beyond their reach.

The field $\mathbb{R}^{sh}$ is an analog of the field of standard elements in the Robinson's NSA, where the element $\alpha^{sh} \in \mathbb{R}^{sh}$ is called the standard part or a shadow (sic!) of α. In what follows the elements of $\mathbb{R}^{sh}$ we call sharp reals.

The analogy between standard reals and sharp reals can be extended to an (incomplete) analogy between standard elements of the NSA and sharp classes.

Definition 5. *The quasi-sets superstructure $\mathcal{S}^{qs}(\mathbb{R})$ over $\mathbb{R}$ is the minimal family of sets that contains $\mathbb{R}$ as an element, is closed under operations defined in Proposition 6 and contains sub quasi-sets of these sets.*

This definition differs from the definition of the standard superstructure $\mathcal{S}(\mathbb{R})$, which includes only sets, but not quasi-sets.

Proposition 9 allows to define the sharp superstructure $\mathcal{S}^{sh}(\mathbb{R}^{sh})$. The definition repeats Definition 5 with replacement of quasi-sets by sharp classes.

Remark 5. *A. Robinson developed his NSA in some proper (nonstandard extension ${}^{*}\mathcal{S}(\mathbb{R})$ of the standard superstructure $\mathcal{S}(\mathbb{R})$ (see e.g. [12]). In our approach $\mathcal{S}^{sh}(\mathbb{R}^{sh})$ is an analog of the standard superstructure $\mathcal{S}(\mathbb{R})$, while $\mathcal{S}^{qs}(\mathbb{R})$ is an analog of its nonstandard extension ${}^{*}\mathcal{S}(\mathbb{R})$.*

A natural question arises how to describe the shadow x^{sh} of an arbitrary set $x \in \mathcal{S}(\mathbb{R})$. The size limitations of the article do not allow me to explore this issue in detail here. Only the simplest examples are discussed below.

By Proposition 10 it is natural to define $\mathbb{R}^{sh}$ as the shadow of the field $\mathbb{R}$. Similarly, for any $\xi, \eta \in \mathbb{R}_b$ it is naturally to define $[\xi, \eta]^{sh}$ as $[\xi^{sh}, \eta^{sh}]^{sh}$. For an arbitrary set $a \subseteq \mathbb{R}$ let $a^{sh} := \{\xi^{sh} \mid \xi \in a\}$. Then the following Proposition is true.

Proposition 12. *If $a \subseteq \mathbb{R}$, then $a^{sh} \subseteq \mathbb{R}^{sh}$ is a closed set.*

▷ Suppose that $r \in \mathbb{R}^{sh}$ is a limit point of a^{sh}. Then for every $n \in \mathbf{S}(\mathbb{N})$ the set $w_n = \{\xi \in a\} \mid 0 < |r - \xi| < n^{-1}$. The intersection $\bigcap\limits_{n \in \mathbf{S}(\mathbb{N})} w_n \neq \emptyset$ by the Axiom of Compactness. So, there exists $\eta \in a$ such that $\eta \in \mathcal{M}(r)$. Since $\eta^{sh} \in \mathbb{R}^{sh}, \eta^{sh} = r \in a$. ◁

Let a function $f \subseteq \mathbb{R}_b \times \mathbb{R}_b$ then it is natural to define f^{sh} by the following conditions:

$$\mathrm{dom}(f^{sh}) = (\mathrm{dom}(f))^{sh}; \ \forall \xi \in \mathrm{dom}(f) \ f^{sh}(\xi^{sh}) = (f(\xi))^{sh} \qquad (4)$$

Proposition 13. f^{sh} *is continuous on its domain if and only if*

$$\forall \xi, \eta \in dom(f) \ \xi \approx \eta \Longrightarrow f(\xi) \approx f(\eta) \tag{5}$$

If the condition (5) hold a function f is said to be S-continuous.

Proposition 14. *Let* $\mathbf{x} = \{x_n \mid n \in \mathbb{N}\}$ *be a sequence of reals that is a set. If* $\mathbf{x} \upharpoonright \mathbf{S}(\mathbb{N}) \subseteq \mathbb{R}_b$. *Define the shadow* $\mathbf{x}^{sh} := \{x_n^{sh} \mid n \in \mathbf{S}(\mathbb{N})\}$. *Then*

$$\lim_{n \in \mathbf{S}(\mathbb{N})} x_n^{sh} = l \in \mathbb{R}^{sh} \Longleftrightarrow \exists \, \omega \in \mathbb{N}_\infty \, \forall n \in \mathbb{N}_\infty \, (n < \omega \Longrightarrow x_n \approx l).$$

Propositions 13 and 14 are formulated and proved in the article [19] (see also the book [18]) in the framework of Robinson's NSA . It can be easily adjusted to **TQS** .

This article is devoted to the study of dynamical systems on hyperfinite measure spaces and their applications to approximations of dynamical systems on Lebesgue spaces by finite dynamical systems. A simple example from [19] is discussed here, which demonstrates some effects arising in the study of the behavior of ergodic means of dynamical systems on very large probability spaces. These effects which easily formulated and proved in the framework of **TQS** , but even their formulation in the standard mathematics seems practically unreadable. It also demonstrates how standard mathematics is used to prove such effects.

Example. Let us demonstrate the application of the theory **TQS** by the example of studying the relationship between the simplest continuous dynamical system and its hyperfinite approximation. As a continuous dynamic system consider the shift by $\tau \in (0, 1)$ modulo 1 of the standard segment $[0, 1]$ (see Remark 5) with the identified ends. The cases a) $\tau \in \mathbb{Q}$ and b) $\tau \notin \mathbb{Q}$ should be considered separately. Obviously for a continuous function f on $[0, 1]$ the ergodic mean

$$A([0, 1], \tau, f, x) = \lim_{n \to \infty} \frac{1}{n} \sum_{k=0}^{n-1} f(\tau^k(x))$$

$$= \begin{cases} \frac{1}{b} \sum_{m=0}^{b-1} f(x + \frac{m}{b}), & \tau = a/b, \ \gcd(a, b) = 1 \\ \int_0^1 f(x) dx, & \tau - \text{ standard irrational number} \end{cases} \tag{6}$$

Define approximating hyperfinite dynamical system as follows. Consider a hyperfinite set $[0, 1]_H = \{k/H \mid k = 0, 1, ..., H - 1\}$, where $H \in \mathbb{N}_\infty$ with a uniform normalized measure p_H as a probability space. The hyperfinite probability space $[0, 1]_H, \mu$ approximate the standard probability space $([0, 1], dx)$ in the following sense. $[0, 1] = [0, 1]^{sh}$ and for any standard continuous function $f \in \mathcal{S}^{sh}(\mathbb{R}^{sh})$ one has

$$\int_0^1 f(x) dx = \left(\frac{1}{H} \sum_{k=0}^{H-1} \varphi(k/H) \right)^{sh}, \tag{7}$$

where $f = \varphi^{sh}$, $\varphi \in \mathcal{S}^{qs}(\mathbb{R})$ see Remark 5 and Proposition 13.

The measure preserving transformation of the hyperfinite probability space $([0,1]_H, p)$ is a permutation $\tau_H : [0,1]_H \to [0,1]_H$. Without loss of generality, assume that τ_H is a cycle of length H. We say that τ_H approximates τ, if $\forall \frac{k}{H} \in [0,1]_H (\tau_H(k/H))^{sh} = \tau((k/H)^{sh})$. For the case a) of a rational shift τ in $[0,1]$ (6) set $H = bM + 1$, where $M \in \mathbb{N}_\infty$ is such that $\gcd(a, H) = 1$. Then $\gcd(aM, H) = 1$ and the shift $\tau_H = \frac{aM}{H}$ in $[0,1]_H$ is a cycle of the length H. For the case b) of an irrational shift τ let $\frac{p_K}{q_K}$ for $K \in \mathbb{N}_\infty$ be the K-th convergent of the continued fraction for τ. Set $H = q_k$. Since $\gcd(p_k, q_k) = 1$ the shift τ_H in $[0,1]_H$ is also a cycle of the length H. It is easy to see that in both cases τ_H approximates τ. So, in both cases the hyperfinite dynamical system $HS = ([0,1]_H, p_H, \tau_H)$ approximates the continuous dynamical system $CS = ([0,1], dx, \tau)$. Let us discuss the relationship between ergodic averages of CS and DS. Let an arbitrary S-continuous function $f : [0,1] \to \mathbb{R}$, be such that $f(0) = f(1)$ (recall that the ends of the segment $[0,1]$ are glued) and $n, m \in \mathbb{N}$.
Put $av(f,n,m) = \frac{1}{n} \sum\limits_{k=0}^{n-1} f(\tau_H^k(m/H))$.

Theorem 4. *For every* $m \in \{0, \ldots, H-1\}$

1. *If* $n_1/H \approx n_2/N \approx a \in \mathbb{R}^{sh}$, $a > 0$, *then* $av(f, n_1, m) \approx av(f, n_2, m)$. *If* $a = 1$, *then* $av(f, n_1, m) \approx \int\limits_0^1 f^{sh}(x)dx$,

2. $\exists L \in \mathbb{N}_\infty \forall n \in \mathbb{N}_\infty$, $n < L$ $(n/H \approx 0 \implies av(f,n,m) \approx A([0,1], \tau, f^{sh}, \xi)$, *where* $\xi = (m/H)^{sh}$,

3. *If* $n/H \approx a$, *then the functional* φ_a *on* $C[0,1]$ *such that* $\varphi_a(f^{sh}) = (av(f,n,m)^{sh}$ *is an invariant mean on* $[0,1]$

4. *If* τ *is irrational, then* $av(f,n,m) \approx \int\limits_0^1 f^{sh}(x)dx$ *for every* $n \in \mathbb{N}_\infty$ *such that*
$n/H \in \mathbb{R}_b$.

All statements of this theorem are formulated and proved in [19] for arbitrary Lebesgue probability spaces within the framework of Robinson's NSA. Statements 1 and 3 can be reformulated and proved in standard terms, but these proofs are unnatural for applied mathematicians. The statement 2 assert that ergodic means $av(f,n,m)$ of the systems HS approximate the ergodic mean (6) of the CS system only on the initial segment of infinitely large moments n of the discrete time that are significantly less than H, i.e. $n/H \approx 0$. The formulation of the statement 2) in standard terms is practically unreadable, to say nothing of its proof. The theorem 4 is illustrated by computer experiments in [19]. The results of the theorem are observed even when in numerical experiments as relatively moderate numbers. e.g. of the order 10^6 are taken for the "infinitely large" H.

Aknowledgements. I am very grateful to my colleagues Dr. Petr Andreev, Professors Lev Glebsky, Ward Henson, Peter Loeb, Paul Schupp and Pavol Zlatos for many years of friendship and joint work on the issues related to this article.

References

1. Zeilberger, D.: Real Analysis is a degenerate case of discrete analysis. In: Proceedings of the Sixth International Conference on Difference Equations, pp. 25–32. CRS, Boca Raton, Fl. (2004)
2. Kolmogoroff, A.-N.: Grudbegriffe der Wahrscheinlichkeitsrechnung. Springe, Heidelberg (1933). https://doi.org/10.1007/978-3-642-49888-6
3. Nelson, E.: Radically Elementary Probability Theory. Princeton University Press, Princeton NJ (1987)
4. Dirac, P.: The Principles of the Quantum Mechanics, 4th edn. The Calderon Press, Oxford AT (1958)
5. Berezin, F.-A.: Lectures on Statistical Physics. MCCME, Moscow (2008)
6. Von Neumann, J.: Mathematical Foundations of Quantum Mechanics. Princeton University Press, Princeton, NJ (1932)
7. Reed, M., Simon, B.: Methods of Modern Mathematical Physics. I. Functional Analysis. Academic Press, New York, London (1973)
8. Halmos, P.: Naïve the Set Theory. D. Van Nostrand Company, Princeton, NJ (1960). Reprinted by Springer-Verlag, New York (1974)
9. Cohen, P.: The Set Theory and the Continuum Hypothesis. Dover Publications, Inc., Mineola, NY (2008)
10. Gödel, K.: Consistency of Continuum Hypothesis, Princeton University Press, Princeton, NJ (1940)
11. Robinson, A.: Nonstandard Analysis, Revised Edition. Princeton University Press, Princeton NJ (1996)
12. Gordon, E.-I.: Some remarks about nonstandard methods in analysis I. Vladikavkaz Math. J. $21(4)$, 25–41 (2019)
13. Nelson, E.: Internal the set theory. A new approach to nonstandard analysis. Bull. AMS $83(6)$, 1165–1198 (1977)
14. Andreev, P.-V., Gordon, E.-I.: An axiomatic for nonstandard the set theory based on von Neumann- Bernays-Gödel theory. J. Symbol. Logic $66(3)$, 1321–1341 (2001)
15. Andreev, P.-V., Gordon, E.-I.: A theory of hyperfinite sets. Ann. Pure Appl. Logic. 143, 3–19 (2006)
16. Vopenka, P.: Mathematics in the Alternative the Set Theory. Teubneg, Leipzig (1979)
17. Kanovei, V., Reeken, M.: Nonstandard Analysis, Axiomatically. Springer-Verlag, Berlin (2004)
18. Loeb, P.A., Wolff, M. Nonstandard Analysis for the Working Mathematician. Kluwer Academic Publishers, 293 p. Dordrecht-Boston-London (2000)
19. Gordon, E.I., Glebsky, L.Y., Henson, C.W.: Nonstandard analysis of the Behavior of ergodic means of dynamical systems on very big finite probability spaces. In: González-Aguilar, H., Ugalde, E. (eds.) Nonlinear Dynamics New Directions. NSC, vol. 11, pp. 115–151. Springer, Cham (2015). https://doi.org/10.1007/978-3-319-09867-8_6

Algorithms and Scenarios of Online Decision Support System for Selection of the Distribution Laws of positive Random Variables

Ivleva Anna[(⊠)] and Smirnov Sergey

Samara Federal Research Scientific Center RAS, Institute for the Control of Complex Systems RAS, 443020 Samara, Russia
borovik@iccs.ru
http://www.iccs.ru/contact.html

Abstract. The problem of making online decision support system for selecting the two-parametric distribution law of a continuous positive random variable with a finite second moment is considered. Algorithms and scenarios of the system operation are given. The main priority is a practical application-oriented approach based on minimal data sets and user's criteria to select the most appropriate type of distribution law out of a finite set of models. The proposed algorithm for selecting the distribution law of random variables is based on a natural factorization of the space of empirical characteristics – the average and MSD. Differential entropy; graphs of densities, distribution functions, and failure rates; scatter plots and a table of metrics are applied to visually compare the distribution laws. To rank the distribution laws, the following simple and intuitive criteria, focused on solving practical problems, are proposed for a user (engineer, economist, biologist, etc.): maximization of the differential entropy, fitting empirical quantiles, fitting empirical moments of higher orders, availability of analytically defined restoration function, behaviour of the intensity function, and the possibility of decomposition into exponential phases. Online decision support system is being developed in Python using the Dash framework. The following Python libraries are used for calculations: math, numpy, random2, scipy, sympy, and plotly module for graphical visualization. The decision support system can be applied in combination with other existing methods for selecting the distribution law of a random variable.

Keywords: Decision support system · Distribution law · Multi-criteria selection · Method of moments

Supported by Ministry of Science and Higher Education of the Russian Federation, R&D registration number AAAA-A19-119030190053-2.

D. Balandin et al. (Eds.): MMST 2020, CCIS 1413, pp. 19–32, 2021.
https://doi.org/10.1007/978-3-030-78759-2_2

1   Introduction

Modeling of different systems requires description of the random factors that affect their behavior [1,2]. A number of publications relating to various branches of science and technology, for instance [3–8], confirm the necessity of selecting theoretical distribution laws for random variables (RV) on the basis of experimental data. The analytical model of a random factor is important for describing various types of systems and processes, since it allows us to determine, predict and optimize important characteristics. Besides, getting analytical models we process experimental data (including smoothing and adjusting). It helps to preserve significant fundamental features of the phenomenon observed. In addition, the distribution laws are the input data for the simulation models. Currently, there are many methods of selecting RV distribution laws. Some reviews on the approaches can be found in [9,10]. The main disadvantages of the methods are:

1. Ambiguity and subjectivity of the result for several reasons:
   – Several distributions may be consistent with the null hypothesis, so that some additional research is necessary to select one distribution law only.
   – The result of distribution law selection depends on the type of goodness-of-fit test.
   – Sensitivity of the result to the selected length of partial intervals.
2. The need for sufficiently big data samples or some a priori information on the distribution shape.
3. Difficulties for engineers, economists, biologists, etc. in adequate selection and assessment of the effectiveness of statistical data processing methods.

In this paper, we consider the problem of making online DSS (decision support system) for selecting a two-parameter distribution law for a continuous, positive random variable with a finite second moment. Such random variables are very important to describe distances, time intervals, deviations, reliability characteristics, economic indicators, mechanical properties etc. The main point of the study is the approach focused on practical application and user's criteria. Hopefully, it will help to select an appropriate distribution law by using less information. There are no formal restrictions and requirements for the dataset size. The point is the bigger dataset is, the more precise calculations of the moments the user takes into account are. The DSS is to be based on simple and intuitive for the user quantitative and qualitative criteria. The degree of criteria significance can be adjusted by the user. And one can take into account a compromise between the accuracy of the model and its computability. The DSS can be applied to a set of distributions previously selected by other methods, or can precede them.

2   DSS Scenarios and Algorithms for Selecting RV Distribution Laws

The DSS is being developed in Python using the Dash framework, which is a bundle of Flask, React.Js, HTML and CSS. The following Python libraries were used for calculations: math, numpy, random2, scipy, sympy. And plotly module was used for graphical visualization. The basic algorithm for comparing and selecting the distribution laws of positive continuous random variables is shown in Fig. 1. The scenario of the DSS operation is as follows.

Actor: user.

Objective: to compare and rank the distribution laws on the basis of user's statistical data.

Prerequisites: The user has opened the DSS web page.

The main sequence:

1. The user uploads a file with statistical data.
2. The DSS calculates the average and mean standard deviation (MSD) with their confidence intervals; a list of distributions, their parameters, and the values of differential entropies; graphs of distribution functions, distribution densities, distribution intensities, and restoration functions; scattering diagrams with the option of selecting the distribution law in a drop-down list on each axis; a table of distribution metrics; a form for selecting groups of criteria importance and filling in additional parameters.
3. The user selects a pair of distribution laws on the axes of a scatter plot.
4. The system automatically outputs a scatter plot for the selected distributions.
5. The user selects a significance group in the drop-down list for each of the 6 criteria and inputs necessary additional data.
6. The DSS provides ranking of RV distribution laws and a reference on its genesis.

Alternative sequence:

If the input data correspond to exponential distribution, the DSS outputs the average and variation with their confidence intervals; a list of distributions, their parameters, and the values of differential entropies; graphs of distribution functions, distribution densities, distribution intensities, and restoration functions; a scattering diagram; a recommendation to use exponential distribution; no multi-criteria selection form is needed in this case.

In Fig. 1 Output 1 is:

- exponential distribution;
- a parameter of the distribution;
- the value of differential entropy;
- graphs of the distribution function, distribution density, distribution intensity, and restoration function;
- recommendation to use the exponential distribution.

Output 2:

- list of distributions;
- parameters of distributions;
- the values of differential entropy;
- graphs of distribution functions, distribution density, distribution intensity, and restoration function;
- table of metrics.

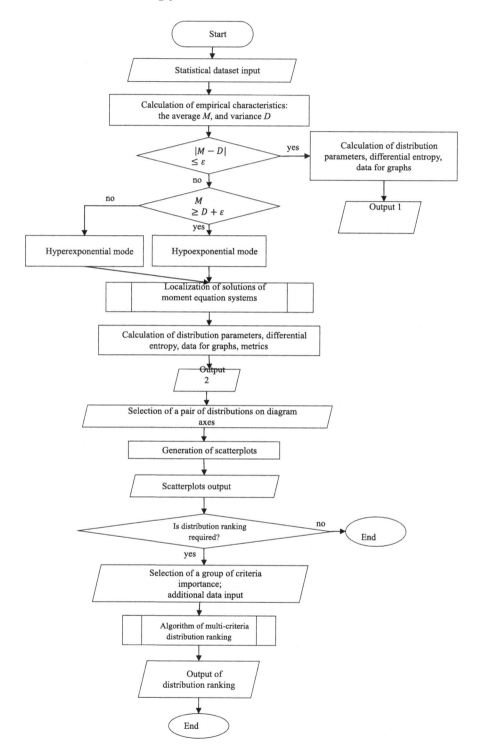

Fig. 1. The algorithm for comparing and ranking the distribution laws

The algorithm for localization, calculation, and selection of distribution parameters is given in Fig. 2. The algorithm for multi-criteria selection of the distribution law is described below.

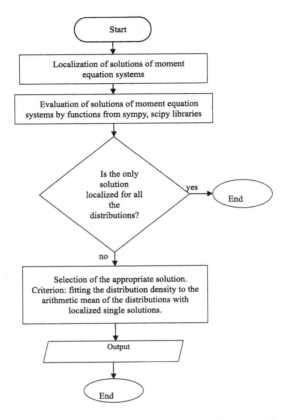

Fig. 2. The algorithm for localization, calculation, and selection of distribution parameters

3   The DSS Basic Elements

The input data block is simple and intuitive for the user. It enables a user to upload a file containing statistical data in .txt or .xlsx formats. The output block consists of several elements.

1. The output form including names of appropriate distribution laws, as well as the view of distribution densities, parameter values, and differential entropies. The set of RV distributions is given in accordance with the natural factorization of the space of empirical characteristics – the average $M > 0$ and MSD $D > 0$. According to this factorization, three main sets of distribution laws are proposed.

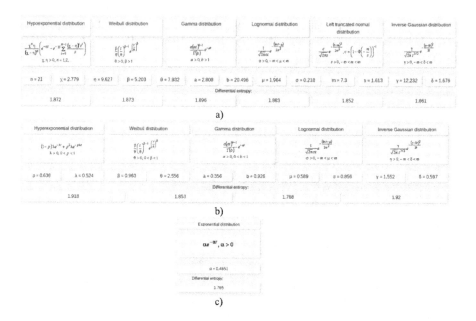

Fig. 3. Output forms with RV distributions and characteristics: densities, parameters, entropies a) hypoexponential mode at $M = 7,3; D^2 = 2,6$; b) hyperexponential mode at $M = 1,2; D = 2,1$; c) exponential mode at $M = 2,15; D^2 = 2,1$.

(a) Exponential mode. The case when the average and MSD are close $|M - D| < \epsilon$ is modelled by an exponential distribution. By default, $\epsilon = 0,1$.

(b) Hyperexponential mode. In the case when the MSD is greater than the average, i.e. $D > M + \epsilon$, the following distributions are proposed as models of a random variable:
   - hyperexponential distribution of a special type;
   - Weibull distribution;
   - gamma distribution with density;
   - lognormal distribution with density;
   - inverse Gaussian distribution.

These distributions are described in more detail in [11]. The two-phase hyperexponential distribution of a special type was first proposed by one of the co-authors of this article in [12] as a method for approximating two-parameter distributions of positive random variables. Its properties were studied in [13,14]. The advantages of this distribution are:
   - requirement of two parameters only (in contrast to the general hyperexponential distribution well studied in [15,16], which requires at least three parameters to be applied);
   - existence and uniqueness of the solution of the system of moment equations for hyperexponential mode [12];

- the meaning of the RV distributed according to this law is the sojourn time of the system in two parallel connected states, each being distributed according to the exponential law;
- explicit analytical form of the restoration function [11].

(c) Hypoexponential mode, M is not divisible by D. The case when the average is greater than the MSD, i.e. $M > D + \epsilon$, the following distributions are offered as models of a random variable:
  - hypoexponential distribution of nth order;
  - Erlang distribution;
  - Weibull distribution;
  - gamma distribution;
  - lognormal distribution;
  - left truncated normal distribution;
  - inverse Gaussian distribution.

The distributions are described in more detail in [17–19]. Hypoexponential distribution of nth order with two types of phases is convenient for approximating two-parameter distributions of positive random variables. The meaning of the RV is the sojourn time in a series-coupled state, each of them is distributed according to the exponential law. The distribution has three parameters determined by two empirical characteristics.

(d) Hypoexponential mode, M is divisible by D. The hypo-exponential distribution is applicable if M is not divisible by D. Otherwise, Erlang distribution of nth order with one type of phases is used. In this case, the rest of distributions are kept, except for the gamma distribution, which degenerates into a special case of a discrete parameter and coincides with the Erlang distribution.

The parameter values for each distribution are determined by the numerical solution of a system of two moment equations using the functions from sympy and scipy libraries. To effectively search for solutions of systems of equations, the initial parameter values are selected adaptively using the equation root localization procedure.

The differential entropy [20] of RV distributions can take any real values. For exponential and lognormal distributions, the entropy is calculated using analytical formulas. For other distributions, numerical calculations are applied.

Examples of output forms with RV distributions, parameter and characteristics for hypo-, hyper- and exponential modes are shown in Fig. 3.

2. Graphs of densities, distribution functions, and failure rates. The user can display some of the curves by choice, change the scale, select an area, and save the image in a graphical format. Examples of graphs for the hyperexponential mode are given in Fig. 4.

3. Scattering diagram is another tool for comparing RV distributions presented in the system (Fig. 5). It allows us to statistically visualize distributions based on random number generators from the random2 library and random sequences from scipy.stats. The following functions were used to generate distributions:

- exponential distribution → random2.expovariate();
- hyperexponential distribution → random2.expovariate()+ random2.expovariate();
- Weibull distribution → random2.weibullvariate();
- gamma distribution → random2.gammavariate();
- lognormal distribution → random2.lognormvariate();
- left truncated normal distribution → random2.normalvariate();
- inverse Gaussian → scipy.stats.invgauss.rvs();
- Erlang distribution → random2.gammavariate().

The left truncated normal distribution can be generated using the algorithm described in [21]. It consists of only positive random numbers obtained by the generator of normal distribution in the sequence. The user can select a pair of distributions to compare, or the same distribution on both axes, as well as the number of points on the chart. The main visual comparison criterion is the symmetry of the scattering diagram over the range of RV values and the form of their concentration. For more information about scattering diagrams and their application to compare distributions of positive RV, see [22,23].

4. An example of table of metrics for pairs of distributions is given in Fig. 6. The metrics presented in [24] are used. A uniform metrics defines the maximum deviation of two distribution functions, while the average metric is an integrally averaged indicator of deviations of distribution functions over the entire positive half-axis. For more information about these metrics application, see [25].

5. Block of multi-criteria ranking of distribution laws.

All of the above elements of the online DSS are tools for comparing the distribution laws, allowing to evaluate the relative differences of distributions only. To rank the distribution laws, we need criteria of "better" or "worse" distributions. The following practical criteria are offered for the user to choose:

1. Maximization of differential entropy
   The principle of maximum differential entropy [26] or entropy coefficient (the ratio of differential entropy to the variation coefficient) is often applied to select the distribution laws. In this case, the model allowing the highest degree of uncertainty for the value of a random variable is selected. We propose to rank the distribution laws by the value of the differential entropy which can take any real values.

2. Fitting higher-order moments
   The first two moments are the basis for determining an appropriate set distribution laws and their parameters. But the user can optionally specify the moments of higher orders. It is relevant if the dataset is big enough to account for higher moments and/or in case such distribution properties as asymmetry (the third moment is responsible for it) and peakedness (the fourth moment) are significant. Then the better the distribution theoretical moments fit (in terms of the sum of squared deviations) corresponding empirical moments, the higher it is ranked.

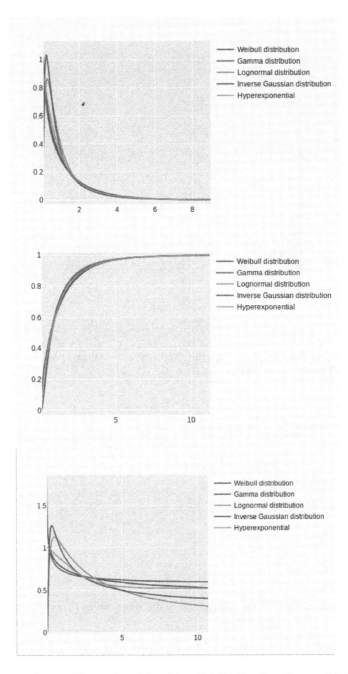

Fig. 4. Output forms with graphs of densities, distribution functions and failure rates: hyperexponential mode at $M = 1, 2; D = 2, 1.$

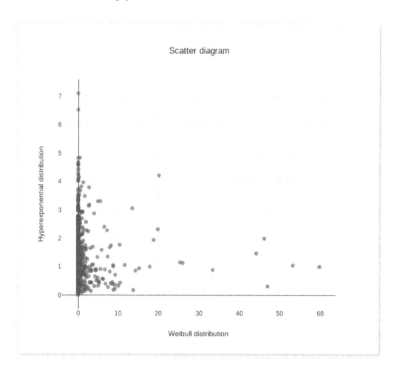

Fig. 5. Scattering diagram for comparing hyperexponential and Weibull distribution at $M = 1, 3; D = 10, 4; n = 1000$

Distributions	Hyperexponential distribution	Weibull distribution	Gamma distribution	Lognormal distribution	Inverse Gaussian distribution
Hyperexponential distribution					
Weibull distribution	p=0.547 ℓ=0.67				
Gamma distribution	p=0.861 ℓ=0.765	p=0.405 ℓ=0.275			
Lognormal distribution	p=0.283 ℓ=0.525	p=0.385 ℓ=0.228	p=0.789 ℓ=0.489		
Inverse Gaussian distribution	p=0.525 ℓ=0.624	p=0.315 ℓ=0.13	p=0.72 ℓ=0.191	p=0.258 ℓ=0.303	

Fig. 6. The table of metrics for comparison of distributions in hyperexponential mode at $M = 1, 3; D = 10, 4; n = 1000$

3. Fitting the given quantiles

The user can use this criterion to fit some quantiles. This criterion is relevant in case of special importance of a certain range of RV values (for example, small or large ones). The ranking is carried out in the same way: the better the distribution fits (in terms of the sum of squared deviations) empirical quantiles, the higher it is ranked.

4. Availability of the restoration function in analytical form

The restoration function is one of the primary concepts in the restoration theory [19]. It is used to calculate important system reliability characteristics. For many well-known distribution laws, the restoration function cannot be obtained in analytical from. Then we need to approximate the restoration function by table values or estimating it by definition. It makes difficulties for further analytical operations. Elementary restoration functions in analytical form are defined for the exponential, hyperexponential, and Erlang distribution. For the latter, the form of the restoration function becomes much more complicated with the growth of the distribution order.

5. The behavior of the failure rate function: monotonic or not.

If we are aware of the failure rate behavior, it is possible to include this criterion and select monotonic or non-monotonic behavior of the failure rate function. It is constant for exponential distribution only. The failure rates of the lognormal and inverse Gaussian distribution are not monotonic functions with break-in [11]. A fairly short period of increasing failure rate and then break-in is also accompanied by a steep decline in the number of failures (the system quickly enters a period of normal operation).

6. The possibility of decomposition into exponential phases.

If the user selects this criterion, the rank of the hyperexponential and hypoexponential distributions will increase.

In the DSS, multi-criteria problem of selecting the distribution is solved by means of the approach developed by S. A. Piyavsky [26]. The significance of these six criteria is determined by the user by assigning them to one of the four groups of significance: very important, important, slightly important, ignore. Quantification of these fuzzy assessments into criteria weight coefficients can be done by the method in [27]. Then the multi-criteria problem is to be solved by reducing it to a linear convolution with the criteria weight coefficients obtained. Herewith, each distribution is assigned the number of points (from 1 to N, where N is the size of distribution set according to the natural factorization of the space of empirical characteristics) due to the rules from Table 1 (column "Distribution ranking method"). Additional data to be specified by user are described in the last column of Table 1 gives a summary of all heading levels.

Table 1. Method of multi-criteria selection of RV distribution law.

	Criterion	Distribution ranking method	Additional data to be specified by user
1	Maximization of differential entropy	From 1 to N points in ascending order of the differential entropy value	
2	Fitting empirical higher-order moments	From 1 to N points in descending order of difference between the empirical and theoretical values of higher-order moments	The number of higher-order moments
3	Fitting empirical quantiles	From 1 to N points in descending order of difference between the empirical and theoretical values of quantiles	The number of quantiles
4.	Availability of the restoration function in analytical form	1 point – no restoration function in analytical form; N/2 points – complex form of restoration function; N points – restoration function in analytical form.	
5.	The behavior of the failure rate function	In case of selecting the monotonic intensity function: 1 point - distributions with non-monotonic intensity rate; N points - distributions with a monotonic intensity function. In case of selecting a non-monotonic intensity function with break-in: vice versa. In case of selecting a different behavior of the intensity function: N/2 points – for all distributions.	
6.	Possibility of decomposition into exponential phases	N points - if it is possible to decompose into exponential phases; 1 point – if it is impossible to decompose into exponential phases.	

4   Conclusions

The above algorithms and scenarios have been implemented in the online DSS for choosing the distribution law of positive random variables. It can become an effective and convenient tool for users (engineers, economists, biologists, etc.) to model random variables, their distributions and characteristics. Advantages of the DSS are intuitive and interactive interface; online accessibility without installing desktop software; ability to save graphic elements as files; light requirements for input dataset; possibility of visual adaptive comparison of the distribution laws of random variables according to different user's criteria. To compare the distribution laws the user can apply: graphs of densities, distribution functions, intensities, and restoration functions; metric tables; and scatter plots. The distributions ranking is based on quantitative and qualitative criteria. The degree of criteria importance can be adjusted by the user with regard to investigation purposes and a compromise between the accuracy of the model and its computability. The prospects of research are to test the system on experimental datasets, to compare it with other methods of distribution laws selection, to add the option of getting advice on how many higher moments one should take into account with regard to the given input dataset.

References

1. Bol'shakov, V.I., Dubov, Y.I.: Accounting for the influence of random factors on the adequacy of the model of a complex system. Visnyk PDABA **5**(158), 8–15 (2011). (in Russian)
2. Ventsel, E.S., Ovcharov, L.A.: Probability Theory and Its Engineering Applications, 3rd edn. Academia, Moscow (2003).(in Russian)
3. Baranova, A.A., Selyaninov, A.A., Vikhareva, E.V.: Reliability criterion for selection of the distribution law when modeling biological destruction processes. Perm Nat. Res. Polytech. Univ. Edn. **1**, 512–514 (2015). (in Russian)
4. Ponomarev, V.P., Beloglazov, I.Y.: On the choice of the distribution law of repair time of oil wells. Probl. Econ. Manage. **5**(45), 140–143 (2015). (in Russian)
5. Duplyakin, V.M., Knyazheva, Y.V.: The choice of the distribution law of the input of the queuing system of a commercial enterprise. Bull. SSAU **6**(37), 102–109 (2012). (in Russian)
6. Pushkareva, L.A., Pushkareva, T.A.: Selection and verification of the analytical distribution law of wind speed for designing a wind turbine in the Udmurt Republic. In: International Scientific and Practical Conference "Theoretical and Practical Aspects of the Development of Scientific Thought in the Modern World", Samara, 8 October 2017, pp. 67–71. LLC "Aeterna", Ufa (2017). (in Russian)
7. Panov, K.V.: The choice of the distribution law of applications input and idle time of locomotives in service in depot. In: Innovative Projects and Technologies In Education, Industry and Transport: Proceedings of the Scientific Conference, Omsk, 8 February 2019, pp. 227–233. Publishing House of the Omsk State University of Railway Transport, Omsk (2019). (in Russian)
8. Linets, G.I., Nikulin, V.I., Melnikov, S.V.: Selection of the identification criterion for the distribution law of random variables of transionospheric communication channels. XLIV Academic readings on cosmonautics, dedicated to the memory of Academician S.P. Korolev and other outstanding Russian scientists-pioneers of space exploration: Collection of theses, Moscow, 28–31 January 2020, pp. 254–256. Publishing House of the Bauman Moscow State Technical University, Moscow (2020). (in Russian)
9. Akimov, S.S.: The problem of selecting a method for obtaining the law of probability distribution. Int. Sci. Res. J. **1–3**(20), 5–8 (2014). (in Russian)

10. Tyrsin, A.N.: Method of selecting the best distribution law of a continuous random variable based on inverse mapping. Bull. SUSU Ser. Math. Mech. Phys. **1**, 31–38 (2017). (in Russian)
11. Bayhelt, F., Franken, P.: Reliability and maintenance. Mathematical approach. Radio and communication, Moscow (1988). (in Russian)
12. Smirnov, S.V.: Modeling of "super irregular" random variables from experimental data. Automation of scientific research: collection of scientific works. KuAI, Kuibyshev, pp. 52–57 (1988). (in Russian)
13. Kovalenko, A.I., Smirnov, S.V.: Comparison of hyperexponential distribution with other models of positive random variables. Infocommunication technologies, Samara **17**(1), 9–16 (2019). (in Russian)
14. Ivleva, A., Smirnov, S.: Comparison of models of positively defined random variables. In: XXI International Conference Complex Systems: Control and Modeling Problems (CSCMP), Samara, pp. 449–454 (2019). (in Russian)
15. Bladt, M., Nielsen, B.F.: Matrix-Exponential Distributions in Applied Probability (Probability Theory and Stochastic Modelling), vol. 81. Springer (2017). https://doi.org/10.1007/978-1-4939-7049-0
16. Ryzhikov, Y.I., Ulanov, A.V.: Application of hyper-exponential approximation in problems of calculation of non-Markov queuing systems. Vestnik Tomsk State Univ. Manage. Comput. Eng. Inform. **3**, 60–65 (2016). (in Russian)
17. Vaynshteyn, I.I., Fedotova, I.M., Vaynshteyn, Y.V., Tsibul'sky, G.M.: Renewal function and optimization of strategies of operation of technical systems if operating time is a mixture of distributions. Siberian J. Sci. Technol. **1**, 15–24 (2017). (in Russian)
18. Kalashnikov, V.V., Rachev, S.T.: Mathematical Methods of Construction of Stochastic Service Models. Nauka, Moscow (1988).(in Russian)
19. Tarakanov, K.V., Ovcharov, L.A., Tyryshkin, A.N.: Analytical methods of systems research. Soviet Radio, Moscow (1974).(in Russian)
20. Guzairov, M.B., Gvozdev, V.E., Ilyasov, B.G., Kolodenkova, A.E. Statistical research of territorial systems. Mechanical Engineering, Moscow (2008). (in Russian)
21. Snetkov, N.N.: Simulation of economic processes: educational and practical guide. Press Center EAOI, Moscow (2008). (in Russian)
22. Khimenko, V.I.: Scatterplots in the analysis of random flows of events. Inf. Control Syst. **83**(4), 85–93 (2016). (in Russian)
23. Khimenko, V.I.: Random data: structure and analysis. TECHNOSPHERE, Moscow (2018).(in Russian)
24. Kalashnikov, V.V.: Quantitative estimates in the theory of reliability. Znaniye, Moscow (1989).(in Russian)
25. Kovalenko, A.I., Smirnov, S.V. Analysis of approximation properties of the hyperexponential distribution of a special type. In: Proceedings of the XX International Conference Problems of Control and Modeling in Complex Systems, 3–6 September 2018, pp. 138–145. LLC Ofort, Samara (2018). (in Russian)
26. Piyavsky, S.A.: Method of universal coefficients for the multi-criterial decision making. Ontol. Design. **8**(3), 449–468 (2018). https://doi.org/10.18287/2223-9537-2018-8-3-449-468. (in Russian)
27. Piyavsky, S.A.: How do we digitize the concept of "more important". Ontol. Design. **6**(4), 414–435 (2016). https://doi.org/10.18287/2223-9537-2016-6-4-414-435. (in Russian)

Application of the Entropic Tilt Limiter to Solve the Gas Dynamics Equations Using the Implicit Scheme of the Discontinuous Galerkin Method

Victor F. Masyagin$^{(\boxtimes)}$ (iD)

National Research Mordovia State University, Saransk, Russia

Abstract. The paper proposes an implicit scheme of the discontinuous Galerkin method for solving two-dimensional equations of gas dynamics on unstructured grids. The implicit scheme is based on the representation of a system of grid equations in the "delta form". To solve the SLAE obtained during the approximation of the initial equations, solvers from the NVIDIA AmgX library are used. To ensure the monotonicity of the solution, a tilt limiter was proposed, which ensures the fulfillment of a discrete analogue of the entropy inequality. The numerical algorithm was successfully verified using three model problems.

Keywords: Gas dynamics equations · Implicit scheme · Discontinuous Galerkin method · Entropic inequality · Tilt limiter · AMGX library

Introduction

Today the discontinuous Galerkin method is actively used to solve gas dynamics problems. This method has a high order of accuracy of the obtained solution, adapts well to unstructured grids, and at the same time has a compact computational template. With all the advantages listed above, the discontinuous Galerkin method requires significant computational costs, which, when using explicit schemes, leads to significant computational time costs. One of the promising areas of research today is the development of effective implicit schemes for the discontinuous Galerkin method on unstructured grids. However, this approach, despite the removal of significant restrictions on the time step, requires significant resources to work with SLAEs of huge dimensions, therefore, the question arises of the most efficient use of all the capabilities of computing technology [1, 2].

Recently, entropy stable methods and algorithms for solving gas dynamics problems have been actively developed. This trend is associated with the desire to improve the quality of numerical solutions, including in numerical algorithms, in addition to the traditionally taken into account conservation laws, the second

The research was supported by RSF (project No. 19-71-00131).

D. Balandin et al. (Eds.): MMST 2020, CCIS 1413, pp. 33–48, 2021.
https://doi.org/10.1007/978-3-030-78759-2_3

law of thermodynamics, which is quantitatively expressed by entropy inequality. The numerical solution algorithm with this approach is reduced to solving one system of linear equations at each time step. In this work, the NVIDIA AmgX library, written in the CUDA C language, is used to solve the SLAE. The advantages of the library include support for parallelism both at the level of several graphics processors and at the level of several computational clusters, which is provided through the support of MPI technology. Also, the AmgX library provides a flexible configuration system, and thanks to this, it becomes possible to create a hierarchy of decision algorithms with an arbitrary depth, in which the external decision the algorithm will use internal ones as preprocessors and preconditioners, which themselves can be processed by other methods. This approach allows the user to quickly experiment with different schemes [3].

At the moment, the library finds more and more widespread use in modern industrial and scientific numerical analysis. In particular, AmgX is part of the commercial computing software ANSYS Fluent [4]. An indicator of the relevance and efficiency of the library used is the fact that at the moment it is used as a standard for comparing the efficiency and speed of new numerical algorithms for solving SLAEs, along with such powerful tools as the HYPRE library [4].

Earlier, the authors in the works [5,6] developed a numerical technique for the implicit scheme of the discontinuous Galerkin method as applied to the solution of inviscid stationary problems of gas dynamics. To solve the SLAE, the HIPRE library was used. This work is devoted to the development of an implicit scheme of the discontinuous Galerkin method for solving two-dimensional gas dynamics equations on unstructured grids. A special entropic tilt limiter is proposed, which ensures the fulfillment of a discrete analog of the entropic inequality. Numerical calculations of some model problems are carried out, allowing to evaluate the effectiveness of the method. A comparison is made of the numerical results obtained using the entropic tilt limiter and using the Barth-Jespersen and Cockburn tilt limiters. The results are presented in the work.

1   Mathematical Model and Computational Algorithm

Consider a two-dimensional system of Navier-Stokes equations written in a conservative form:

$$\frac{\partial \rho}{\partial t} + \frac{\partial (\rho u)}{\partial x} + \frac{\partial (\rho v)}{\partial y} = 0, \tag{1}$$

$$\frac{(\partial \rho u)}{\partial t} + \frac{\partial (\rho u^2 + p)}{\partial x} + \frac{\partial (\rho u v)}{\partial y} = 0, \tag{2}$$

$$\frac{\partial (\rho v)}{\partial t} + \frac{\partial (\rho u v)}{\partial x} + \frac{\partial (\rho v^2 + p)}{\partial y} = 0, \tag{3}$$

$$\frac{\partial (\rho E)}{\partial t} + \frac{\partial ((\rho E + p)u)}{\partial x} + \frac{\partial ((\rho E + p)v)}{\partial y} = 0, \tag{4}$$

where ρ is the density, $\mathbf{v} = (u, v)$ is the velocity vector, p is the pressure, $E = e + \frac{u^2 + v^2}{2}$ is the specific total energy, e is the internal energy.

The system of equations is closed by the equation of state $p = \rho e \, (\gamma - 1)$, where γ is the adiabatic exponent. These equations must be supplemented with initial and boundary conditions, the form of which depends on the specific problem, and will be specified further.

Introduce the notation

$$
\mathbf{U} = \begin{pmatrix} U_1 \\ U_2 \\ U_3 \\ U_4 \end{pmatrix} = \begin{pmatrix} \rho \\ \rho u \\ \rho v \\ \rho E \end{pmatrix}, \mathbf{F}^{(1)}(\mathbf{U}) = \begin{pmatrix} F_1^{(1)} \\ F_2^{(1)} \\ F_3^{(1)} \\ F_4^{(1)} \end{pmatrix} = \begin{pmatrix} \rho u \\ \rho u^2 + p \\ \rho u v \\ (\rho E + p) u \end{pmatrix},
$$

$$
\mathbf{F}^{(2)}(\mathbf{U}) = \begin{pmatrix} F_1^{(2)} \\ F_2^{(2)} \\ F_3^{(2)} \\ F_4^{(2)} \end{pmatrix} = \begin{pmatrix} \rho v \\ \rho u v \\ \rho v^2 + p \\ (\rho E + p) v \end{pmatrix}.
$$

Taking into account the introduced notations, the system (1)–(4) will be rewritten as

$$
\frac{\partial \mathbf{U}}{\partial t} + \nabla \cdot \mathbf{F}(\mathbf{U}) = 0. \tag{5}
$$

To approximate the equations, we cover the region Ω, on which the solution is sought, with an unstructured triangular mesh $K_h : \Omega = \cup K_i$, $i = 0, ..., ..., N_h$. All triangles K_i have nonzero area and intersect at most along the edges or vertices that form them. Each inner edge of one cell is the entire edge of another cell.

As basis functions on each element of K_i, we choose all possible polynomials of the form

$$
\varphi_{il} = \left(\frac{x - x_{ci}}{\Delta x_i} \right)^{\alpha_{K_i l}} \cdot \left(\frac{y - y_{ci}}{\Delta y_i} \right)^{\beta_{K_i l}}, \quad l = 0, \ldots, N,
$$

such that the sum of the powers $\alpha_{K_i l} + \beta_{K_i l}$ does not exceed some given number p. Here x_{ci}, y_{ci} are the coordinates of the center of mass of the cell, and $\Delta x_i, \Delta y_i$ are the characteristic dimensions of the K_i cell.

We obtain a discrete analogue of the (5) system, assuming that inside each mesh element K_i the approximate solution $\mathbf{U}_{ih}$ is represented as $\mathbf{U}_{ih}(t, x, y) = \sum_{k=0}^{N} \mathbf{U}_{ik}(t) \varphi_{ik}(x, y)$.

Let us multiply the equations of the discrete analog of system (5) by the test functions taken from the space of basis functions, and integrate over each element of the grid [7,8]. As a result, we get the system

$$
\int_{K_i} \sum_{k=0}^{N} \frac{d\mathbf{U}_{ik}}{dt} \varphi_{ik} \varphi_{il} dS + \oint_{\partial K_i} \left(\mathbf{n} \cdot \hat{\mathbf{F}}^{\sigma} \right) \varphi_{il} d\sigma - \int_{K_i} \mathbf{F}(\mathbf{U}_{ih}) \cdot \nabla \varphi_{il} dS = 0, l = 0, \ldots, N,
$$

where $\mathbf{n}$ – normal to the boundary ∂K_i, $\hat{\mathbf{F}}_n^{\sigma} = \hat{\mathbf{F}}^{\sigma} \cdot \mathbf{n}$ – flux function to be defined later.

We replace the derivative $\frac{dU_{ik}}{dt}$ with a discrete analogue and, taking into account the time step t, we rewrite the system in the form

$$\int_{K_i} \sum_{k=0}^{N} \frac{\mathbf{U}_{ik}^{m+1} - \mathbf{U}_{ik}^{m}}{\Delta t} \varphi_{ik} \varphi_{il} dS + \oint_{\partial K_i} (\hat{\mathbf{F}}_n^{\sigma})^{m+1} \varphi_{il} d\sigma$$

$$- \int_{K_i} \left(\mathbf{F}^{(1)}(\mathbf{U}_{ih}^{m+1}) \frac{\partial \varphi_{il}}{\partial x} + \mathbf{F}^{(2)}(\mathbf{U}_{ih}^{m+1}) \frac{\partial \varphi_{il}}{\partial y} \right) dS = 0, \qquad (6)$$

where the superscripts indicate the time step at which the value of the corresponding field is taken.

Next, consider finding the elements of the resulting matrix in the system (6).

$$\mathbf{F}^{(\alpha)}(\mathbf{U}_{ih}^{m+1}) = \mathbf{F}^{(\alpha)}(\mathbf{U}_{ih}^{m}) + \left(\frac{\partial \mathbf{F}^{(\alpha)}}{\partial \mathbf{U}} \right)^m (\mathbf{U}_{ih}^{m+1} - \mathbf{U}_{ih}^{m}), \ \alpha = \overline{1,2}.$$

Let's introduce the notation:

$$\mathbf{A} = \left(\frac{\partial \mathbf{F}_n}{\partial \mathbf{U}} \right)^m \Big|_{\mathbf{U} = \mathbf{U}_{avg}^{\sigma}}, \ \mathbf{A} = \mathbf{L}\mathbf{\Lambda}\mathbf{R},$$

$$\mathbf{\Lambda} = \mathbf{\Lambda}^- + \mathbf{\Lambda}^+, \ \mathbf{\Lambda}^- = \tfrac{1}{2}(\mathbf{\Lambda} - |\mathbf{\Lambda}|), \ \mathbf{\Lambda}^+ = \tfrac{1}{2}(\mathbf{\Lambda} + |\mathbf{\Lambda}|), \ \mathbf{A} = \mathbf{A}^- + \mathbf{A}^+,$$

$$\mathbf{A}^- = \mathbf{L}\mathbf{\Lambda}^-\mathbf{R}, \ \mathbf{A}^+ = \mathbf{L}\mathbf{\Lambda}^+\mathbf{R}, \ \mathbf{A}^{(1)} = \mathbf{A}|_{\mathbf{n}=(1,0)}, \ \mathbf{A}^{(2)} = \mathbf{A}|_{\mathbf{n}=(0,1)},$$

where $\mathbf{R}, \mathbf{L}$ are matrices composed of the right and left eigenvectors of the matrix $\mathbf{A}$, $\mathbf{\Lambda}$ – is a diagonal matrix composed of the eigenvalues of the matrix $\mathbf{A}$, $\mathbf{U}_{avg}^{\sigma}$ computed using Godunov's flux function [9].

The flux values, taking into account the previously introduced designations, are in the form

$$(\hat{\mathbf{F}}_n^{\sigma})^{m+1} = (\hat{\mathbf{F}}_n^{\sigma})^m + \mathbf{A}^+(\mathbf{U}_{ih}^{m+1} - \mathbf{U}_{ih}^{m}) + \mathbf{A}^-(\mathbf{U}_{jh}^{m+1} - \mathbf{U}_{jh}^{m}).$$

To find the value of $(\hat{\mathbf{F}}_n^{\sigma})^m$, the Godunov flux function is used. The normal $\mathbf{n}$ to the edge σ is directed from the cell with index i to the cell with index j.

Let's denote by Γ_{ij} the border between cells i and j and by $\Delta \mathbf{U}_{ih}^{m+1}$ the increment per step in time from the solution $\mathbf{U}_{ih}^{m+1}$, i.e. $\Delta \mathbf{U}_{ih}^{m+1} = \mathbf{U}_{ih}^{m+1} - \mathbf{U}_{ih}^{m}$.

We will look for the increments of the required functions in the same space of basis functions as the functions themselves: $\Delta \mathbf{U}_{ih}^{m+1} = \sum_{k=0}^{N} \Delta \mathbf{U}_{ik}^{m+1} \varphi_{ik}$.

Finally, let's rewrite the system in delta form [10]:

$$\sum_{k=0}^{N} \frac{\Delta \mathbf{U}_{ik}^{m+1}}{\Delta t} \int_{K_i} \varphi_{il} \varphi_{ik} dS + \sum_j \oint_{\Gamma_{ij}} \mathbf{A}^+ \sum_{k=0}^{N} \Delta \mathbf{U}_{ik}^{m+1} \varphi_{ik} \varphi_{il} d\sigma$$

$$+ \sum_j \oint_{\Gamma_{ij}} \mathbf{A}^- \sum_{k=0}^{N} \Delta \mathbf{U}_{jk}^{m+1} \varphi_{jk} \varphi_{il} d\sigma + \int_{K_i} \left(\mathbf{A}^{(1)} \sum_{k=0}^{N} \Delta \mathbf{U}_{ik}^{m+1} \varphi_{ik} \frac{\partial \varphi_{il}}{\partial x} \right) dS$$

$$+ \int_{K_i} \left(\mathbf{A}^{(2)} \sum_{k=0}^{N} \Delta \mathbf{U}_{ik}^{m+1} \varphi_{ik} \frac{\partial \varphi_{il}}{\partial y} \right) dS$$

$$= \int_{K_i} \left(\mathbf{F}^{(1)} (\Delta \mathbf{U}_{ih}^m) \frac{\partial \varphi_{il}}{\partial x} + \mathbf{F}^{(2)} (\Delta \mathbf{U}_{ih}^m) \frac{\partial \varphi_{il}}{\partial y} \right) dS - \oint_{\partial K_i} (\hat{\mathbf{F}}_n^\sigma)^m \varphi_{il} d\sigma.$$

We solve the obtained SLAE using solvers from the NVIDIA AmgX library.

2   Entropic Tilt Limiter

According to modern concepts, a realistic model of a gas dynamic flow should, in a generalized sense, satisfy the entropy inequality [11]:

$$\frac{\partial (\rho s)}{\partial t} + \frac{\partial (\rho s u)}{\partial x} + \frac{\partial (\rho s v)}{\partial y} \geq 0, \tag{7}$$

in which the dimensionless specific entropy is defined as $s = \ln \frac{p}{p_*} - \gamma \ln \frac{\rho}{\rho_*}$, where as standard values p_* and ρ_* you can choose any constant positive quantities with dimensions of pressure and density.

Using the notation $S(\mathbf{U}) = \rho s$, $H_x(\mathbf{U}) = \rho s u$, $H_y(\mathbf{U}) = \rho s v$, entropy inequality (7) can be rewritten in differential form

$$\frac{\partial S(\mathbf{U})}{\partial t} + \frac{\partial H_x(\mathbf{U})}{\partial x} + \frac{\partial H_y(\mathbf{U})}{\partial y} = \nabla_{\mathbf{U}} S(\mathbf{U}) \cdot \frac{\partial \mathbf{U}}{\partial t} + \frac{\partial H(\mathbf{U})}{\partial x} + \frac{\partial H(\mathbf{U})}{\partial y} \geq 0, \tag{8}$$

where $\nabla_{\mathbf{U}} S(\mathbf{U}) \cdot \frac{\partial \mathbf{U}}{\partial t} = \sum_{i=1}^{4} \frac{\partial S(\mathbf{U})}{\partial U_i} \frac{\partial U_i}{\partial t}$.

For each cell K_i, the discrete analogue of the entropy inequality (8) can be written as:

$$\frac{d}{dt} \int_{K_i} S(\mathbf{U}(x,y,t)) dV + \oint_{\partial K_i} \left(H_x(\mathbf{U}_G(x,y,t)) n_x \right.$$

$$\left. + H_y(\mathbf{U}_G(x,y,t)) n_y \right) dS \geq 0, \tag{9}$$

where $\mathbf{U}_G(x,y,t)$ computed using Godunov's flux function.

Let's integrate (9) over the time interval $[t, t + \tau]$ and divide the result by τ:

$$\frac{1}{\tau} \int_{K_i} [S\left(\mathbf{U}\left(x, y, t + \tau\right)\right) - S\left(\mathbf{U}\left(x, y, t\right)\right)] dV$$

$$+ \frac{1}{\tau} \int_{t}^{t+\tau} \left[\oint_{\partial K_i} \left(H_x\left(\mathbf{U}_G\left(x, y, t'\right)\right) n_x + H_y\left(\mathbf{U}_G\left(x, y, t'\right)\right) n_y\right) dS\right] dt' \geq 0. \quad (10)$$

The resulting integral form (10) of the discrete analogue of the entropy inequality can be rewritten as $\int_{K_i} S\left(\mathbf{U}\left(x, y, z, t + \tau\right)\right) dV \geq C_s$, where

$$C_s = \int_{K_i} S\left(\mathbf{U}\left(x, y, t\right)\right) dV + \int_{t}^{t+\tau} \left[\oint_{\partial K_i} \left(H_x\left(\mathbf{U}_G\left(x, y, t'\right)\right) n_x\right.\right.$$

$$\left.\left. + H_y\left(\mathbf{U}_G\left(x, y, t'\right)\right) n_y\right) dS\right] dt'. \quad (11)$$

The second integral on the right-hand side (11) can be approximated by different ways. In the case of using the Euler scheme, we obtain the following expression for C_s:

$$C_s = \int_{K_i} S\left(\mathbf{U}\left(x, y, t\right)\right) dV + \tau \oint_{\partial K_i} \left(H_x\left(\mathbf{U}_G\left(x, y, t\right)\right) n_x + H_y\left(\mathbf{U}_G\left(x, y, t\right)\right) n_y\right) dS.$$

Consider one cell K_i and keep the notation $\mathbf{U}_{i0}$, $\mathbf{U}_{i1}$, $\mathbf{U}_{i2}$ for the values of the coefficients of the discontinuous Galerkin method corresponding to the found point of the conditional minimum. Let's calculate the value of four functions

$$\Phi_{jk}\left(\lambda\right) = p\left(\mathbf{U}_{i0} + (-1)^j \lambda \mathbf{U}_{i1}\varphi_{i1} + (-1)^k \lambda \mathbf{U}_{i2}\varphi_{i2}\right), \ j, k = \{0, 1\},$$

where $p(\mathbf{U}) = (\gamma - 1)\left(E - \frac{u^2 + v^2}{2\rho}\right)$ – pressure in conservative variables, with a value of $\lambda = 1$. If $\min\left[\Phi_1\left(1\right), \Phi_2\left(1\right), \ldots, \Phi_4\left(1\right)\right] \geq p_{min}$ ($p_{min} > 0$ is a preselected small positive number), then we set $\lambda_p = 1$. Otherwise, we take as λ_p the root of the equation closest to 1 from the left

$$\min\left[\Phi_1\left(\lambda\right), \Phi_2\left(\lambda\right), \ldots, \Phi_4\left(\lambda\right)\right] = p_{min}. \quad (12)$$

Next, we calculate the function

$$\Psi\left(\mu\right) = \int_{K_i} S\left(\mathbf{U}_{i0} + \lambda_p \mu \mathbf{U}_{i1}\varphi_{i1}(x, y) + \lambda_p \mu \mathbf{U}_{i2}\varphi_{i2}(x, y)\right) dV$$

for the value $\mu = 1$. If $\Psi\left(1\right) \geq C_s$, then we set $\mu_s = 1$. Otherwise, we take as μ_s the root of the equation closest to 1 from the left

$$\Psi\left(\mu\right) = C_s. \quad (13)$$

After determining the values of the numeric parameters λ_p and μ_s as described above by way of setting the values of the coefficients of the discontinuous Galerkin method in the cell K_i equal: $\mathbf{U}_{i0}^* = \mathbf{U}_{i0}, \quad \mathbf{U}_{i1}^* = \lambda_p\mu_s\mathbf{U}_{i1}, \quad \mathbf{U}_{i2}^* = \lambda_p\mu_s\mathbf{U}_{i2}.$

This procedure excludes the possibility of obtaining negative pressures in any cell on a new time layer and ensures the fulfillment of the discrete analogue of the entropy inequality in the conservative form [12,13]. Due to the upward convexity of the functions $\Phi_j(\lambda), j = 1,\ldots,4$ and $\Psi(\mu)$, the solution of Eqs. (12) and (13) was found using Newton's method, with an initial argument value of 1.

3   Test Results

3.1   Computing Tools and Libraries Used

To carry out computational experiments, we used a computer with an Intel Core i5-8265U processor and an NVIDIA GeForce MX250 video card. Below are the results obtained using the PBICGSTAB solver from AmgX library version 2.1.0.131-opensource. Calculations were made using the entropic limiter, the Cockburn limiter [14] and the Barth-Jespersen limiter [15].

3.2   The Flow of an Inviscid Compressible Gas in a Flat Channel with a Wedge

The calculation of the flow of an inviscid compressible gas in a flat channel with a wedge at $M = 2$ and Courant number equal to 5. The angle of the wedge in the channel is $10°$. The configurations of the system of shock waves arising from the flow around the initial wedge and multiple reflection of the initial shock from the channel walls were simulated under the conditions: $p_\infty = 10^5$ Pa, $T_\infty = 300$ K, $c_v = 724.4$ J/(kg·K) [16]. The computational grid consisted of 17096 cells. Figure 1 shows the geometry of the problem. In Fig. 2 shows the pictures of the Mach number distribution in the computational domain when using various limiters. Figure 3 demonstrates the pictures of the total pressure distribution in the computational domain when using various limiters. In Fig. 4 shows the dependence of the total pressure, referred to the total pressure of the unperturbed flow, on the x coordinate along the section $y = 850$.

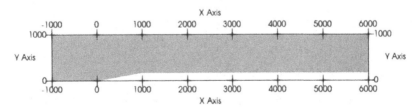

Fig. 1. Problem geometry

The results are in good agreement with the results obtained using the finite volume method on block-structured [16] and unstructured grids [17]. The results

obtained show the development of the shock-wave structure of the flow in a
channel with a wedge, namely, the initiation of a shock wave, its development,
reflection from the channel walls, and interaction with a fan of rarefaction waves.
It should be noted that the solution obtained using the entropic limiter shows
a clearer flow structure, which is in better agreement with the known numeric
solutions of this problem [16, 17]. The numerical solution with the entropy limiter
converged in 4000 time steps, with the Cockburn limiter - for 4500, with the
Barth-Jespersen limiter - for 5000. Table 1 shows the average time to complete
one time step for the task under consideration when using different limiters.

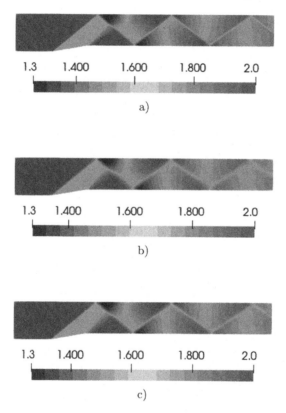

Fig. 2. Mach number distribution: a) entropic limiter; b) Cockburn limiter; c) Barth-Jespersen limiter

3.3   Prandtl-Meyer Expansion Fan

Figure 5 shows a schematic of an expansion fan flow, depicting the relative location of the expansion fan and its centering at the corner, the boundary conditions,
the two regions of uniform flow, and the wall angle.

The problem was solved in a setting taken from the work [18]. We take
$M_1 = 2.0$, $\delta = -10°$, $M_2 = 2.383$. The relationship between the properties
of the expansion fan is set as follows: $p_2/p_1 = 0.5471$, $\rho_2/\rho_1 = 0.6500$ and
$v_2/v_1 = 1.0931$.

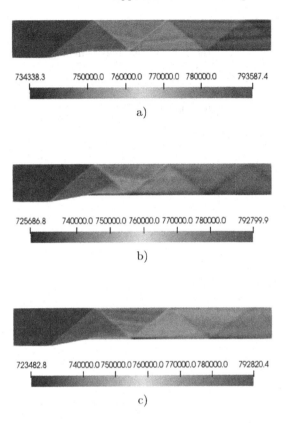

Fig. 3. Total pressure distribution: a) entropic limiter; b) Cockburn limiter; c) Barth-Jespersen limiter

An unstructured triangular mesh consisting of 8998 cells was used in the calculation.

Figure 6 shows the distribution of the Mach number in the computational domain in the case of using different limiters. Figure 7 shows the pressure distribution along the streamline. The presented figures show that when using the entropic limiter, the thickness of the flow layer, which is influenced by the numerically generated entropy, is smaller compared to other limiters. Thus, the entropic limiter gives a numerical solution that is closest to the exact solution.

3.4   Steady-State Oblique Shock Wave

Figure 8 shows a schematic of an oblique shock flow, depicting the shock-wave, the boundary conditions, the two regions of uniform flow, the wall angle, and the shock angle.

The problem was solved in a setting taken from the work [18]. We take $M_1 = 3.0$, $\delta = -15°$, $\gamma = 1.4$, $M_2 = 2.383$, $\varepsilon = 32.24$. The relationship between the properties of the expansion fan is set as follows: $p_2/p_1 = 2.822$, $\rho_2/\rho_1 = 2.0342$ and $v_2/v_1 = 0.888$.

Table 1. Average time to complete one time step.

Limiter	Average time, s
Entropic	2.99
Cockburn	2.93
Barth-Jespersen	3.74

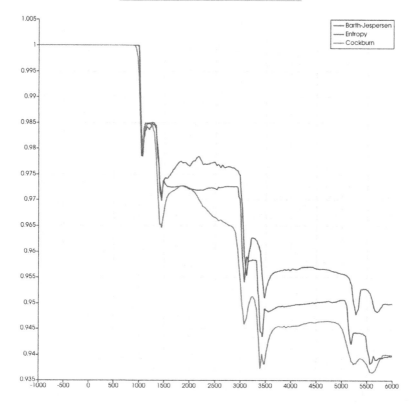

Fig. 4. Dependence of the total pressure related to the total pressure of the undisturbed flow on the x coordinate on the side wall in the section $y = 850$

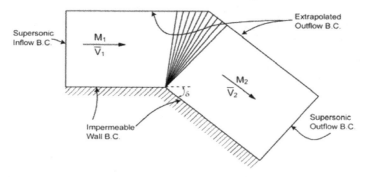

Fig. 5. Schematic of an expansion fan flow

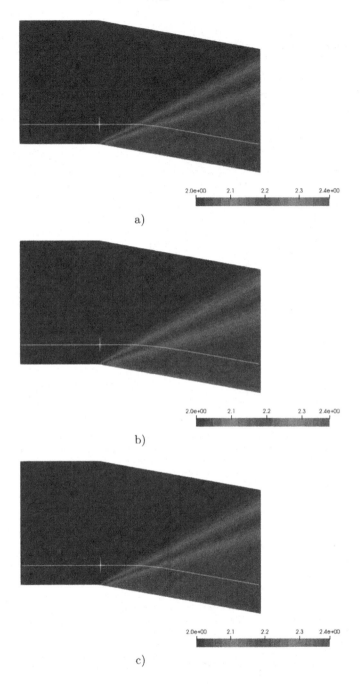

Fig. 6. Mach number distribution: a) entropic limiter; b) Cockburn limiter; c) Barth-Jespersen limiter

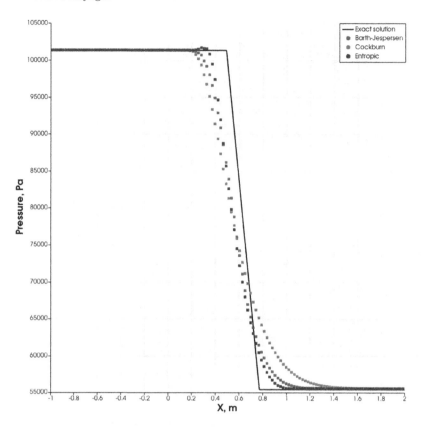

Fig. 7. Pressure distributions along the streamline: exact solution (exact), entropic limiter (entropic), Cockburn limiter (Cockburn), Barth-Jespersen limiter (Barth-Jespersen)

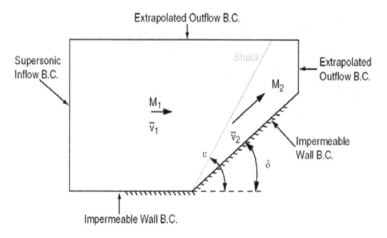

Fig. 8. Schematic of an oblique shock flow

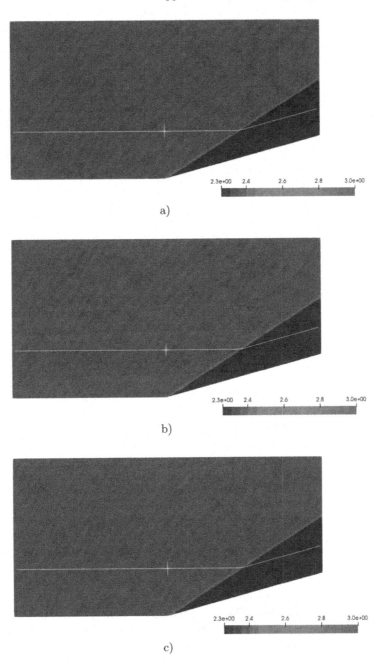

Fig. 9. Mach number distribution: a) entropic limiter; b) Cockburn limiter; c) Barth-Jespersen limiter

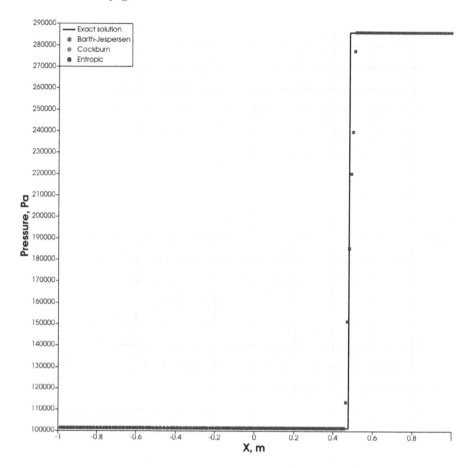

Fig. 10. Pressure distributions along the streamline: exact solution (exact), entropic limiter (entropic), Cockburn limiter (Cockburn), Barth-Jespersen limiter (Barth-Jespersen)

An unstructured triangular mesh consisting of 9785 cells was used in the calculation.

Figure 9 shows the distribution of the Mach number in the computational domain in the case of using different limiters. Figure 10 shows the pressure distribution along the streamline. The presented figures demonstrate approximately the same numerical results obtained using the considered limiters. All limiters show good agreement with the exact solution.

4   Conclusion

As a result, a numerical technique was created based on the implicit scheme of the discontinuous Galerkin method for solving gas dynamics equations using the

NVIDIA AmgX library. The obtained numerical results show good agreement with the known numerical solutions of the problems under consideration and their exact solutions. It shows the possibility of using the numerical scheme and the entropic limiter for solving the gas dynamics equations. Investigation of the efficiency of using the NVIDIA AmgX library solvers is beyond the scope of this work.

References

1. Krasnov, M.M., Kuchugov, P.A., Ladonkina, M.E., Tishkin, V.F.: Discontinuous Galerkin method on three-dimensional tetrahedral grids: using the operator programming method. Math. Models Comput. Simul. **9**(5), 529–543 (2017). https://doi.org/10.1134/S2070048217050064
2. Bogdanov, P.B., Gorobets, A.V., Sukov, S.A.: Adaptation and optimization of basic operations for an unstructured mesh CFD algorithm for computation on massively parallel accelerators. Comput. Math. Math. Phys. **53**, 1183–1194 (2013). https://doi.org/10.1134/S0965542513080046
3. Simoncini, V., Szvld, D.B.: Flexible inner-outer Krylov subspace methods. SIAM J. Numer. Anal. **40**(6), 2219–2239 (2003). https://doi.org/10.1137/S0036142902401074
4. Naumov, M., et al.: AmgX: a library for GPU accelerated algebraic multigrid and preconditioned iterative methods. SIAM J. Sci. Comput. **37**(5), 602–626 (2015). https://doi.org/10.1137/140980260
5. Zhalnin, R.V., et al.: Research of the order of accuracy of an implicit discontinuous Galerkin method for solving problems of gas dynamics. Middle Volga Math. Soc. J. **17**(1), 48–54 (2015)
6. Zhalnin, R.V., et al.: About the use of WENO-limiter in the implicit scheme for the discontinuous Galerkin method. Middle Volga Math. Soc. J. **17**(3), 75–81 (2015)
7. Cockburn, B., Shu, C.-W.: The local discontinuous Galerkin method for time-dependent convection-diffusion systems. SIAM J. Numer. Anal. **35**(6), 2440–2463 (1998). https://doi.org/10.1137/S0036142997316712
8. Bassi, F.A., Rebay, S.: A high-order accurate discontinuous finite element method for the numerical solution of the compressible Navier-Stokes equations. J. Comput. Phys. **131**(2), 267–279 (1997). https://doi.org/10.1006/jcph.1996.5572
9. Godunov, S.K.: A difference scheme for numerical solution of discontinuous solution of hydrodynamic equations. Math. Sbornik. **47**(89), 271–306 (1959)
10. Anderson, D., Tannehill, J., Pletcher, R.: Computational Fluid Mechanics and Heat Transfer. Mir, Moscow (1990)
11. Godunov, S.K., Zabrodin, A.V., Ivanov, M.Ya., Kraiko, A.N., Prokopov, G.P.: Numerical solution of multidimensional problems of gas dynamics. Nauka, Moscow (1976)
12. Kriksin, Y.A., Tishkin, V.F.: Variational entropic regularization of discontinuous Galerkin method for gas dynamics equations. Math. Models Comput. Simul. **31**(5), 69–84 (2019). https://doi.org/10.1134/S0234087919050058
13. Bragin, M.D., Kriksin, Yu.A., Tishkin, V.F.: Ensuring the entropy stability of the discontinuous Galerkin method in gas-dynamics problems. Keldysh Institute preprints no. 51, pp. 1–22 (2019). https://doi.org/10.20948/prepr-2019-51

14. Cockburn, B.: An introduction to the discontinuous Galerkin method for convection-dominated problems. In: Quarteroni, A. (ed.) Advanced Numerical Approximation of Nonlinear Hyperbolic Equations. LNM, vol. 1697, pp. 150–268. Springer, Heidelberg (1998). https://doi.org/10.1007/BFb0096353
15. Barth, T.J., Jespersen, D.C.: The design and application of upwind schemes on unstructured meshes. In: 27th Aerospace Sciences Meeting (1989). https://doi.org/10.2514/6.1989-366
16. Zhalnin, R.V., et al.: Software package LOGOS. The high order of accuracy method on block-structured meshes with WENO reconstruction. Modern Probl. Sci. Educ. Surg. **6** (2012). http://science-education.ru/ru/article/view?id=7329
17. Volkov, K.N., et al.: Implementation of parallel calculations on graphics processor units in the LOGOS computational fluid dynamics package. Numer. Methods Program. **14**(1), 334–342 (2013)
18. Ghia, U., et al.: The AIAA code verification project - test cases for CFD code verification. In: 48th AIAA Aerospace Sciences Meeting Including the New Horizons Forum and Aerospace Exposition (2010). https://doi.org/10.2514/6.2010-125

Methods for Simulation the Nonlinear Dynamics of Gyrotrons

Evgeny Semenov$^{(\boxtimes)}$, Vladimir Zapevalov , and Andrey Zuev

Federal Research Center Institute of Applied Physics of the Russian Academy of
Sciences (IAP RAS), Nizhny Novgorod, Russia
`semes@ipfran.ru`

Abstract. Determining the efficiency of high-power gyrotrons and
studying operating regimes require solving self-consistent problems of
the electron-wave interaction in the gyrotron cavity. Two mathematical models are considered: the stationary model that reduces to the
Sturm—Liouville problem for the inhomogeneous string equation, and
the time-depended model that combines the Schrödinger equation and
equations of electron motion. The algorithms described in this paper
are implemented in the code-package ANGEL (Analyzer of a Gyrating
Electrons), which is used at IAP RAS and GYCOM Ltd. to analyze an
electron-wave interaction in gyrotrons. The impact of ohmic losses in
a terahertz gyrotron cavity on the efficiency and a stable single-mode
operation is investigated. It is shown that a high fraction of ohmic losses
can lead to disruption of the stable single-mode generation of terahertz
gyrotrons operating at a high cyclotron harmonic.

Keywords: Inhomogeneous string equation · Schrödinger equation ·
Crank—Nicolson method · Electron-wave interaction · Gyrotron

1 Introduction

Nowadays, one of the most promising high-powered vacuum tube is the gyrotron.
To improve the gyrotron efficiency, complex models of an electron-wave interaction are required. In particular, it is important to take into account the influence of an electron beam on the structure of the electromagnetic field in the
cavity, which leads to the need to solve self-consistent problems with a non-
fixed field structure. In this paper, two models are considered: stationary and
time-depended. The stationary model [1,2] allows to quickly find the generation
efficiency and the output radiation power under the assumption of a single-mode
operation. Gyrotrons typically operate in a one TE_{mp} mode of a cylindrical cavity. Various transition processes are possible in a gyrotron, especially in cases
of operating in modes with high azimuthal and radial indices. In this case, the

This work was supported by a subsidy for Russian Federation State Assignment (Topic
No. 0030-2019-0019).

D. Balandin et al. (Eds.): MMST 2020, CCIS 1413, pp. 49–62, 2021.
https://doi.org/10.1007/978-3-030-78759-2_4

problem of mode competition from neighboring modes or modes operating at lower cyclotron harmonics plays a significant role. Of interest is the analysis of the gyrotron startup scenarios (taking into account the dependence of voltage, current, and other parameters of the device on time). Different regimes of mode interaction, self-modulation and stochastic regimes are also possible in gyrotrons. In all these cases, it is necessary to use complicated time-depended model [1,3]. At present, there is a sufficient number of code-packages for modeling an electron-wave interaction in gyrotrons. Examples of such programs are the code package GYRO1-3 developed by Borie and Dumbrajs [4], the Maryland Gyrotron (MAGY) code [5], the code-package EURIDICE developed at the National Technical University of Athens [6] and the gyrotron design toolbox GYROCOMPU developed at Wuhan National High Magnetic Field Center (China) [7]. Papers [4–7] are mainly devoted to the description of mathematical models, but aren't fully revealing numerical methods which were used. This paper is devoted to the description of the methods implemented in the code-package ANGEL developed at IAP RAS for calculating an electron-wave interaction and analyzing the competition of modes in gyrotrons.

2 Stationary Model

In the stationary and the time-depended models, we assume that the cavity is formed by segments of circular (azimuthally symmetric) weakly irregular waveguides, the radius of which is close to the cutoff radius of the operating mode. A typical weakly irregular cavity profile of gyrotron is shown in Fig. 2a. An electron-wave interaction is considered in following both models on the interval $[z_{in}, z_{out}]$.

The equations describing an interaction of the electron beam with the eigenmodes of a cylindrical cavity can take into account the inhomogeneity of the static magnetic field, the spread of oscillatory velocities, and the spread of the leading centers of the electron trajectories. Electrons in the beam can be classified into several groups: N_θ initial electron phases of entering the cavity, N_g groups with different oscillatory velocities (but with constant initial beam energy), N_R fractions with the radii of the leading centers of Larmor orbits (for simulation the misalignment of the beam and the magnetic field axis, wide beams, multi-beam systems etc.)

In the stationary model $N_a = N_\theta \times N_g \times N_R$ pairs of equations of motion (for the transverse and longitudinal components of the electron momentum)

$$\frac{dp_c}{dz} = f_{p_c}(z, p_c, p_\parallel, F, F'), \qquad \frac{dp_\parallel}{dz} = f_{p_\parallel}(z, p_c, p_\parallel) \tag{1}$$

and the equation for the electromagnetic field

$$\frac{d^2 F}{dz} + \kappa_\parallel^2(z) \cdot F = \mathcal{J}(z, \{p_c, p_\parallel\}) \tag{2}$$

are jointly solved with initial conditions

$$F(z_{in}) = F_{in}, \quad p_{\parallel}(z_{in}) = p_{\parallel,in}\{g\}, \quad p_c(z_{in}) = e^{i\theta_0} \cdot p_{\perp,in}\{g\}, \qquad (3)$$

$$g = \frac{p_{\perp}}{p_{\parallel}}, \quad \theta_0 = 2\pi \frac{j_\theta}{N_\theta}, \quad j_\theta = \overline{1, N_\theta},$$

and boundary conditions

$$\frac{dF}{dz} - i\,\kappa_{\parallel}F = 0, \quad z = z_{in}; \qquad \frac{dF}{dz} + i\,\kappa_{\parallel}F = 0, \quad z = z_{out}. \qquad (4)$$

Here F is the normalized high-frequency field, p_c and p_z are the parameters describing the electron beam, z is the axial coordinate of the cavity. The values F and p_c are complex, p_z is real. The function $\kappa_{\parallel}(z)$ is determined by the geometry of the cavity $R(z)$. Equations (1)–(2) are self-consistently integrated, Eq. (1) describes the influence of a high-frequency field on the parameters of the electron beam, Eq. (2) describes the longitudinal structure of the mode, which depends on the electron beam. The integration of the equations of motion together with the inhomogeneous string equation is carried out by the Runge-Kutta method of the 4th order. As a result of solving the Sturm—Liouville problem (1)–(4), the longitudinal structure of the field $F(z)$ is obtained for the found values of the frequency ω and the initial amplitude F_{in}. The search for the eigenvalues (ω, F_{in}) is carried out similarly to the method described in using the argument principle and further refinement of the solution by the two-dimensional Newton method. The parameters of the electron beam p_c, $p_{\parallel}$, the right-hand sides of Eqs. (1)–(2) f_{p_c}, $f_{p_{\parallel}}$, $\mathcal{J}$ are defined in Sect. 4.

3   Multimode Time-Depended Model

3.1   Mathematical Formulation of the Problem

In the time-depended model $N_a = N_\psi \times N_\theta \times N_g \times N_R$ pairs of equations of motion

$$\frac{\partial p_c}{\partial z} = f_{p_c}(z, p_c, p_{\parallel}, \{F_s\}), \qquad \frac{\partial p_{\parallel}}{\partial z} = f_{p_{\parallel}}(z, p_c, p_{\parallel}), \qquad (5)$$

and N_s equations for the field

$$\frac{\partial^2 F_s}{\partial z^2} + C_0\frac{\partial F_s}{\partial \tau} + \kappa_{\parallel,s}^2(z) \cdot F_s = \mathcal{J}_s(z, \{p_c, p_{\parallel}\}) \qquad (6)$$

are jointly solved with initial particle momenta (3), nonzero initial field $F_s(z, 0) = F_{in,s}(z)$ and the nonreflective radiation boundary conditions at the bounds for each s-th mode

$$F(z_{in}, \tau) - C_1 \cdot \int\limits_0^\tau \frac{\partial F(z_{in}, \partial\tilde{\tau})}{\partial z} \cdot \frac{e^{C_L \cdot (\tau - \tilde{\tau})}}{\sqrt{\tau - \tilde{\tau}}} \cdot d\tilde{\tau} = 0, \qquad (7)$$

$$F(z_{out}, \tau) + C_1 \cdot \int\limits_0^\tau \frac{\partial F(z_{out}, \partial\tilde\tau)}{\partial z} \cdot \frac{e^{C_R \cdot (\tau - \tilde\tau)}}{\sqrt{\tau - \tilde\tau}} \cdot d\tilde\tau = 0. \tag{8}$$

The system of Eq. (5)–(6) with boundary conditions (7)–(8) is solved in the rectangular region $[z_{in}, z_{out}] \times [0, T_{fin}]$. In (6)–(8) τ is the normalized time. The use of equations in general form (5)–(8) is convenient for describing numerical methods. The specific expressions for the right parts f_{P_c}, $f_{P_{\|}}$, $\mathcal{J}_s$, the constants C_0, C_1, C_L, C_R and other parameters are defined in Sect. 4.

3.2   Implicit Crank—Nicolson Scheme

To solve the system of Eqs. (5)–(6) by the finite difference method, the implicit Crank—Nicolson scheme [8, pp. 192–193] is used, which provides the second order of accuracy, but requires repeated solution of equations of motion at each time step, as well as solving SLAE by the tridiagonal matrix algorithm to find the field amplitude at each time layer. This numerical scheme uses the 3 + 3 stencil, that is, 3 nodes from the previous time layer and 3 nodes from the current layer (see Fig. 1).

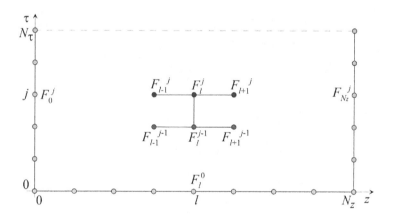

Fig. 1. The Crank—Nicolson stencil (3+3).

It is convenient to use a uniform grid in space and time

$$\{z_l, \tau^j\} \equiv \{z_{in} + \delta_z \cdot l, \ \delta_\tau \cdot j\}, \quad l = \overline{0, N_z}, \quad j = \overline{0, N_\tau};$$

$$\delta_z = \frac{(z_{out} - z_{in})}{N_z}, \quad \delta_\tau = \frac{T_{fin}}{N_\tau}.$$

The complex field amplitudes at grid nodes are denoted as $F_l^j \equiv F_s(z_l, \tau^j)$. The right-hand sides of the equation are denoted as $f_l^j \equiv \mathcal{J}_s(z_l, \tau^j) - \kappa_{\|,s}^2(z_l) \cdot F_l^j$. Approximation of derivatives:

$$\frac{\partial F}{\partial \tau}(z_l, \tau^j) \approx \frac{1}{\delta_\tau} \cdot \left(F_l^j - F_l^{j-1} \right), \tag{9}$$

$$\frac{\partial^2 F}{\partial z^2}(z_l, \tau^j) \approx \frac{1}{\delta_z^2} \cdot \left(F_{l-1}^j - 2 F_l^j + F_{l+1}^j \right), \tag{10}$$

$$\frac{\partial F}{\partial z}(z_l, \tau^j) \approx \frac{1}{\delta_z} \cdot \left(F_l^j - F_{l-1}^j \right) \approx \frac{1}{\delta_z} \cdot \left(F_{l+1}^j - F_l^j \right). \tag{11}$$

For internal nodes ($z_{in} < z < z_{out}$) the approximation of Eq. (6) takes the form

$$\frac{C_0}{\delta_\tau} \cdot \left(F_l^j - F_l^{j-1} \right) = \frac{1}{2} \cdot \left(f_l^j + f_l^{j-1} \right) - \frac{1}{2\,\delta_z^2} \cdot \left(F_{l-1}^j \right.$$
$$\left. -2 F_l^j + F_{l+1}^j + F_{l-1}^{j-1} - 2 F_l^{j-1} + F_{l+1}^{j-1} \right), \; l = \overline{1, N_z - 1}. \tag{12}$$

The link of boundary nodes with internal nodes is determined by boundary conditions for z_{in} and z_{out}. The result is a tridiagonal SLAE matrix

$$\mathcal{A} \cdot \mathbf{F} = \mathbf{d}, \tag{13}$$

$$\mathcal{A} = \begin{pmatrix} a_0 & b_0 & 0 & \cdots & & 0 & 0 \\ c_1 & a_1 & b_1 & 0 & \cdots & & 0 \\ 0 & c_2 & a_2 & b_2 & 0 & \cdots & 0 \\ \vdots & & & \ddots & & & \vdots \\ 0 & \cdots & & 0 & c_{N_z-2} & a_{N_z-2} & b_{N_z-2} & 0 \\ 0 & \cdots & & & 0 & c_{N_z-1} & a_{N_z-1} & b_{N_z-1} \\ 0 & 0 & \cdots & & & 0 & c_{N_z} & a_{N_z} \end{pmatrix},$$

$$\mathbf{F} = \{F_l\}, \quad \mathbf{d} = \{d_l\}, \quad l = \overline{0, N_z}.$$

To bring to this form, it's suitable to rewrite the Eq. (12):

$$F_{l-1}^j + \left(2 C_0 \tfrac{\delta_z^2}{\delta_\tau} - 2 \right) \cdot F_l^j + F_{l+1}^j = \delta_z^2 \cdot \left(f_F^j + f_F^{j-1} \right) +$$
$$+ \left(2 C_0 \tfrac{\delta_z^2}{\delta_\tau} + 2 \right) \cdot F_l^{j-1} - \left(F_{l-1}^{j-1} + F_{l+1}^{j-1} \right), \quad l = \overline{1, N_z - 1}.$$

It is followed that for all columns of the matrix except the zero and the last, the elements are constant

$$a_l = 2 C_0 \frac{\delta_z^2}{\delta_\tau} - 2 = -2 \left(1 - C_0 \frac{\delta_z^2}{\delta_\tau} \right), \quad b_l = c_l = 1, \quad l = \overline{1, N_z - 1},$$

and the corresponding elements of the right-hand side of the Eq. (13) have the form

$$d_l = \delta_z^2 \cdot \left(f_F^j + f_F^{j-1} \right) + 2 \left(1 + C_0 \frac{\delta_z^2}{\delta_\tau} \right) \cdot F_l^{j-1} - \left(F_{l-1}^{j-1} + F_{l+1}^{j-1} \right).$$

The elements of the zero and last columns are determined by the boundary conditions.

If we approximate the derivative in the boundary conditions (7)–(8) using formula (11), the solution accuracy drops to $O(\delta_z)$, but using information about the structure of Eq. (6), we can increase the order of approximation of the derivative. Let's introduce the notation:

$$\sigma_l^j = \frac{1}{6} \cdot \frac{\partial^2 F}{\partial z^2}(z_l, \tau^j),$$

and write down the finite difference for Eq. (6) according to the Crank–Nicolson scheme only in time (9) similarly to Eq. (12), but keeping the second derivative unchanged:

$$\frac{C_0}{\delta_\tau}\left(F_l^j - F_l^{j-1}\right) = \frac{1}{2}\left(f_l^j + f_l^{j-1}\right) - 3\left(\sigma_l^j + \sigma_l^{j-1}\right). \tag{14}$$

Next, it is used the spline representation of the first derivative from [9]. For the left and right bounds it has the form:

$$\frac{\partial F}{\partial z}(z_{in}, \tau^j) = \frac{1}{\delta_z} \cdot \left(F_1^j - F_0^j\right) - \delta_z \cdot \left(2\sigma_0^j + \sigma_1^j\right), \tag{15}$$

$$\frac{\partial F}{\partial z}(z_{out}, \tau^j) = \frac{1}{\delta_z} \cdot \left(F_{N_z}^j - F_{N_z-1}^j\right) + \delta_z \cdot \left(2\sigma_{N_z}^j + \sigma_{N_z-1}^j\right).$$

Using the relation (14) for $l = 0$ and $l = 1$ the sum $\left(2\sigma_0^j + \sigma_1^j\right)$ is expressed and substituted in (15a). The first derivative at the left bound for the left column of the SLAE matrix takes the form:

$$\frac{\partial F}{\partial z}(z_{in}, \tau^j) = W_0 \cdot F_0^j + W_1 \cdot F_1^j + W_L,$$

$$W_0 = \frac{2}{3} C_0 \frac{\delta_z}{\delta_\tau} - \frac{1}{\delta_z}, \qquad W_1 = \frac{C_0}{3} \frac{\delta_z}{\delta_\tau} + \frac{1}{\delta_z},$$

$$W_L = -\delta_z \cdot \left\{ \frac{C_0}{3\delta_\tau}\left(2F_0^{j-1} + F_1^{j-1}\right) - \left(2\sigma_0^{j-1} + \sigma_1^{j-1}\right) \right.$$
$$\left. + \frac{1}{6}\left(2f_{F,0}^{j-1} + f_{F,1}^{j-1} + 2f_{F,0}^j + f_{F,1}^j\right) \right\}.$$

Similarly, one finds the expression for $\left(2\sigma_{N_z}^j + \sigma_{N_z-1}^j\right)$ in (15b) using the relation (14) with $l = N_z$ and $l = N_z - 1$ and gets the first derivative at the right bound for the right column in the form:

$$\frac{\partial F}{\partial z}(z_{out}, \tau^j) = W_{N_z-1} \cdot F_{N_z-1}^j + W_{N_z} \cdot F_{N_z}^j + W_R, \tag{16}$$

$$W_{N_z-1} = -\left(\frac{1}{\delta_z} + \frac{C_0}{3}\frac{\delta_z}{\delta_\tau}\right), \qquad W_{N_z} = \left(\frac{1}{\delta_z} - \frac{2}{3}C_0\frac{\delta_z}{\delta_\tau}\right),$$

$$W_R = \delta_z \cdot \left\{ \frac{C_0}{3\delta_\tau}\left(2F_{N_z}^{j-1} + F_{N_z-1}^{j-1}\right) - \left(2\sigma_{N_z}^{j-1} + \sigma_{N_z-1}^{j-1}\right) \right.$$
$$\left. + \frac{1}{6}\left(2f_{N_z}^{j-1} + f_{N_z-1}^{j-1} + 2f_{N_z}^j + f_{N_z-1}^j\right) \right\}.$$

3.3   Approximation of the Boundary Conditions

An additional difficulty is added by the singularity contained in the boundary conditions (7)–(8). The integral is splited into the sum of integrals corresponding to the time steps:

$$\int_0^\tau \frac{\partial F(z_{in}, \tilde{\tau})}{\partial z} \cdot \frac{e^{C_L \cdot (\tau - \tilde{\tau})}}{\sqrt{\tau - \tilde{\tau}}} \cdot d\tilde{\tau} = I_L + I_j(z_{in}),$$

$$\int_0^\tau \frac{\partial F(z_{out}, \tilde{\tau})}{\partial z} \cdot \frac{e^{C_R \cdot (\tau - \tilde{\tau})}}{\sqrt{\tau - \tilde{\tau}}} \cdot d\tilde{\tau} = I_R + I_j(z_{out}),$$

$$I_L = \sum_{\tilde{j}=1}^{j-1} I_{\tilde{j}}(z_{in}), \qquad I_R = \sum_{\tilde{j}=1}^{j-1} I_{\tilde{j}}(z_{out}),$$

$$I_{\tilde{j}}(z_q) = \int_{(\tilde{j}-1)\cdot\delta_\tau}^{\tilde{j}\cdot\delta_\tau} \frac{\partial F(z_q, \tilde{\tau})}{\partial z} \cdot \frac{e^{C_q \cdot (j\cdot\delta_\tau - \tilde{\tau})}}{\sqrt{j\cdot\delta_\tau - \tilde{\tau}}} \cdot d\tilde{\tau},$$

$$C_q = C_L, \ z_q = z_{in}, \qquad C_q = C_R, \ z_q = z_{out}.$$

The following analytical calculations are given for the boundary condition (8). Similar calculations can be repeated for the boundary condition (7). The integrand of each $I_{\tilde{j}}$ is linearized in $\tilde{\tau}$ only the derivative and the exponent; the integrals of $(\tau - \tilde{\tau})^{-1/2}$ and $\tilde{\tau}/\sqrt{\tau - \tilde{\tau}}$ are taken analytically:

$$I_{\tilde{j}} \approx \tilde{I}_{\tilde{j}} = a_{\tilde{j}} I_{\tilde{j}}^a + b_{\tilde{j}} I_{\tilde{j}}^b,$$

$$a_{\tilde{j}} = \frac{1}{\delta_\tau} \left(\tilde{\tau}_{\tilde{j}} u_{\tilde{j}-1} - \tilde{\tau}_{\tilde{j}-1} u_{\tilde{j}} \right) = \left(\tilde{j} \cdot u_{\tilde{j}-1} - (\tilde{j}-1) \cdot u_{\tilde{j}} \right),$$

$$b_{\tilde{j}} = \frac{u_{\tilde{j}} - u_{\tilde{j}-1}}{\delta_\tau}, \qquad u_{\tilde{j}} = \frac{\partial F(z_{out}, \tilde{\tau})}{\partial z} \cdot e^{C_R \cdot (\tau - \tilde{\tau})},$$

$$I_{\tilde{j}}^a = \int_{\tau^{\tilde{j}} - \delta_\tau}^{\tau^{\tilde{j}}} \frac{d\tilde{\tau}}{\sqrt{\tau - \tilde{\tau}}} = \int_{(\tilde{j}-1)\cdot\delta_\tau}^{\tilde{j}\cdot\delta_\tau} \frac{d\tilde{\tau}}{\sqrt{j\cdot\delta_\tau - \tilde{\tau}}}$$

$$= -2\sqrt{\tau - \tilde{\tau}} \Big|_{(\tilde{j}-1)\cdot\delta_\tau}^{\tilde{j}\cdot\delta_\tau} = -2\sqrt{\delta_\tau} \left\{ \sqrt{j - \tilde{j}} - \sqrt{j - \tilde{j} + 1} \right\},$$

$$I_{\tilde{j}}^b = \int_{\tau^{\tilde{j}} - \delta_\tau}^{\tau^{\tilde{j}}} \frac{\tilde{\tau}\, d\tilde{\tau}}{\sqrt{\tau - \tilde{\tau}}} = \int_{(\tilde{j}-1)\cdot\delta_\tau}^{\tilde{j}\cdot\delta_\tau} \frac{\tilde{\tau}\, d\tilde{\tau}}{\sqrt{j\cdot\delta_\tau - \tilde{\tau}}} = \left\{ \frac{2}{3}(\tau - \tilde{\tau})^{3/2} - 2\tau\sqrt{\tau - \tilde{\tau}} \right\} \Big|_{(\tilde{j}-1)\cdot\delta_\tau}^{\tilde{j}\cdot\delta_\tau}$$

$$= \frac{2}{3}\delta_\tau^{3/2}\left\{(j-\tilde{j})^{3/2}-(j-\tilde{j}+1)^{3/2}\right\}-2j\,\delta_\tau^{3/2}\left\{\sqrt{j-\tilde{j}}-\sqrt{j-\tilde{j}+1}\right\}.$$

As a result, we obtain

$$\tilde{I}_{\tilde{j}}=\frac{4}{3}\sqrt{\delta_\tau}\left\{u_{\tilde{j}-1}\cdot\left[(j-\tilde{j})\sqrt{j-\tilde{j}}-\left(j-\tilde{j}-\frac{1}{2}\right)\sqrt{j-\tilde{j}+1}\right]\right.$$

$$\left.+u_{\tilde{j}}\cdot\left[(j-\tilde{j}+1)\sqrt{j-\tilde{j}+1}-\left(j-\tilde{j}+\frac{3}{2}\right)\sqrt{j-\tilde{j}}\right]\right\}.$$

Special cases:

$$I_R\equiv 0,\qquad j=0,1.$$

For $j=2$, the case is already quite regular, but the sum contains only one term:

$$I_R=\tilde{I}_1=\frac{4}{3}\sqrt{\delta_\tau}\left\{u_0\cdot\left(1-\frac{\sqrt{2}}{2}\right)+u_1\cdot\left(2\sqrt{2}-\frac{5}{2}\right)\right\}.$$

To speed up the summation, it is better to reorder the terms to minimize the access to the tables $\frac{\partial F(z_{out},\tilde{\tau})}{\partial z}$ and $e^{C_R\cdot(\tau-\tilde{\tau})}$. Square roots of integers in the range from 2 to the final time step number N_τ are calculated once per calculation and are also kept in the tables $\frac{\partial F(z_{out},\tilde{\tau})}{\partial z}$ and $e^{C_R\cdot(\tau-\tilde{\tau})}$. For $j>2$, we obtain

$$I_R=\frac{4}{3}\sqrt{\delta_\tau}\left\{u_0\cdot\left[(j-1)^{3/2}-\left(j-\frac{3}{2}\right)\sqrt{j}\right]+u_{j-1}\cdot\left[2\sqrt{2}-\frac{5}{2}\right]\right.$$

$$\left.+\sum_{\tilde{j}=1}^{j-2}u_{\tilde{j}}\cdot\left[(j-\tilde{j}-1)^{3/2}-2\,(j-\tilde{j})^{3/2}+(j-\tilde{j}+1)^{3/2}\right]\right\}.$$

It is possible to further increase the accuracy of approximation of the integral $I_{\tilde{j}}$ (or increase the integration step δ_τ, while maintaining accuracy) using the Fresnel integrals.

Using the refined representation of the derivative (16) and the method of eliminating the singularity in the integral I_j described above, one obtains the boundary conditions (7)–(8) in the form

$$F_0^j-C_1\cdot\left(I_L+\frac{4}{3}\sqrt{\delta_\tau}\frac{\partial F}{\partial z}(z_{in},\tau)+\frac{2}{3}\delta_\tau\frac{\partial F}{\partial z}(z_{in},\tau-\delta_\tau)\cdot\frac{e^{C_L\cdot\delta_\tau}}{\sqrt{\delta_\tau}}\right)=0,$$

$$F_{N_z}^j+C_1\cdot\left(I_R+\frac{4}{3}\sqrt{\delta_\tau}\frac{\partial F}{\partial z}(z_{out},\tau)+\frac{2}{3}\delta_\tau\frac{\partial F}{\partial z}(z_{out},\tau-\delta_\tau)\cdot\frac{e^{C_R\cdot\delta_\tau}}{\sqrt{\delta_\tau}}\right)=0,$$

The elements of the SLAE matrix corresponding to the boundary conditions take the following form

$$a_0=a_{N_z}=1-\frac{4}{3}C_1W_0\sqrt{\delta_\tau},\qquad b_0=c_{N_z}=-\frac{4}{3}C_1W_1\sqrt{\delta_\tau},\qquad(17)$$

$$d_0 = C_1 \cdot \left(I_L + \frac{4}{3} W_L \sqrt{\delta_\tau} + \frac{2}{3} \delta_\tau \frac{\partial F}{\partial z}(z_{in}, \delta_\tau \cdot (j-1)) \cdot \frac{\exp(C_L \cdot \delta_\tau)}{\sqrt{\delta_\tau}} \right),$$

$$d_{N_z} = -C_1 \cdot \left(I_R + \frac{4}{3} W_R \sqrt{\delta_\tau} + \frac{2}{3} \delta_\tau \frac{\partial F}{\partial z}(z_{out}, \delta_\tau \cdot (j-1)) \cdot \frac{\exp(C_R \cdot \delta_\tau)}{\sqrt{\delta_\tau}} \right).$$

The result is a tridiagonal SLAE matrix, the right side of which contains combinations of field values from the previous time layer F^{j-1}. Solving this SLAE by the tridiagonal matrix algorithm, it is possible in $O(N_z)$ operations to find the values of the nodes F_l^j, $l = \overline{0, N_z}$.

The implicit dependences between the field and particle momenta can be resolved at each time step in several iterations. First, the equations of motion are solved in the field on the previous layer and $\mathcal{J}[F^{j-1}]$ is found, after this, using (12)–(17), the SLAE is composed and solved, and thus the non-self-consistent field is found on the current layer $\tilde{F}_l^j$. Then, the equations of motion are again solved and $\mathcal{J}[\tilde{F}^j]$ is obtained, and already with such a right-hand side, a self-consistent field is found in the same way on the current layer F_l^j.

4 Examples of Mode Competition in Gyrotrons Operating in Different Frequency Ranges

The algorithms described in this article are implemented in the code-package ANGEL (Analyzer of a Gyrating Electrons), which is written in FORTRAN-90 and used at IAP RAS and GYCOM Ltd. Previously, this code-package was used to simulate electron-optical systems of gyrotrons [10]. Implementation described above complex algorithms in the code-package ANGEL allows to simulate the electron-wave interaction in gyrotrons. The stationary model affords to determine the efficiency and output power in the single-mode operation. The multimode time-depended self-consistent model allows to analyze transition processes, to investigate various dynamic and self-modulation regimes of generation. The described calculation methods can be used to investigate various effects in the gyrotron.

For modeling an electron-wave interaction in gyrotrons by the stationary model, the specific expressions for the right parts f_{p_c}, $f_{p_\parallel}$, $\mathcal{J}_s$ of the Eq. (1)–(2) take the form:

$$f_{p_c} = -i \frac{p_c}{p_\parallel} \cdot \left(\gamma \frac{\kappa}{n} - \frac{\omega_{H_0}}{c} \right)$$

$$+ \kappa_\perp \cdot J_{m-n}(\kappa_\perp R_b) \cdot \frac{i \gamma F}{p_\parallel} \cdot (p_c^*)^{n-1} \cdot \left(\frac{J_{n-1}(\xi)}{2 p_\perp^{n-1}} \right) + \frac{p_c}{2 B_0} \cdot \frac{d B_0}{dz},$$

$$f_{p_\parallel} = -\frac{p_\perp^2}{p_\parallel \cdot 2 B_0} \cdot \frac{d B_0}{dz},$$

$$\mathcal{J} = I \cdot \kappa \cdot \kappa_\perp \cdot \left\langle\!\!\!\left\langle J_{m-n}(\kappa_\perp R_b) \cdot \frac{p_c^n}{p_\parallel} \cdot \left(\frac{J_n'(\xi)}{p_\perp^{n-1}} \right) \right\rangle\!\!\!\right\rangle.$$

Here $\gamma(z) = \sqrt{1 + p_\perp^2 + p_\parallel^2}$ is the Lorentz factor, $\gamma_0 = \gamma(z_{in})$, $p_\perp = \gamma \dfrac{v_\perp}{c}$, $p_\parallel = \gamma \dfrac{v_\parallel}{c}$ are the normalized transverse and longitudinal momenta of electrons in a helical beam, n is the number of the cyclotron harmonic, κ is the wave number, $\kappa_\perp$ and $\kappa_\parallel$ are the transverse and longitudinal wave numbers, ω_{H_0} is the nonrelativistic gyrofrequency, B_0 is the external magnetic field, R_b is the average radius of the leading centers of electron orbits, J_m is the Bessel function, $\xi(z) = \dfrac{\kappa_\perp c}{\omega_{H_0}} p_\perp$, $I = \dfrac{C_I \cdot I_b}{(\nu_{m,p}^2 - m^2) \cdot J_m^2(\nu_{m,p})}$ is the normalized beam current, $C_I = \dfrac{8 \, e_0}{c \, m_0} \cdot 10^{-7} \approx 0.4693 \cdot 10^{-3}$, I_b is the beam current in amperes.

In the case of the multimode time-depended model, the expressions for the right parts of the equations different and take the form:

$$f_{p_c} = -i\frac{p_c}{p_\parallel} \cdot \left(\gamma \frac{\omega_p}{c} - \frac{\omega_{H_0}}{c}\right) + \frac{p_c}{2\,B_0} \cdot \frac{dB_0}{dz}$$
$$+ i \sum_{s=1}^{N_s} \left\{ J_{m_s - n_s}(\kappa_{\perp,s} R_b) \cdot \frac{\gamma \, \kappa_{\perp,s}}{p_\parallel} \cdot (p_c^*)^{n_s - 1} \cdot \left(\frac{J_{n_s - 1}(\xi)}{2\,p_\perp^{n_s - 1}}\right) \cdot F_s \, e^{i\,(\Delta_s \tilde{z} + \psi_s)} \right\},$$

$$J_s = I_s \cdot \kappa_s \cdot \kappa_{\perp,s} \cdot \left\langle\!\!\!\left\langle J_{m_s - n_s}(\kappa_{\perp,s} R_b) \cdot \frac{p_c^{n_s}}{p_\parallel} \cdot \left(\frac{J_{n_s}'(\xi)}{p_\perp^{n_s - 1}}\right) \cdot e^{-i\,(\Delta_s \tilde{z} + \psi_s)} \right\rangle\!\!\!\right\rangle,$$

$$\Delta_s = \frac{(\omega_s - n_s \, \omega_p)}{c \, \beta_\parallel}, \qquad \psi_s = \left(\frac{m_1}{n_1} \cdot n_s - m_s\right) \cdot \psi_0, \qquad \psi_0 = [0, 2\pi),$$

$$\tau = \frac{t \cdot c^2}{2 \cdot \omega_p}, \qquad \kappa_{\perp,s} = \frac{\nu_{m_s, p_s}}{R_r(z)}, \qquad \kappa_{\parallel,s}^2 = \kappa_s^2 - \kappa_{\perp,s}^2 \cdot \left(1 + \frac{i}{Q_{ohm,s}}\right),$$

$$\tilde{z} = z - z_{in}, \qquad \kappa_s = \frac{\omega_s}{c}, \qquad C_0 = -i\,n_s.$$

Here ω_p is the averaging frequency, usually defined as ω_1 / n_1 ($s = 1$ is index of the operating mode), ω_s is the reference frequency of s-th mode. The sign $\langle\!\langle ... \rangle\!\rangle$ means averaging in initial phases θ_0, in angles ψ_0, in groups with different pitch factor g and in fractions with the radii of the leading centers of Larmor orbits R_b.

The constants C_1, C_L, C_R in the nonreflective radiation boundary conditions (7)–(8) are following

$$C_1 = \frac{e^{-i\pi/4}}{\sqrt{\pi \cdot n_s}}, \qquad C_L = -\frac{i}{n_s} \cdot \kappa_{\parallel,s}^2(z_{in}), \qquad C_R = -\frac{i}{n_s} \cdot \kappa_{\parallel,s}^2(z_{out}).$$

As an example of using these algorithms, we illustrate the impact of ohmic losses on the efficiency and stability of the gyrotron operation at high cyclotron harmonics under the condition of strong competition from spurious modes. The transition to operation at high cyclotron harmonics allows to reduce the external magnetic field by n times. It expands a number of gyrotron applications

[11]. Two versions of gyrotron operating at 30 GHz (the low-frequency version) and 500 GHz (the high-frequency version) are considered. They have the same operating parameters (accelerating voltage 15 kV and current 0.5 A) and scalable dimensions of the electrodynamic system. For example, the $TE_{28,3}$ mode at the second cyclotron harmonic was selected as an operating mode. In this case the nearest spurious mode is the $TE_{-10,3}$ mode, operating at the fundamental cyclotron resonance. It should be noted that the version of a gyrotron with the operating mode at 30 GHz is practically infeasible due to the large size of the electrodynamic system. Nevertheless, the chosen mode is typical in the case of terahertz gyrotrons and has the problem of mode competition, which is specific issue of gyrotrons of this range.

The ohmic quality factor (ohmic Q-factor) is introduced into the system of equations through the longitudinal wave number $\kappa_{\parallel}$. For each gyrotron versions, the length of interaction space and the magnetic field were optimized to achieve the maximum efficiency of the operating mode. Figure 2a shows the calculated distributions of the high-frequency field along the interaction space of the operating mode and the spurious mode. The $TE_{28,3}$ mode has a similar to gaussian structure with a high diffraction Q-factor, corresponding to the optimal electron-wave interaction; then the $TE_{-10,3}$ mode is traveling wave and has a longitudinal structure with several variations of the RF field and respectively a lower diffraction Q-factor. Figure 2b shows the typical dependence of the efficiency on the generation frequency of these modes. With an increase in the generation frequency of the operating mode from 30 GHz to 1000 GHz, the calculated efficiency decreases from 11.35% to 1.77%; while ohmic losses practically do not affect the efficiency of the spurious mode.

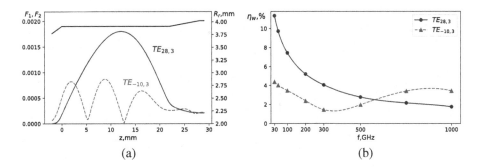

Fig. 2. (a) Example of a cavity profile and longitudinal structures of modes. (b) The gyrotron efficiency as a function of the generation frequency of the $TE_{28,3}$ mode.

The multimode time-depended self-consistent model allows to analyze the stability of $TE_{28,3}$ and $TE_{-10,3}$ modes in this frequency range. Figure 3 shows the time dependence of the mode amplitudes with the optimal initial operating mode amplitude and small amplitude of spurious mode. Figure 4 shows the planes of dimensionless amplitudes of the fields of the operating and spurious modes for the low-frequency (Fig. 4a) and the high-frequency (Fig. 4b) gyrotron versions. In the case of the low-frequency version the operating mode has stable generation. With an increase in the generation frequency and, accordingly, the fraction of ohmic losses in the cavity, the "saddle point" come up to the equilibrium point "stable node" corresponding to the single-mode operation. In the case of a high-frequency gyrotron, the operating mode turns unstable due to the bifurcation of equilibrium points. The generation of radiation on the spurious mode remains stable in the entire frequency range.

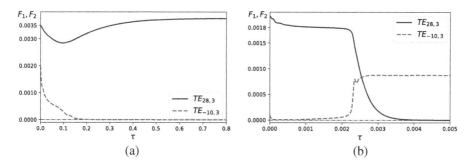

Fig. 3. Time dependence of the amplitudes F_1 and F_2 of the $TE_{28,3}$ and the $TE_{-10,3}$ modes respectively, obtained with the multimode time-depended self-consistent model. (a) The low-frequency version. (b) The high-frequency version.

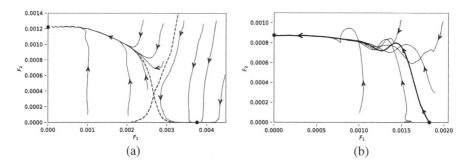

Fig. 4. Planes of amplitudes of the operating and the spurious modes. (a) The low-frequency version. (b) The high-frequency version.

The ratio of the total Q-factor of the spurious mode to the total Q-factor of the operating mode is essential for the stability of the gyrotron operation. In the case of the low-frequency gyrotron version the contribution of the ohmic Q-factor to the total Q-factor is negligible, then the parameter is determined mainly by the ratio of the diffraction Q-factors. In the case of the high-frequency gyrotron version, due to an increase in the fraction of ohmic losses, the parameter increases by 2.3 times relative to the low-frequency gyrotron version, which leads to a significant decrease in the stability of the operating mode.

In the calculations, the number of points along the axis z was equal to 200, the number of initial phases N_θ was equal to 57, the number groups with different oscillatory velocities N_g was equal to 37, the number of azimuthal angle N_ψ was equal to 37. Only one fraction with the optimal radii of the leading centers was considered. This corresponds to calculating the gyrotron efficiency with stationary model for a split second and calculating the mode competition using the time-depended model for several hours (the number of time steps was equal to about 20000).

5   Conclusion

The paper presents the stationary and the multimode time-depended models designed to calculate the electron-wave interaction in gyrotron cavity and implemented in the code-package ANGEL. Used numerical methods are described in details. The approach of refined representation of the first derivative by using information about the structure of the Schrödinger equation is applied to increase the solution accuracy. The method of eliminating the singularity in the integral of the nonreflective radiation boundary condition is shown.

The calculations showed that the main problem of terahertz gyrotrons is the problem of realizing the stable single-mode operation at high cyclotron harmonics, which, as a rule, is solved by choosing the operating mode, the parameters of the electron beam and the electrodynamic system, as well as using various methods of additional selection of the operating mode.

Acknowledgments. The authors are grateful to prof. A.S. Sergeev for his valuable pieces of advice. This work was supported by a subsidy for Russian Federation State Assignment (Topic No. 0030-2019-0019).

References

1. Moiseev, M.A., Nemirovskaya, L.L., Zapevalov, V.E., Zavolsky, N.A.: Numerical simulation of mode interaction in 170 GHz/1 MW gyrotrons for ITER. Int. J. Infrared Millimeter Waves **18**, 2117–2128 (1997). https://doi.org/10.1007/BF02678254
2. Moiseev, M.A., Zapevalov, V.E., Zavolsky, N.A.: Efficiency enhancement of the relativistic gyrotron. Int. J. Infrared Millimeter Waves **22**, 813–833 (2001). https://doi.org/10.1023/A:1014954012067
3. Ginzburg, N.S., Zavol'skii, N.A., Nusinovich, G.S., Sergeev, A.S.: Self-oscillation in UHF generators with diffraction radiation output. Radiophys. Quantum Electron. **29**, 89–97 (1986). https://doi.org/10.1007/BF01034008
4. Dumbrajs, O., Borie, E.: Calculation of eigenmodes of tapered gyrotron resonators. Int. J. Electron. **60**, 143–154 (1986). https://doi.org/10.1080/00207218608920768

5. Botton, M., Antonsen, T.M., Levush, B., Nguyen, K.T., Vlasov, A.N.: MAGY: a time-dependent code for simulation of slow and fast microwave sources. IEEE Trans. Plasma Sci. **26**, 882–892 (1998). https://doi.org/10.1109/27.700860

6. Avramides, K.A., Pagonakis, I.Gr., Iatrou, C.T., Vomvoridis, J.L.: EURIDICE: a code-package for gyrotron interaction simulations and cavity design. In: EPJ Web of Conferences, vol. 32, p. 04016 (2012). https://doi.org/10.1051/epjconf/20123204016

7. Wang, P., Chen, X., Xiao, H., Dumbrajs, O., Qi, X., Li, L.: GYROCOMPU: toolbox designed for the analysis of gyrotron resonators. IEEE Trans. Plasma Sci. **48**, 3007–3016 (2020). https://doi.org/10.1109/TPS.2020.3013299

8. Richtmyer, R.D., Morton, K.W.: Difference Methods for Initial-Value Problems, 2nd edn. Interscience Publishers, New York (1967)

9. Forsythe, G.E., Malcolm, M.A., Moler, C.B.: Computer Methods for Mathematical Computations. Prentice Hall, Englewood Cliffs (1977)

10. Plankin, O.P., Semenov, E.S.: Trajector analysis of the electronic-optical system technological gyrotron. Vestnik Novosib. Gos. Univ. Ser. Fiz. **8**, 44–54 (2013). [in Russian]

11. Glyavin, M.Yu., Idehara, T., Sabchevski, S.P: Development of THz gyrotrons at IAP RAS and FIR UF and their applications in physical research and high-power THz technologies. IEEE Trans. Terahertz Sci. Technol. **5**, 788–797 (2015). https://doi.org/10.1109/TTHZ.2015.2442836

Mathematical Modeling of Multidimensional Strongly Nonlinear Dynamic Systems

Irina V. Nikiforova$^{(\boxtimes)}$ ⓘ, Vladimir S. Metrikin ⓘ, and Leonid A. Igumnov ⓘ

Lobachevsky University, Nizhny Novgorod, Gagarin Avenue, 23 bld,
Nizhny Novgorod, Russia

Abstract. The paper investigates the dynamics of the systems with a crank vibration exciter designed for processing of various media, taking into account the properties of processing medium. The proposed scheme of vibration mechanisms can be described as follows: the rotary motion of a shaft with a constant frequency is converted into a reciprocating body motion relative anvil block with the help of a crank mechanism. In this case, the sliders-strikers alternately strike the corresponding anvils and the excited vibro-impact effect is transmitted to the processing medium through an anvil block. Due to the choice of the phase shift values between cranks, the values of eccentricities length and anvil heights, the required (may be optimal) impact interaction of the sliders-strikers with an anvil is ensured at such phase crankshaft positions at which the sliders-strikers exhibit the maximum velocity and, in addition, stable operating mode is being achieved. It should be noted that the presence of several sliders-strikers contributes to a longer exposure of the processing medium to loads and, consequently, to a more efficient process of plastic deformation. The research on dynamics of mechanisms is reduced to studying the properties of the point mapping of a two-dimensional non-analytical Poincaré surface into itself. Analytical relations have been obtained that allow to determine in the parameter space the boundaries of existence and stability regions of periodic operating motion modes. Numerical calculations were performed using a software package developed in C++.

Keywords: Soil · Sliders-strikers · Bifurcation · Sustainability

1 Introduction

Currently, there are a number of different types of vibration and vibration impact machines with various purposes, in particular, along with shock - vibration mechanisms with an unbalanced vibration exciter, eccentric shock - vibration mechanisms with a crank vibration exciter have found wide application in construction

This work was supported by the Ministry of Science and Higher Education of the Russian Federation, agreement No 0729-2020-0054.

D. Balandin et al. (Eds.): MMST 2020, CCIS 1413, pp. 63–76, 2021.
https://doi.org/10.1007/978-3-030-78759-2_5

[1,2]. The study of the nonlinear dynamics of devices with a crank-connecting rod exciter of vibrations still attracts the attention of many researchers, in particular [3]. The new constructive solution was based on the principle of "inverted vibrator", in which the working body, being an unbalance, is hinged on the eccentric shaft and balanced during rotation by the unbalance [4]. The force impulse transmitted to the surface (soil, piles, etc.) arises both due to the thrust with the eccentric shaft shoulder, and due to the kinetic energy of the fall of the working body. The efficiency of compacting and immersion machines essentially depends not so much on the amount of energy transferred to the processed medium, but on the nature of the transfer of this energy - the "shape" of the pulse, which should be changed due to the redistribution of individual dynamic factors of a single loading cycle. Obviously, a dense and at the same time strong soil structure is achieved only if, during the compaction process, the specific pressure on the contact surface of the working body with the soil increases gradually, the lower boundary of which is determined by the physical properties of the soil in the initial state with respect to the compaction process, and the upper one is the ultimate strength of the soil or technological conditions. In this regard, the parameters of such machines and mechanisms should be determined from conditions close to quasi-plastic interaction. In this case, the pulse repetition rate in each single cycle should be such as to exclude the possibility of developing an elastic aftereffect of the treated medium in the intervals between pulses. Such a multi-pulse loading method can be implemented using multi-hammer mechanisms with a crank-connecting rod vibration exciter, the design of which makes it possible to quite simply regulate the operating modes by changing the geometry of the kinematic links and solve the problems of soil compaction in the cramped conditions of industrial and civil construction.

In this work, a mathematical model of a shock-vibration mechanism with a crank-connecting rod vibration exciter with a different number of pistons-strikers (PS) has been constructed, both with and without (with a fixed limiter) the medium being processed.

2   Problem Setting

The operation of the vibration mechanisms under consideration can be described as follows Fig. 1: the rotating movement of the shaft with a constant ω is converted by a crank mechanism into a reciprocating movement of the body of mass M (1) relative to the anvil block (2) In this case, the striking sliders strike (3) the anvils of height h_i (4) and the shock-vibration effect created in this way is transmitted through the anvil block to the processed environment. Due to the choice of phase shift values φ between cranks of eccentricities r_i, heights of the anvils can be provided the required (can be optimal in a sense) impact interaction of the pistons-strikers on the slabs, in which the pistons-strikers have the maximum speed, ensuring the stability of the periodic operation. It should be noted that the presence of several PSs contributes to a longer exposure to the medium being processed, and, consequently, to an effective plastic deformation process.

Equations describing the shock - vibrational motion of a mechanism with a fixed limiter in dimensionless coordinates under the condition $r_i \ll l$

$$
\begin{cases}
\dfrac{d^2x}{dt^2} = -p, (x > f(\tau)) \\[3mm]
\dfrac{dx}{dt}\Big|_{+} = -R\dfrac{dx}{dt}\Big|_{-} + (1+R)\dfrac{df}{dt}.(x = f(\tau), \dot{x} - \dfrac{df}{dt} < 0)
\end{cases}
\tag{1}
$$

where in (1): $x = \dfrac{y - s_2 - l}{l}, \tau = \omega t, \mu = r_1/l, \gamma_i = r_i/r_1, \varepsilon_i = (s_i - s_2)/l, p = g/\omega^2 l, f(\tau) = \max_{\tau}(f_1(\tau), f_2(\tau), ..., f_N(\tau)), f_i(\tau) = \varepsilon_i - \mu\gamma_i \cos(\tau - \varphi_i),$

$\dfrac{dx}{dt}\Big|_{+} = \dot{x}_{+}, \dfrac{dx}{dt}\Big|_{-} = \dot{x}_{-}$ are velocities of the ith PS immediately before and after the impact interaction, respectively.

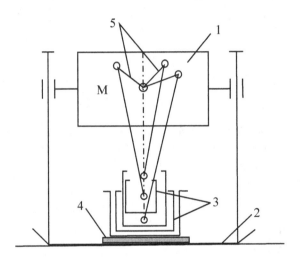

Fig. 1. Scheme shock-vibration mechanism.

3   Solution Method

Phase space for system $\Phi(x \geq f(\tau), \dot{x} < \infty)$ in coordinates $x, \dot{x}, \tau$ is reduced with respect to x. The surface $S(x = f(\tau))$ is a cylindrical surface formed by an intersection of N surfaces $S_i(x = f_i(\tau)), i = 1, 2, ..., N$. All phase trajectories are located either on the surface S or above it. If $x > f(\tau)$ it corresponds to free (impactless) motion of the mechanism; if $x = f(\tau)$, it corresponds to an impact interaction of one of the pistons with the stop (anvil block). Phase trajectories are shown on Fig. 2 in a qualitative form.

Obviously, the kind of the surface S, shown in Fig. 2, is preserved only when provided that two subsequent surfaces $f_i(\tau), f_{i+1}(\tau)$ intersect in pairs.

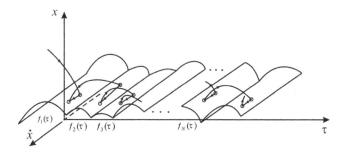

Fig. 2. The qualitative view of the phase space.

Thus, from the description of the structure of the phase space and the behavior of phase trajectories in it, it follows that the surface S can be taken as the section plane and the dynamics can be investigated by the method of point mappings [5]. In this case, this means mapping the section plane S on to itself, which can be written in the form

$$\begin{cases} f_{k+2}(\tau_{k+1}) = \Delta\tau_{k+1}(\dot{x}_k - p\Delta\tau_{k+1}/2) + f_{k+1}(\tau_k), \\ \dot{x}_{k+1} = R(p\Delta\tau_{k+1} - \dot{x}_k) + (1+R)\dfrac{df(\tau_{k+1})}{d\tau} \\ (\Delta\tau_{k+1} = \tau_{k+1} - \tau_k, k = 0, 1, \dots N-1) \end{cases} \qquad (2)$$

The coordinates of fixed points $M^*(\tau^*, X^*)$, corresponding to periodic modes of motion with alternate collision of each piston-striker (main mode), are determined from the system of $2(N+1)$ equations (2) supplemented by the conditions of periodicity $\dot{X}_N = \dot{X}_0 = \dot{X}^*, \tau_N = \tau_0 + nT = \tau^*$, having the form

$$\begin{cases} \dot{X}^* = \dfrac{b_N - R^N \sum_{k=1}^N (-1)^{k+1} b_{N-k}}{1 + (-1)^{N-1} R^N} \\ \dot{X}_{k+1} = R^k[(-1)^{k+1}\dot{X}^* + \sum_{i=0}^N (-1)^i b_{k-i}] \end{cases} \qquad (3)$$

where the components of the N-dimensional vector are functions $b(b_1, \dots, b_N)$ do not depend on $\dot{X}^*, \dot{X}_{k+1}$, but are functions of τ^*, τ_{k+1} and system parameters

$$b_j = Rp(\tau_{j+1} - \tau_j) + (1+R)\frac{df_{j+1}(\tau_{j+1})}{d\tau}$$

The times of motion along individual sections of the phase trajectory are determined by solving a system of nonlinear equations of the form

$$\begin{cases} f_{j+1}(\tau^*_{j+1}) + \Delta\tau^*_{j+1}(p\Delta\tau_j/2 - \dot{X}^*_j) = f_j(\tau^*_0) \\ f_1(\tau^*) - f_N(\tau_N) - (\tau^* - \tau^*_N)[p(\tau^* - \tau^*_N)/2 - \dot{X}^*_N] = 0 \\ (j = 1, 2, \dots, N-1) \end{cases} \qquad (4)$$

The stability in a small of a fixed point of the mapping (Lyapunov stability) corresponding to the main periodic regime of motion, as is known [6], is determined by the magnitude of the roots of the characteristic equation

$$\chi(z) = 0, \tag{5}$$

namely,

$$\begin{vmatrix} A(z) & B \\ C(z) & D(z) \end{vmatrix} \tag{6}$$

where $A(z), B, C(z), D(z)$ are square matrices, nonzero elements of which have the form

$$a_{i,j+1} = (-1)^j (p\Delta\tau_{i+1} - \dot{X}_i^*) + \frac{df_{i+j}(\tau_{i+j}^*)}{d\tau};$$

$$a_{N,1} = z[p(\tau_n^1 - \tau^*) + \dot{X}_N^* - \frac{df_1(\tau^*)}{d\tau}];$$

$$c_{i,i+j} = -R(-1)^j p + (1+R)j\frac{d^2 f_{i+1}(\tau_{i+1}^*)}{d\tau^2};$$

$$b_{i,j} = \Delta\tau_{i+1}^*; d_{i,i+j} = (j-1)R - j; d_{N,1} = -z; j = 0, 1; i = 1, 2, ..., N.$$

A periodic solution is stable if all roots (2) are located inside the unit circle, i.e. the inequality $|z| < 1$ holds. The violation of the stability conditions for a fixed point occurs either when the absolute value of one of the roots of the characteristic equation becomes equal to $z = \pm 1$, or when $z = \exp(\pm j\varphi), 0 \le \varphi \le \pi$. The boundaries of the stability region of the considered periodic motion, denoted by N_+, N_-, N_φ, satisfy the equations

$$\begin{cases} \chi(1) = 0, \\ \chi(-1) = 0, \\ \chi(e^{\pm j\varphi}) = 0, 0 \le \varphi \le \pi. \end{cases} \tag{7}$$

In addition, it is known [5] that with a continuous change in the parameters, the periodic mode of motion of interest to us disappears either due to violations of the stability conditions, or due to the exit of the phase trajectory from the domain of definition of the corresponding point transformation. The exit of the phase trajectory from the domain of definition of the point mapping, determined by the bifurcation surface N_C, is associated with the tangency of the trajectory at some points in time τ_i^* to the surface S Fig. 2. Thus, the region of existence of stable periodic motions of the system under consideration is limited by the surfaces N_+, N_-, N_φ and the surface N_C.

4   Numerical Study of the Dynamics of a Two-Piston Vibroimpact Mechanism

Since the mode of operation of the system with alternate impacts of the launcher is of particular interest, then, using the equations of the Poincaré surface S,

relations were obtained between the parameters of the system, at which this
mode is possible (excluding stands), namely

$$\varepsilon/\mu = \sqrt{1 - 2\gamma\cos\varphi + \gamma^2}, \tag{8}$$

$(\varepsilon_1 = \varepsilon, \gamma_2 = \gamma)$

Figure 3 in the plane $(\gamma, \varepsilon/\mu)$ shows the boundaries (8) at different values of
the phase shift φ with no stands. Only for the values of the parameters lying
above the curves Γ_i the main mode with alternate impact of the PSs is possible.

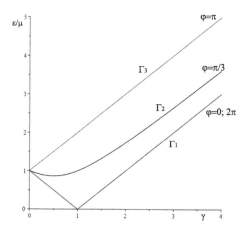

Fig. 3. Boundary regions of motion modes with alternate impacts of each PS.

Below are given the domains of existence and stable motions in the plane
(p, R), denoted by $D(m_1, m_2)$ [7] for different values of the parameters φ, γ, ε,
Δk, k.

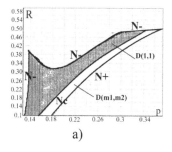

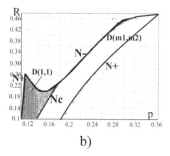

Fig. 4. Regions $D(1, 1)$ of existence of stable periodic motions of alternate PS impacts
$\varepsilon = 0.02\,(a), \varepsilon = 0.15\,(b)$.

Comparing Fig. 4a and Fig. 4b, it can be seen that an increase in parameter
ε leads to a significant decrease in the size of the existence and stability region

of the periodic motion modes of the mechanism and to the displacement of the region itself in the direction of decreasing the frequency parameter p. Hence it is better to connect the rods to PS at equal distances from PS bases.

Figure 5 shows power supply and stability for the following parameter values $\mu = 0.1, \varepsilon = 0.02, \gamma = 4, k_1 = 0, \Delta k = 0$ and two values $\varphi = 0.3$ Fig. 5a and $\varphi = 2$ Fig. 5b.

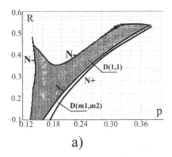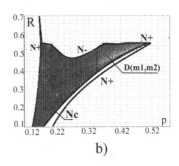

a)   b)

Fig. 5. Regions $D(1,1)$ of existence of stable periodic motions of alternate PS impacts $\varphi = 0.3\,(a), \varphi = 2\,(b)$.

Figure 6 and 7 shows bifurcation diagrams for the frequency parameter p (the abscissa shows the values of the parameter p, and the ordinates show the values of the post-impact velocities) with the same set of parameters as in Figs. 5a, b respectively, but with $R = 0.4$.

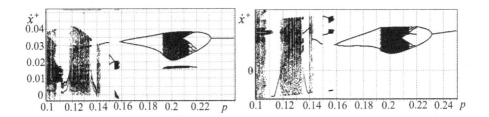

Fig. 6. Bifurcation diagrams with frequency parameter p for $\varphi = 0.3$.

An analysis of the presented stability regions and bifurcation diagrams showed that an increase in φ (phase shift between eccentricities) leads to an increase in the regions of existence and stability of periodic regimes, i.e. the range of values of the frequency parameter p, at which there are periodic motions, becomes wider.

The numerical calculations of stability regions with a wide range of parameters, carried out using a program complex developed in C++, showed that taking

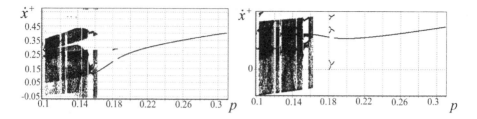

Fig. 7. Bifurcation diagrams with frequency parameter p for $\varphi = 2$.

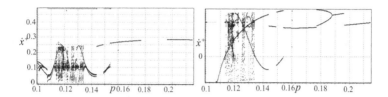

Fig. 8. Bifurcation diagrams with frequency parameter p for $k_1 = 0, \Delta k = 0$.

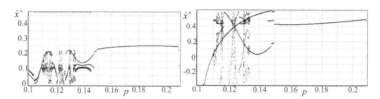

Fig. 9. Bifurcation diagrams with frequency parameter p for $k_1 = 0.1, \Delta k = 0.1$.

into account the bias in the mechanism scheme leads to an increase in the size of the stability region and its shift towards lower frequencies.

The analysis of the presented diagrams and stability regions showed that taking into account the heights of the anvils supports $k_1, k_1 + \Delta k$ leads to an increase in the regions of existence of periodic modes of motion [7]. So in Fig. 8 that the periodic regime for the selected values of the parameters exists at $0.195 \leq p \leq 0.215$ (excluding the heights of the anvils), and in Fig. 9 the periodic regime exists at $0.15 \leq p \leq 0.21$.

The numerical experiments carried out allowed us to conclude that the parameters $\varepsilon, \varphi, k_1, \Delta k$ affect the dynamics of the mechanism most significantly.

5   Numerical Study of the Dynamics of a Three-Piston Vibroimpact Mechanism

The equation of the bifurcation surface in the parameter space of the vibroimpact mechanism (9), which distinguishes the range of parameter values at which the main mode is possible, has the form

$$\begin{cases} \varepsilon_1^2/\mu^2 = (\gamma_2 - \cos\varphi_2)^2 + \sin^2\varphi_2, \\ \varepsilon_3^2/\mu^2 = (\sqrt{(\varepsilon_1^2/\mu^2 - \sin^2\varphi_2)} + \cos\varphi_2 - \gamma_3\cos(\varphi_3 - \varphi_2))^2 + \\ +(\gamma_3\sin(\varphi_3 - \varphi_2))^2. \end{cases} \qquad (9)$$

So, in Fig. 10 for $\gamma_3 = 3, \varphi_3 = 1.1$ and different angle values φ_2: $\varphi_2 = 0.3$ (upper surface), 0.7 and 1 (lower surface) shows the type of surfaces that highlight the range of Ω parameters, according to equations (9), in which modes of movement of the mechanism with alternate impact by each piston are possible - hammer on the anvil. These values, undoubtedly, can serve as a fairly convenient adjustment of the mechanism to the required mode of movement. It is shown that an increase in the size of the Ω region occurs with the growth of γ_3, φ_3 when the inequality is satisfied

$$\gamma_3 > \frac{\cos\varphi_2 + \sqrt{(\varepsilon_1/\mu)^2 - \sin^2\varphi_2}}{\cos(\varphi_3 - \varphi_2)} \qquad (10)$$

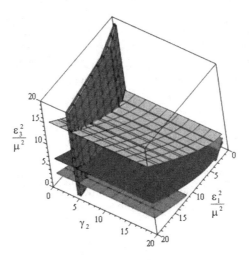

Fig. 10. Range of parameter values at which motion modes with alternate collision of each launcher with an anvil are possible.

In Fig. 11 shows bifurcation diagrams for the frequency parameter p of the three-piston mechanism, where the abscissa shows the values of the parameter, and the ordinates show the values of the post-impact velocities, built for the following sets of parameters $\mu = 0.12, \varepsilon_1 = 0.018, \varepsilon_3 = 0.02, \gamma_2 = 3, \gamma_3 = 3, \varphi = 0.2, \varphi_3 = 1.1, \lambda_1 = 0.1, \lambda_2 = 0.2, \lambda_3 = 0.3$ and different R equal to 0.2 (Fig. 11a) and R=0.4 (Fig. 11b).

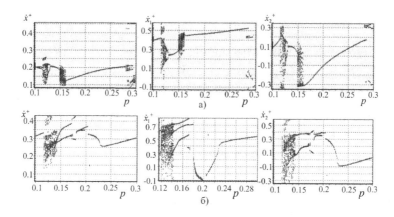

Fig. 11. Bifurcation diagrams with frequency parameter p for $R = 0.2$ (a) and $R = 0.4$ (b).

Figure 11a shows that at frequencies $0.13 \leq p \leq 0.15$ and $0.16 \leq p \leq 0.28$, there is a periodic motion with $m_i = 1, i = 1, 2, 3$; whereas at frequencies $0.11 \leq p \leq 0.12$ and $0.15 \leq p \leq 0.16$, a chaotic regime of motion is observed. It is also seen that the value $p = 0.13$ is the bifurcation value of the parameter at which the process of doubling the number of strokes by the first, second, and third pistons is observed. It follows from Fig. 11b that at frequencies $0.2 \leq p \leq 0.3$, the main mode of movement is observed ($m_i = 1, i = 1, 2, 3$); and with a decrease in the frequency parameter, there are periodic movements with a different number of strokes by each of the PSs. At $0.11 \leq p \leq 0.14$, a chaotic regime of motion is observed.

Thus, comparing Fig. 11a and Fig. 11b, we can conclude that the periodic mode of motion $D(1, 1, 1)$ with an increase in R shifts towards large values frequency parameter p.

In conclusion, it should be noted that the obtained results of computer simulation at various values of the parameters make it possible to clearly trace the processes of reconstruction from periodic motions of arbitrary multiplicity to stochastic ones [8,9].

6   Dynamics of the Vibroimpact Mechanism Taking into Account the Influence of the Processed Medium

The first model of a two-piston vibro-shock mechanism, taking into account the processed medium, is shown in Fig. 12. The treated medium (soil, pile, etc.) is presented in the form of an elastic mass M_1 with a coefficient of elasticity k. Energy losses in the medium are taken into account as viscous friction with a damping factor b.

Figure 13 shows bifurcation diagrams for parameters $\mu_0 = 5, \mu = 0.1, \gamma = 4, \phi = 0.6$, and $h = 0.06$ (Fig. 13a) and $h = 0.04$ (Fig. 13b), and in Fig. 14 -

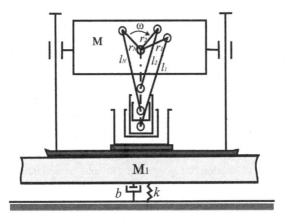

Fig. 12. Diagram of a two-piston vibro-shock mechanism taking into account the influence of the processed medium.

oscillograms of motion at the same values of parameters as in Fig. 13a, but in Fig. 14a $p = 0.02$, and in Fig. 14b $p = 0.025$ (red color in Fig. 14a, b indicates the oscillogram movement of the first piston, green - the second).

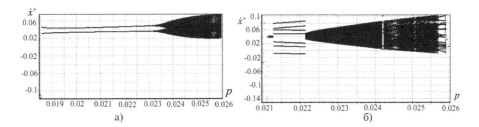

Fig. 13. Bifurcation diagrams with frequency parameter p for different values of coefficient h (dimensionless damping coefficient) $h = 0.06$ (a) and $h = 0.04$ (b).

From these figures it can be seen that at $0.0185 \le p \le 0.0235(h = 0.06)$ there is a periodic motion with $m_1 = 1, m_2 = 1$; at $0.0235 \le p \le 0.026$, a chaotic regime of motion is observed. At $p \ge 0.0235$, the process of doubling the number of impacts of the launcher is observed. The studies have shown that the main mode of motion exists and is stable at lower frequencies with a decrease in the damping coefficient.

The second model of the mechanism with one PS, taking into account the medium being processed (an analogue of the fundamental work of Yu.I. Neimark [10]) in the form of a plug, on which a constant force Q and dry friction acts F_0, is shown in Fig. 15. Similar models of interaction of the working body with the base were later used in many works and applications (see [11,12] and the literature cited therein).

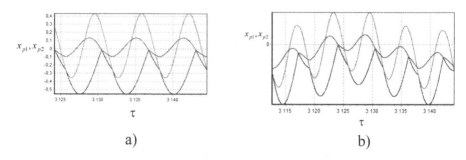

Fig. 14. Oscillograms of motion for $p = 0.02$ (a) and $p = 0.025$ (b). (Color figure online)

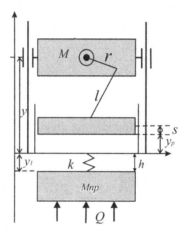

Fig. 15. Diagram of a single-piston vibro-impact mechanism taking into account the properties of the medium being processed.

Oscillograms of the piston and plug movement are plotted. So, in Fig. 16 shows one of the oscillograms with the following set of parameters $p = 0.1; \lambda_0 = 120; \lambda = 140, \mu = 0.1; \mu_0 = 0.1; \eta = 0.1; f^* = 10; q = 1$, where $\lambda_0^2 = \dfrac{k}{M\omega^2}, \lambda^2 = \dfrac{k}{M_{np}\omega^2}, \eta = h/l, f^* = \dfrac{F_0}{M_{np}l\omega^2}, q = \dfrac{Q}{M_{np}l\omega^2}$.

It follows from Fig. 16 that when there is no interaction between the piston and the plug, its coordinate remains constant, and when interacting with the piston, the plug begins to move together with it and sinks to a certain depth.

The third model of soil compaction with two PSs is shown in Fig. 17.

In this model it is assumed that when the medium is compacted, the resistance force depends on the value of the previous soil settlement. Analysis of the results of numerical experiments showed that the compaction process is most effective for soils with a lower stiffness index. So in Fig. 18a,b, oscillograms of movement are shown, which represent the process of soil compaction for various

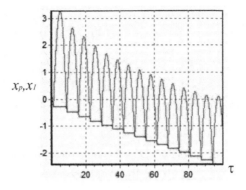

Fig. 16. Oscillogram of piston and plug movement.

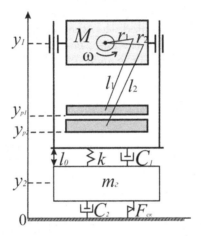

Fig. 17. Model of the soil compaction process.

values model parameters characterizing properties of the soil. Figure 18a is presented for the following set of parameters $p = 0.1, \varepsilon = 0.018, \mu = 0.1, \gamma = 4, \lambda_1 = 40, \lambda_2 = 40, \lambda_3 = 30, h_2 = 150, \alpha = 5, \beta = 2, \eta = 0.05$. Figure 18b presents the oscillogram of movement with the same set of parameters, as in Fig. 18a, but for $h_2 = 7$. Dimensionless parameters characterize the properties of the processed medium.

The main results of the present work are as follows:

– a new mathematical model of a two-piston vibro-impact mechanism with a crank vibration exciter is presented;
– analytical relations for the mechanism parameters are given, with the help of which one can indicate regions where mechanism motions are possible after impacts either by one PS or alternately by two PSs;
– bifurcation diagrams for all parameters of the mechanism allowed us to identify existence regions of various motion modes, including chaotic ones;

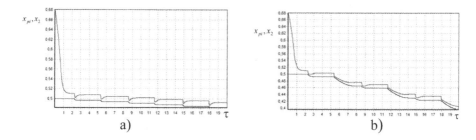

Fig. 18. Model of the soil compaction process.

- two periodic motion modes are proved to exist in the stability regions, formed by surfaces N_+, N_-, N_φ and N_C with one set of parameters, one periodic motion being always stable, the other unstable;
- it has been numerically computed, that the parameters ε (distance from the connecting point of the rod to the lower base of PS), φ (phase shift between eccentricities) and the height of the anvils $k_1, k_1 + \Delta k$ are the most significant parameters that affect the dynamics of the mechanism.

References

1. Babitsky, V.I.: Theory of Vibro-Impact Systems and Applications, Springer, Heidelberg (2013). https://doi.org/10.1007/978-3-540-69635-3
2. Pavlovskaia, E., Hendry, D.C., Wiercigroch, M.: Modelling of high frequency vibro-impact drilling. Int. J. Mech. Sci. **91**, 110–119 (2015)
3. Korendiy, V., Lanets, O., Kachur, O., Dmyterko, P., Kachmar, R.: Determination of inertia-stiffness parameters and motion modelling of three-mass vibratory system with crank excitation mechanism. Vibroeng. PROCEDIA **369**, 7–12 (2021)
4. Igumnov, L.A., Metrikin, V.S., Nikiforova, I.V.: The dynamics of eccentric vibration mechanism (Part 1). JVE J. Vibroeng. **19**, 4816–5656 (2017)
5. Neimark, Yu. I. The point mapping method in the theory of nonlinear oscillations. LIBROKOM, Moscow, p. 472 (2010). (in Russian)
6. Feigin, M.I.: Forced oscillations of systems with discontinuous nonlinearities. Nauka, Moscow (1994). (in Russian)
7. Igumnov, L., Metrikin, V., Nikiforova, I., Fevral'skikh, L.: The Dynamics of Eccentric Vibration Mechanism (Part 2): Dynamics. Strength of Materials and Durability in Multiscale Mechanics, vol. 137, pp. 173–190 (2021)
8. Guo, Y.: Bifurcation and Chaos of nonlinear vibro-impact systems. Ph.D. thesis (2013)
9. Shuster, H.G.: Deterministic Chaos. Physik-Verlag, Weinheim, p. 248 (1984)
10. Neimark, Y.I.: Vibration immersion and vibration pull theory. Eng. Collect. USSR Acad. Sci. **2**, 14–48 (1953)
11. Pavlovskaia, E., Wiercigroch, M.: Periodic solution finder for an impact oscillator with a drift. J. Sound Vibr. **267**, 893–911 (2003)
12. Lai, Z.H., Thomson, G., Yurchenco, D., Val, D.V., Rodgers, E.: On energy harvesting from a vibro-impact oscillator with dielectric membranes. J. Mech. Syst. Signal Process. **107**, 105–121 (2018)

Hierarchy of Models of Quasi-stationary Electromagnetic Fields

Aleksey Kalinin[1,2] and Alla Tyukhtina[1(✉)]

[1] N.I. Lobachevsky State University, 23 Gagarin Avenue,
603950 Nizhny Novgorod, Russia
[2] Institute of Applied Physics, Russian Academy of Sciences, 46 Ulyanova Street,
603950 Nizhny Novgorod, Russia

Abstract. The hierarchy of quasi-stationary models for the system of Maxwell's equations in homogeneous and inhomogeneous media is studied. The non-relativistic magnetic approximation, the non-relativistic electric approximations and the generalizing quasi-stationary approximation, in which the displacement current contains only a component corresponding to the potential part of the electric field, are considered. The relationship between solutions of initial-boundary value problems for the system of Maxwell's equations in various approximations is established and estimates of the proximity of these solutions are given. The obtained results show that the generalizing quasi-stationary approximation considered in this work has the same accuracy as the non-relativistic magnetic approximation in determining the magnetic field and the transverse component of the electric field and allows more accurate determination of the potential component of the electric field and the volume density of charges. The resulting generalized quasi-stationary approximation thus covers both classical non-relativistic approximations and can be used in modeling electromagnetic processes in substantially inhomogeneous media, in particular, in solving problems of atmospheric electricity.

Keywords: The system of Maxwell's equations · Quasi-stationary approximations · Inhomogeneous media

1 Introduction

The system of Maxwell's equations is written as [1]

$$\operatorname{curl} \boldsymbol{H}(x,t) = \frac{4\pi}{c} \boldsymbol{J}(x,t) + \frac{1}{c}\frac{\partial \boldsymbol{D}(x,t)}{\partial t}, \tag{1}$$

$$\operatorname{curl} \boldsymbol{E}(x,t) = -\frac{1}{c}\frac{\partial \boldsymbol{B}(x,t)}{\partial t}, \tag{2}$$

Supported by the Scientific and Education Mathematical Center "Mathematics for Future Technologies" (Project No. 075-02-2020-1483/1).

D. Balandin et al. (Eds.): MMST 2020, CCIS 1413, pp. 77–92, 2021.
https://doi.org/10.1007/978-3-030-78759-2_6

$$\text{div}\,\boldsymbol{B}(x,t) = 0, \tag{3}$$

$$\text{div}\,\boldsymbol{D}(x,t) = 4\pi\rho(x,t), \tag{4}$$

where $(x,t) \in Q = \Omega \times (0,T)$, $\Omega \subset \mathbb{R}^3$, $T > 0$.

It is assumed, the constitutive relations

$$\boldsymbol{D}(x,t) = \varepsilon\boldsymbol{E}, \qquad \boldsymbol{B}(x,t) = \mu\boldsymbol{H}, \; \boldsymbol{J} = \sigma\boldsymbol{E} + \boldsymbol{J}^{\text{ext}} \tag{5}$$

are valid, where $\boldsymbol{J}^{\text{ext}}$ is the source current density.

We consider the system (1)–(5) under boundary and initial conditions

$$\boldsymbol{E}(x,t) \times \boldsymbol{\nu}(x) = 0, \; (x,t) \in \partial\Omega \times (0,T),$$

where $\boldsymbol{\nu}(x)$ denotes the outward unit normal vector to the boundary at a point $x \in \partial\Omega$,

$$\boldsymbol{H}(x,0) = \boldsymbol{h}(x), \; \boldsymbol{E}(x,0) = \boldsymbol{e}(x), \; x \in \Omega.$$

In applied problems various quasi-stationary approximations [1,2] are often used instead of the system (1)–(5) for modeling sufficiently slow electromagnetic processes. In different physical situations the non-relativistic magnetic and electric approximations can be considered.

The non-relativistic magnetic approximation or eddy current approximation is characteristic of slow processes in media with a sufficiently high conductivity [2–5]. Formally, this approximation consists in ignoring the displacement current, that is, in Eq. (1) we can set $\partial\boldsymbol{D}/c\partial t \approx 0$. In this case, instead of Eq. (1) equation

$$\text{curl}\,\boldsymbol{H}(x,t) = \frac{4\pi}{c}\boldsymbol{J}(x,t) \tag{6}$$

is considered and the system (6), (2)–(5) can be studied under boundary and initial conditions

$$\boldsymbol{E}(x,t) \times \boldsymbol{\nu}(x) = 0, \; (x,t) \in \partial\Omega \times (0,T), \; \boldsymbol{H}(x,0) = \boldsymbol{h}(x), \; x \in \Omega.$$

A quite extensive literature is devoted to the study of various formulations of problems for this approximation and the issues of the numerical implementation of algorithms for their solution, in particular, [4–10]. The justification of the quasi-stationary magnetic approximation is discussed in [3–5].

Another quasi-stationary approximation, called the non-relativistic electrical approximation [2], is used to describe fairly slow processes in media with low conductivity. In particular, it is traditionally used for modeling electromagnetic processes in the lower atmosphere [11–14]. Formally, the approximation consists in neglecting the term $\partial\boldsymbol{B}/c\partial t$ in equality (2), which leads to the potentiality of the electric field in spatial simply connected regions:

$$\text{curl}\,\boldsymbol{E}(x,t) = 0. \tag{7}$$

The system (1), (3)–(5), (7) can be considered in this case under boundary and initial conditions

$$\boldsymbol{E}(x,t) \times \boldsymbol{\nu}(x) = 0, \; (x,t) \in \partial\Omega \times (0,T), \; \boldsymbol{E}(x,0) = \boldsymbol{e}(x), \; x \in \Omega.$$

In [21] for the case of inhomogeneous media a theoretical justification was carried out of the quasi-stationary approximation, in which the displacement current $\frac{1}{c}\frac{\partial}{\partial t}\boldsymbol{D} = \frac{1}{c}\frac{\partial}{\partial t}\varepsilon\boldsymbol{E}$ in (1) can be replaced by $-\frac{1}{c}\frac{\partial}{\partial t}\varepsilon\mathrm{grad}\varphi$, where

$$\boldsymbol{E} = \boldsymbol{\mathcal{E}} - \mathrm{grad}\varphi, \ \mathrm{div}\varepsilon\boldsymbol{\mathcal{E}} = 0. \tag{8}$$

For spatial inhomogeneous regions filled with an inhomogeneous conducting medium with a permittivity $\varepsilon(x)$, the first equation of the system in this approximation takes the form

$$\mathrm{curl}\boldsymbol{H}(x,t) = \frac{4\pi}{c}\boldsymbol{J}(x,t) - \frac{1}{c}\varepsilon(x)\frac{\partial}{\partial t}\mathrm{grad}\varphi(x,t). \tag{9}$$

The system (9), (2)–(5) is considered under boundary and initial conditions

$$\boldsymbol{E}(x,t)\times\boldsymbol{\nu}(x) = 0, \ (x,t) \in \partial\Omega\times(0,T), \ \boldsymbol{H}(x,0) = \boldsymbol{h}(x), \ \varphi(x,0) = \varphi_0(x), \ x \in \Omega. \tag{10}$$

In [15,17–19] problems for this approximation are studied under the assumption that in the system (1)–(4) the volume current density $\boldsymbol{J}$ and the volume charge density ρ are given functions, which formally corresponds to the assumption of a non-conducting medium (in (5) $\sigma \equiv 0$). In this case the considered approximation is called the Darwin approximation. Strict results on the correctness of problems for the linear system of Maxwell's equations in the framework of the Darwin approximation are obtained and an asymptotic relationship between the solutions of problems for the Darwin approximation and the solutions of corresponding problems for the non-stationary Maxwell system are established in terms of the small parameter $\beta = \Delta x/(c\Delta t)$, where Δx is the characteristic spatial scale, Δt is the characteristic time, c is the speed of light. The issues hierarchy of various quasi-stationary approximations are discussed in [2,15,16]. In particular, in [16] it is noted that the Darwin approximation covers the traditional non-relativistic magnetic and electric approximations.

In this paper, we study the quasi-stationary approximation (9) for a system of Maxwell's equations in homogeneous and inhomogeneous conducting media. The condition of inhomogeneity of media leads, in contrast to the works [17–19], to a related system of differential equations for unknown functions $\boldsymbol{H}$, $\boldsymbol{\mathcal{E}}$, $-\mathrm{grad}\varphi$, which is not reduced to classical problems of mathematical physics.

The chapter is organized as follow. In Sect. 2 we introduce the functional spaces necessary for a strict formulation of the initial boundary value problems under consideration. To illustrate the relationship between different approximations for the Maxwell equation system, Sect. 3 provides a preliminary analysis of initial boundary value problems in the case of homogeneous media. In Sect. 4 we present results concerning the correctness of initial boundary value problems for the system of Maxwell equations in different approximations. Section 5 is devoted to estimates of proximity of the solution of these problems.

All inequalities are obtained under the additional condition of coordination the initial data

$$\mathrm{curl}\boldsymbol{h} = \frac{4\pi}{c}\sigma\boldsymbol{e} + \frac{4\pi}{c}\boldsymbol{J}^{\mathrm{ext}}(0), \ \mathrm{curl}\boldsymbol{e} = 0, \tag{11}$$

which makes it possible to avoid the effect of the boundary layer.

2   Functional Spaces

Let $\Omega \subset \mathbb{R}^3$ be an open bounded Lipschitz domain with boundary Γ, homeomorphic to a sphere in $\mathbb{R}^3$. The following Hilbert spaces of vector functions with corresponding scalar products are defined [20]:

$$H(\mathrm{div}; \Omega) = \{\boldsymbol{u} \in \{L_2(\Omega)\}^3 : \mathrm{div}\boldsymbol{u} \in L_2(\Omega)\},$$
$$K(\mathrm{div}; \Omega) = \{\boldsymbol{u} \in \{L_2(\Omega)\}^3 : \mathrm{div}\boldsymbol{u} = 0\},$$
$$(\boldsymbol{u}, \boldsymbol{v})_{\mathrm{div}} = (\boldsymbol{u}, \boldsymbol{v})_{2,\Omega} + (\mathrm{div}\boldsymbol{u}, \mathrm{div}\boldsymbol{v})_{2,\Omega},$$
$$H(\mathrm{curl}; \Omega) = \{\boldsymbol{u} \in \{L_2(\Omega)\}^3 : \mathrm{curl}\boldsymbol{u} \in \{L_2(\Omega)\}^3\},$$
$$K(\mathrm{curl}; \Omega) = \{\boldsymbol{u} \in \{L_2(\Omega)\}^3 : \mathrm{curl}\boldsymbol{u} = 0\},$$
$$(\boldsymbol{u}, \boldsymbol{v})_{\mathrm{curl}} = (\boldsymbol{u}, \boldsymbol{v})_{2,\Omega} + (\mathrm{curl}\boldsymbol{u}, \mathrm{curl}\boldsymbol{v})_{2,\Omega},$$

where $(\cdot, \cdot)_{2,\Omega}$ denotes the scalar product in $L_2(\Omega)$ or in $\{L_2(\Omega)\}^3$. The closures of the set of test vector functions $\{\mathcal{D}(\Omega)\}^3$ in $H(\mathrm{div}; \Omega)$ and $H(\mathrm{curl}; \Omega)$ are denoted by $H_0(\mathrm{div}; \Omega)$ and $H_0(\mathrm{curl}; \Omega)$ respectively, $K_0(\mathrm{curl}; \Omega) = K(\mathrm{curl}; \Omega) \cap H_0(\mathrm{curl}; \Omega)$, $K_0(\mathrm{div}; \Omega) = K(\mathrm{div}; \Omega) \cap H_0(\mathrm{div}; \Omega)$.

The following propositions hold [20].

Lemma 1. *For any function $\boldsymbol{u} \in K(\mathrm{curl}; \Omega)$ there is a function $p \in H^1(\Omega)$ that $\boldsymbol{u} = \mathrm{grad}p$. If $\boldsymbol{u} \in K_0(\mathrm{curl}; \Omega)$, then it possible to choose $p \in H_0^1(\Omega)$.*

Lemma 2 *(Friedrichs inequality). There is a constant $A(\Omega) > 0$ that for all $\varphi \in H_0^1(\Omega)$*

$$\|\varphi\|_{2,\Omega} \le A(\Omega)\|\mathrm{grad}\varphi\|_{2,\Omega}.$$

Let $\eta \in L_\infty(\Omega)$ and there are positive constants η_1, η_2 that for almost all $x \in \Omega$

$$\eta_1 \le \eta(x) \le \eta_2.$$

We denote by $\{L_2(\eta; \Omega)\}^3$ the space $\{L_2(\Omega)\}^3$ provided with the scalar product $(\eta\boldsymbol{u}, \boldsymbol{v})_{2,\Omega}$. We set also

$$K(\mathrm{div}\eta; \Omega) = \{\boldsymbol{u} \in \{L_2(\Omega)\}^3 : \mathrm{div}\eta\boldsymbol{u} = 0\},$$
$$K_0(\mathrm{div}\eta; \Omega) = \{\boldsymbol{u} \in \{L_2(\Omega)\}^3 : \eta\boldsymbol{u} \in K_0(\mathrm{div}; \Omega)\},$$
$$U_1(\eta; \Omega) = K(\mathrm{div}\eta; \Omega) \cap H_0(\mathrm{curl}; \Omega), \; U_2(\eta; \Omega) = K_0(\mathrm{div}\eta; \Omega) \cap H(\mathrm{curl}; \Omega).$$

Lemma 3. *The orthogonal complement to $K(\mathrm{div}\eta; \Omega)$ in $\{L_2(\eta; \Omega)\}^3$ coincides with $K_0(\mathrm{curl}; \Omega)$.*

Lemma 4. *The orthogonal complement to $K_0(\mathrm{div}\eta; \Omega)$ in $\{L_2(\eta; \Omega)\}^3$ coincides with $K(\mathrm{curl}; \Omega)$.*

Lemmas 3, 4 follow from the corresponding statements for $\eta \equiv 1$ proved in [20], since

$$\eta K(\mathrm{div}\eta; \Omega) = K(\Omega), \; \eta K_0(\mathrm{div}\eta; \Omega) = K_0(\mathrm{div}; \Omega).$$

Lemma 5. *There exists a constant $C(\Omega) > 0$ that for all $\boldsymbol{u} \in U_i(\eta; \Omega)$, $i = 1, 2$,*

$$\|\boldsymbol{u}\|_{2,\Omega} \leq C(\Omega)\|\mathrm{curl}\boldsymbol{u}\|_{2,\Omega}. \tag{12}$$

This lemma is proved, for example, in [8].

3   Problems for Maxwell's Equations in Homogeneous Media

We suppose that $\boldsymbol{J}^{\mathrm{ext}} : Q \to \mathbb{R}^3$, $\boldsymbol{h} : \Omega \to \mathbb{R}^3$, $\boldsymbol{e} : \Omega \to \mathbb{R}^3$ are square integrable functions. We set $\boldsymbol{J}^{\mathrm{ext}} = \sigma \boldsymbol{E}^{\mathrm{ext}}$. In this section we consider the case of homogeneous conducting media, that is σ, ε, μ are positive constants. Let Δx is a characteristic spatial scale, Δt is a characteristic time, σ^* is a characteristic value of electrical conductivity (in homogeneous media $\sigma^* = \sigma$), ρ^* is a characteristic value of charge density. We replace x with $\Delta x \cdot x'$ and t with $\Delta t \cdot t'$ and denote

$$\gamma = 4\pi \Delta t \sigma^*, \quad \beta = \frac{\Delta x}{c\Delta t}, \quad \kappa = 4\pi \Delta x \rho^*.$$

The system of Maxwell's equations (1)–(4) with regard the constitutive relations (5) become

$$\mathrm{curl}' \boldsymbol{H} = \beta\gamma \boldsymbol{E} + \beta\gamma \boldsymbol{E}^{\mathrm{ext}} + \beta\frac{\partial}{\partial t'}\varepsilon \boldsymbol{E}, \tag{13}$$

$$\mathrm{curl}' \boldsymbol{E} = -\beta\frac{\partial}{\partial t'}\mu \boldsymbol{H}. \tag{14}$$

System (13), (14) is considered under the boundary conditions

$$\boldsymbol{E}(x', t') \times \boldsymbol{\nu}(x') = 0, \quad (x', t') \in \Gamma' \times (0, T'), \tag{15}$$

and the initial conditions

$$\boldsymbol{H}(x', 0) = \boldsymbol{h}(x'), \quad \boldsymbol{E}(x', 0) = \boldsymbol{e}(x'), \quad x' \in \Omega'. \tag{16}$$

The unknown functions $\boldsymbol{J}$, $\boldsymbol{D}$, $\boldsymbol{B}$ can be found from (5). Equation (4), which becomes

$$\kappa\rho = -\mathrm{div}\varepsilon\mathrm{grad}\varphi,$$

use to define the function ρ. Further for simplicity we will drop the primes for dimensionless variables (x', t') and their areas of change $Q' = \Omega' \times (0, T')$.

From Lemma 3 in the partial case $\eta \equiv 1$ follows the orthogonal decomposition $\{L_2(\Omega)\}^3 = K(\mathrm{div}; \Omega) \oplus K_0(\mathrm{curl}; \Omega)$. Using this decomposition and Lemma 1, we can set $\boldsymbol{E}^{\mathrm{ext}} = \boldsymbol{\mathcal{E}}^{\mathrm{ext}} + \mathrm{grad}\psi^{\mathrm{ext}}$, $\boldsymbol{e} = \boldsymbol{e}_0 - \mathrm{grad}\varphi_0$, where $\boldsymbol{\mathcal{E}}^{\mathrm{ext}} \in L_2(0, T, K(\mathrm{div}; \Omega))$, $\mathrm{grad}\psi^{\mathrm{ext}} \in L_2(0, T, K_0(\mathrm{curl}; \Omega))$, $\boldsymbol{e}_0 \in K(\mathrm{div}; \Omega)$, $\mathrm{grad}\varphi_0 \in K_0(\mathrm{curl}; \Omega)$.

Let $\boldsymbol{E} = \boldsymbol{\mathcal{E}} - \mathrm{grad}\varphi$, where $\boldsymbol{\mathcal{E}}(t) \in K(\mathrm{div}; \Omega)$ and $\mathrm{grad}\varphi(t) \in K_0(\mathrm{curl}; \Omega)$ for $t \in (0, T)$. Projecting (13) to orthogonal subspaces, we obtain that the initial-boundary value problem (13)–(16) for non-stationary Maxwell's equations is divided into the problem of defining a function $\mathrm{grad}\varphi \in L_2(0, T, K_0(\mathrm{curl}; \Omega))$ such that

$$\varepsilon\frac{\partial}{\partial t}\mathrm{grad}\varphi + \gamma\mathrm{grad}\varphi = \gamma\mathrm{grad}\psi^{\mathrm{ext}}, \tag{17}$$

$$\mathrm{grad}\varphi(0) = \mathrm{grad}\varphi_0, \tag{18}$$

and the problem of defining functions $\boldsymbol{H} \in L_2(0, T, H(\mathrm{curl}; \Omega))$, $\boldsymbol{\mathcal{E}} \in L_2(0, T, U_1(1; \Omega))$ such that

$$\mathrm{curl}\boldsymbol{H} = \beta\gamma\boldsymbol{\mathcal{E}} + \beta\gamma\boldsymbol{E}^{\mathrm{ext}} + \beta\frac{\partial}{\partial t}\varepsilon\boldsymbol{\mathcal{E}}, \tag{19}$$

$$\mathrm{curl}\boldsymbol{\mathcal{E}} = -\beta\frac{\partial}{\partial t}\mu\boldsymbol{H}, \tag{20}$$

$$\boldsymbol{H}(0) = \boldsymbol{h}, \ \boldsymbol{\mathcal{E}}(0) = \boldsymbol{e}_0. \tag{21}$$

The system (17), (18) corresponds to the quasi-stationary electrical approximation [2], which is used, in particular, for modeling electrical processes in the atmosphere [11–14]. The mathematical theory for various formulations of problems for this approximation in generally was presented in [13]. Problem (17), (18) allows the following statement: find $\varphi \in L_2(0, T, H_0^1(\Omega))$, that (18) is valid and for all $\psi \in H_0^1(\Omega)$

$$\frac{d}{dt}(\varepsilon\mathrm{grad}\varphi, \mathrm{grad}\psi)_{2,\Omega} + \gamma(\mathrm{grad}\varphi, \mathrm{grad}\psi)_{2,\Omega} = \gamma(\mathrm{grad}\psi^{\mathrm{ext}}, \mathrm{grad}\psi)_{2,\Omega}. \tag{22}$$

Theorem 1. *There exists a unique solution* $\varphi \in L_2(0, T, H_0^1(\Omega))$ *of (17), (18) and* $\mathrm{grad}\varphi \in C(0, T, K_0(\mathrm{curl}; \Omega))$, $\partial/\partial t\mathrm{grad}\varphi \in L_2(0, T, K_0(\mathrm{curl}; \Omega))$.
If $\mathrm{grad}\psi^{\mathrm{ext}} \in H^1(0, T, \{L_2(\Omega)\}^3)$ *and*

$$\mathrm{grad}\varphi_0 = \mathrm{grad}\psi^{\mathrm{ext}}(0), \tag{23}$$

then

$$\|\frac{\partial}{\partial t}\mathrm{grad}\varphi\|_{2,Q} \leq (1 - \exp(-\frac{\gamma}{\varepsilon}T))\|\frac{\partial}{\partial t}\mathrm{grad}\psi^{\mathrm{ext}}\|_{2,Q}. \tag{24}$$

Let $V_1(\Omega) = H(\mathrm{curl}; \Omega) \times U_1(1; \Omega)$, $L(\Omega) = \{L_2(\Omega)\}^3 \times \{L_2(\Omega)\}^3$,

$$(\Phi_1, \Phi_2)_L = (\mu\boldsymbol{u}_1, \boldsymbol{u}_2)_{2,\Omega} + (\varepsilon\boldsymbol{v}_1, \boldsymbol{v}_2)_{2,\Omega}, \ \Phi_i = \{\boldsymbol{u}_i, \boldsymbol{v}_i\} \in L(\Omega), \ i = 1, 2.$$

By introducing the liner operator $A : V_1(\Omega) \to L(\Omega)$,

$$A\Phi = \{\mu^{-1}\mathrm{curl}\boldsymbol{v}, -\varepsilon^{-1}\mathrm{curl}\boldsymbol{u}\}, \ \Phi = \{\boldsymbol{u}, \boldsymbol{v}\} \in V_1(\Omega), \tag{25}$$

we get the following formulation of the problem (19)–(21):
find $\Psi = \{\boldsymbol{H}, \boldsymbol{\mathcal{E}}\} \in L_2(0, T, \{L_2(\Omega)\}^3 \times K(\mathrm{div}; \Omega))$, that $\Psi(0) = \{\boldsymbol{h}, \boldsymbol{e}_0\}$ and for all $\Phi = \{\boldsymbol{u}, \boldsymbol{v}\} \in V_1(\Omega)$

$$\beta\frac{d}{dt}(\Psi, \Phi)_L - (\Psi, A\Phi)_L + \beta\gamma(\boldsymbol{\mathcal{E}}, \boldsymbol{v})_{2,\Omega} = -\beta\gamma(\boldsymbol{\mathcal{E}}^{\mathrm{ext}}, \boldsymbol{v})_{2,\Omega}. \tag{26}$$

Theorem 2. *There exists a unique solution* $\Psi = \{H, \mathcal{E}\}$ *of problem (26).*
If $\Psi_0 \in V_1(\Omega)$, $\mathcal{E}^{\text{ext}} \in H^1(0, T, \{L_2(\Omega)\}^3)$, *then* $\Psi \in L_2(0, T, V_1(\Omega))$, $\partial/\partial t \Psi$
belongs to $L_\infty(0, T, L(\Omega))$ *and (19), (20) hold. If, in addition,*

$$e_0 = 0, \quad \operatorname{curl} h = \beta\gamma\mathcal{E}^{\text{ext}}(0), \tag{27}$$

then

$$\|\partial/\partial t\mathcal{E}\|_{2,Q} \le (1 - \exp(-\gamma T/\varepsilon))\|\partial/\partial t\mathcal{E}^{\text{ext}}\|_{2,Q}. \tag{28}$$

The statements of Theorems 1, 2 follow from the more general Theorem 4 formulated in the next section.

Consider the initial-boundary value problem for the Maxwell's equations in the quasi-stationary magnetic approximation, which has the dimensionless form

$$\operatorname{curl} H = \beta\gamma E + \beta\gamma E^{\text{ext}}, \tag{29}$$

$$\operatorname{curl} E = -\beta\frac{\partial}{\partial t}\mu H, \tag{30}$$

$$E(x, t) \times \nu(x) = 0, \ (x, t) \in \Gamma \times (0, T), \tag{31}$$

$$H(x, 0) = h(x), \ x \in \Omega. \tag{32}$$

Using the method of orthogonal projection according to the Lemma 3 the problem reduced to the problem of defining functions $H \in L_2(0, T, H(\operatorname{curl}; \Omega))$ and $\mathcal{E}$ belongs to $L_2(0, T, U_1(1; \Omega))$ such that

$$\operatorname{curl} H = \beta\gamma\mathcal{E} + \beta\gamma\sigma\mathcal{E}^{\text{ext}}, \tag{33}$$

$$\operatorname{curl}\mathcal{E} = -\beta\frac{\partial}{\partial t}\mu H, \tag{34}$$

$$H(0) = h, \tag{35}$$

and equation

$$\operatorname{grad}\varphi = \operatorname{grad}\psi^{\text{ext}}. \tag{36}$$

The problem (33)–(35) admits the following formulation: find a function $H \in L_2(0, T, U_2(1; \Omega))$, that (35) is valid and for all $v \in U_2(1; \Omega)$

$$\beta^2\gamma\frac{d}{dt}\int_\Omega (\mu H \cdot v)dx + \int_\Omega (\operatorname{curl} H \cdot \operatorname{curl} v)dx = \beta\gamma\int_\Omega (\mathcal{E}^{\text{ext}} \cdot \operatorname{curl} v)dx. \tag{37}$$

Theorem 3. *For any* $h \in K(\operatorname{div}; \Omega)$, $\mathcal{E}^{\text{ext}} \in L_2(0, T, K(\operatorname{div}; \Omega))$ *there exists a unique solution* $H \in L_2(0, T, U_2(1; \Omega))$ *of problem (37).*
If $\mathcal{E}^{\text{ext}} \in H^1(0, T, \{L_2(\Omega)\}^3)$, $h \in U_2(1; \Omega)$ *and* $\operatorname{curl} h = \beta\gamma\mathcal{E}^{\text{ext}}(0)$, *then* $\partial/\partial t H \in L_\infty(0, T, K_0(\operatorname{div}; \Omega))$, *the function*

$$\mathcal{E} = (\beta\gamma)^{-1}\operatorname{curl} H - \mathcal{E}^{\text{ext}}$$

belongs to $L_2(0, T, U_1(1; \Omega))$ *and (34) is valid.*

The statements of the theorem follows from Theorem 6 and Lemma 7 of Sect. 4.

The initial-boundary value problem for Maxwell's equation in generalizing quasi-stationary approximation has the form

$$\text{curl}\boldsymbol{H} = \beta\gamma\boldsymbol{E} + \beta\gamma\boldsymbol{E}^{\text{ext}} - \beta\frac{\partial}{\partial t}\varepsilon\text{grad}\varphi, \tag{38}$$

$$\text{curl}\boldsymbol{E} = -\beta\frac{\partial}{\partial t}\mu\boldsymbol{H}, \tag{39}$$

$$\boldsymbol{E}(x,t) \times \boldsymbol{\nu}(x) = 0, \ (x,t) \in \Gamma \times (0,T), \tag{40}$$

$$\boldsymbol{H}(x,0) = h(x), \ \varphi(x,0) = \varphi_0(x), \ x \in \Omega. \tag{41}$$

This problem is divided into problem (17), (18) of defining a function $\text{grad}\varphi$ from $L_2(0,T,K_0(\text{curl};\Omega))$ and problem (33)–(35) for $\boldsymbol{H} \in L_2(0,T,H(\text{curl};\Omega))$, $\boldsymbol{\mathcal{E}} \in L_2(0,T,U_1(1;\Omega))$.

Thus, using the indices n, d and m for solutions of initial-boundary value problems for the non-stationary system of Maxwell's equations and the Maxwell's equations in the quasi-stationary and in the magnetic approximations respectively, we obtain when the initial functions coincide

$$\text{grad}\varphi^n = \text{grad}\varphi^d, \ \boldsymbol{H}^m = \boldsymbol{H}^d, \ \boldsymbol{\mathcal{E}}^m = \boldsymbol{\mathcal{E}}^d.$$

The next estimates were obtained in [21].

Lemma 6. *Let* $\boldsymbol{\mathcal{E}}^{\text{ext}} \in H^1(0,T,\{L_2(\Omega)\}^3)$, $h \in U_2(1;\Omega)$ *and* (27) *holds. Then*

$$\|\boldsymbol{\mathcal{E}}^n - \boldsymbol{\mathcal{E}}^d\|_{2,Q} \le \tfrac{\varepsilon}{\gamma}(1 - \exp(-\tfrac{\gamma}{\varepsilon}T))\|\tfrac{\partial}{\partial t}\boldsymbol{\mathcal{E}}^{\text{ext}}\|_{2,Q},$$

$$\|\boldsymbol{H}^n - \boldsymbol{H}^d\|^2_{L_\infty(0,T,\{L_2(\Omega)\}^3)} \le \tfrac{\varepsilon}{\sqrt{\gamma\mu}}(1 - \exp(-\gamma T/\varepsilon))\|\tfrac{\partial}{\partial t}\boldsymbol{\mathcal{E}}^{\text{ext}}\|_{2,Q},$$

$$\|\boldsymbol{H}^n - \boldsymbol{H}^d\|_{L_2(0,T,H(\text{curl};\Omega))} \le 2(1 + C(\Omega))^{1/2}\varepsilon\beta(1 - \exp(-\gamma T/\varepsilon))\|\tfrac{\partial}{\partial t}\boldsymbol{\mathcal{E}}^{\text{ext}}\|_{2,Q},$$

where $C(\Omega)$ *is the constant from* (12). *If* (23) *is met,*

$$\|\text{grad}\varphi^d - \text{grad}\varphi^m\|_{2,Q} \le \varepsilon\gamma^{-1}(1 - \exp(-\gamma T/\varepsilon))\|\frac{\partial}{\partial t}\text{grad}\psi^{\text{ext}}\|_{2,Q}.$$

The above results show that in the case of homogeneous media, the quasi-stationary approximation does not change the potential component of the electric field. In addition, it follows from (4) that $\rho^d = \rho^n$ and the charge conservation equation

$$\frac{\partial\rho}{\partial t} + \text{div}\boldsymbol{J} = 0$$

remains valid. The magnetic field and the solenoid component of the electric field are the same for the quasi-stationary approximation and for the non-relativistic magnetic approximation.

4   Initial-Boundary Value Problems for Maxwell's Equations in Inhomogeneous Media

4.1   The Non-stationary Maxwell's Equations System

Now we suppose that μ, σ, ε are functions from $L_\infty(\Omega)$ and

$$\varepsilon_1 \leq \varepsilon(x) \leq \varepsilon_2, \ \mu_1 \leq \mu(x) \leq \mu_2, \ \sigma_1 \leq \sigma(x) \leq \sigma_2, \ x \in \Omega,$$

where μ_i, σ_i, ε_i ($i = 1, 2$) are given positive constants.

Passing, as in the previous section, to dimensionless variables, we assume $\sigma = \sigma^* \sigma_0$, $\sigma_{01} \leq \sigma_0 \leq \sigma_{02}$, $\boldsymbol{J}^{\text{ext}} = \sigma^* \sigma_0 \boldsymbol{E}^{\text{ext}}$.

The initial-value problem for system of Maxwell's equations (1)–(4) with regard the constitutive relations (5) become

$$\text{curl}\boldsymbol{H} = \gamma\beta\sigma_0 \boldsymbol{E} + \gamma\beta\sigma_0 \boldsymbol{E}^{\text{ext}} + \beta\frac{\partial}{\partial t}\varepsilon\boldsymbol{E}, \tag{42}$$

$$\text{curl}\boldsymbol{E} = -\beta\frac{\partial}{\partial t}\mu\boldsymbol{H}, \tag{43}$$

$$\boldsymbol{E}(x, t) \times \boldsymbol{\nu}(x) = 0, \ (x, t) \in \Gamma \times (0, T), \tag{44}$$

$$\boldsymbol{H}(x, 0) = \boldsymbol{h}(x), \ \boldsymbol{E}(x, 0) = \boldsymbol{e}(x), \ x \in \Omega. \tag{45}$$

Constrains (11) for initial data takes the form

$$\text{curl}\boldsymbol{h} = \beta\gamma\sigma_0(\boldsymbol{e} + \boldsymbol{E}^{\text{ext}}(0)), \ \text{curl}\boldsymbol{e} = 0. \tag{46}$$

Note, that in the case of homogeneous media (46) implies (27) and (23).

Let $V(\Omega) = H(\text{curl}; \Omega) \times H_0(\text{curl}; \Omega)$ and $A : V(\Omega) \rightarrow L(\Omega)$ is a linear operator defined by (25). Problem (42)–(45) allows the following formulation: find a function $\Psi = \{\boldsymbol{H}, \boldsymbol{E}\} \in L_2(0, T, L(\Omega))$, that for all $\Phi = \{\boldsymbol{u}, \boldsymbol{v}\} \in V(\Omega)$

$$\beta\frac{d}{dt}(\Psi, \Phi)_L - (\Psi, A\Phi)_L + \beta\gamma(\sigma_0 \boldsymbol{E}, \boldsymbol{v})_{2,\Omega} = -\beta\gamma(\sigma_0 \boldsymbol{E}^{\text{ext}}, \boldsymbol{v})_{2,\Omega}, \tag{47}$$

$$\Psi(0) = \Psi_0 = \{\boldsymbol{h}, \boldsymbol{e}\}. \tag{48}$$

Theorem 4. *For any $\Psi_0 \in L(\Omega)$, $\boldsymbol{E}^{\text{ext}} \in \{L_2(Q)\}^3$ there exists a unique solution $\Psi \in L_2(0, T, L(\Omega))$ of problem (47), (48). If $\Psi_0 \in V(\Omega)$, $\boldsymbol{E}^{\text{ext}}$ in $H^1(0, T, \{L_2(\Omega)\}^3)$, then $\Psi \in L_2(0, T, V(\Omega))$, $\partial/\partial t\Psi \in L_\infty(0, T, L(\Omega))$ and Eqs. (42), (43) hold. If, in addition, the constrain (46) is met, then*

$$\left\|\frac{\partial}{\partial t}E\right\|_{2,Q} \leq \frac{\sigma_{02}}{\sigma_{01}}\left(1 - \exp(-\gamma\frac{\sigma_{01}}{\varepsilon_1}T)\right)\left\|\frac{\partial}{\partial t}E^{\text{ext}}\right\|_{2,Q}. \tag{49}$$

The existence of a solution to the problem (42)–(43) is proved in the same way as the corresponding statements in [22] (Theorems 4.1, 5.1 of Chapter VII). The estimate (49) is obtained in [21].

4.2   Quasi-stationary Approximation to Maxwell's Equations

Due to Lemmas 1, 3 we can set

$$\boldsymbol{E}(t) = \boldsymbol{E}(t) - \mathrm{grad}\varphi(t), \ \boldsymbol{\mathcal{E}}(t) \in K(\mathrm{div}\varepsilon; \Omega), \ \mathrm{grad}\varphi(t) \in K_0(\mathrm{curl}; \Omega), \ t \in (0, T).$$

The initial-boundary value problem for the system of Maxwell's equations in the quasi-stationary approximation is written in dimensionless form as

$$\mathrm{curl}\boldsymbol{H} = \beta\gamma\sigma_0\boldsymbol{E} + \beta\gamma\sigma_0\boldsymbol{E}^{\mathrm{ext}} - \beta\frac{\partial}{\partial t}\varepsilon\mathrm{grad}\varphi, \tag{50}$$

$$\mathrm{curl}\boldsymbol{E} = -\beta\frac{\partial}{\partial t}\mu\boldsymbol{H}, \tag{51}$$

$$\boldsymbol{E}(x, t) \times \boldsymbol{\nu}(x) = 0, \ (x, t) \in \Gamma \times (0, T), \tag{52}$$

$$\boldsymbol{H}(x, 0) = \boldsymbol{h}(x), \ \mathrm{grad}\varphi(x, 0) = \mathrm{grad}\varphi_0(x), \ x \in \Omega. \tag{53}$$

We denote $V_0(\Omega) = H(\mathrm{curl}; \Omega) \times K_0(\mathrm{curl}; \Omega)$. The problem (50)–(53) admits the following statement:
find $\Psi = \{\boldsymbol{H}, \mathrm{grad}\varphi\} \in L_2(0, T, V_0(\Omega))$ and $\boldsymbol{\mathcal{E}} \in L_2(0, T, U_1(\varepsilon; \Omega))$ such that for all $\Phi = \{\boldsymbol{u}, \mathrm{grad}\psi\} \in V_0(\Omega), \ \boldsymbol{v} \in U_1(\varepsilon; \Omega)$

$$\beta\frac{d}{dt}(\Psi, \Phi)_L + (\boldsymbol{\mathcal{E}}, \mathrm{curl}\boldsymbol{u})_{2,\Omega} - \beta\gamma(\sigma_0\boldsymbol{\mathcal{E}}, \mathrm{grad}\psi)_{2,\Omega} + \beta\gamma(\sigma_0\mathrm{grad}\varphi, \mathrm{grad}\psi)_{2,\Omega}$$
$$= \beta\gamma(\sigma_0\boldsymbol{E}^{\mathrm{ext}}, \mathrm{grad}\psi)_{2,\Omega}, \tag{54}$$

$$\beta\gamma(\sigma_0\boldsymbol{\mathcal{E}}, \boldsymbol{v})_{2,\Omega} - \beta\gamma(\sigma_0\mathrm{grad}\varphi, \boldsymbol{v})_{2,\Omega} - (\boldsymbol{H}, \mathrm{curl}\boldsymbol{v})_{2,\Omega} = -\beta\gamma(\sigma_0\boldsymbol{E}^{\mathrm{ext}}, \boldsymbol{v})_{2,\Omega} \tag{55}$$

$$\Psi(0) = \Psi_0 = \{\boldsymbol{h}, \mathrm{grad}\varphi_0\}. \tag{56}$$

The condition on the initial data, corresponding to (46), has the form

$$\mathrm{curl}\boldsymbol{h} = \beta\gamma\sigma_0(-\mathrm{grad}\varphi_0 + \boldsymbol{E}^{\mathrm{ext}}(0)). \tag{57}$$

Theorem 5. *Let $\boldsymbol{h} \in H(\mathrm{curl}; \Omega)$, $\boldsymbol{E}^{\mathrm{ext}} \in H^1(0, T, \{L_2(\Omega)\}^3)$. Then problem (54)–(56) has a unique solution $\Psi, \boldsymbol{\mathcal{E}}$. Moreover, $\Psi \in L_\infty(0, T, L(\Omega))$, $\partial/\partial t\Psi \in L_2(0, T, L(\Omega))$ and (50), (51) hold. If (57) is satisfied, then*

$$\left\|\frac{\partial}{\partial t}\boldsymbol{\mathcal{E}}\right\|_{2,Q} \leq \left(\frac{\varepsilon_2\sigma_{02}}{\varepsilon_1\sigma_{01}}\right)^{1/2}\left\|\frac{\partial}{\partial t}\boldsymbol{E}^{\mathrm{ext}}\right\|_{2,Q}, \tag{58}$$

$$\left\|\frac{\partial}{\partial t}\mathrm{grad}\varphi\right\|_{2,Q} \leq \frac{\sigma_{02}}{\sigma_{01}}\left(1 + \frac{\sqrt{\varepsilon_2\sigma_{02}}}{\sqrt{\varepsilon_1\sigma_{01}}}\right)(1 - \exp(-\frac{\gamma\sigma_{01}}{\varepsilon_1}T))\left\|\frac{\partial}{\partial t}\boldsymbol{E}^{\mathrm{ext}}\right\|_{2,Q}^2. \tag{59}$$

Proof. The existence of a solution to the problem and (58) are proved in [21]. Let us prove (59). According to Lemmas 1, 3 we have

$$\varepsilon^{-1}\sigma_0(\boldsymbol{E} + \boldsymbol{E}^{\mathrm{ext}}) = \boldsymbol{v} - \mathrm{grad}\psi,$$

where $v \in H^1(0, T, K(\mathrm{div}\varepsilon; \Omega))$, $\mathrm{grad}\psi \in H^1(0, T, K_0(\mathrm{curl}; \Omega))$. From (50) we get

$$\mathrm{curl}\boldsymbol{H} = \beta\gamma\varepsilon\boldsymbol{v}, \quad \gamma\mathrm{grad}\psi + \partial/\partial t \mathrm{grad}\varphi = 0.$$

Thus $\partial^2/\partial t^2 \mathrm{grad}\varphi \in L_2(0, T, \{L_2(\Omega)\}^3)$, hence $\partial/\partial t \mathrm{grad}\varphi \in C([0, T], \{L_2(\Omega)\}^3)$. In view to (57) $\boldsymbol{\mathcal{E}}(0) = 0$ [21] and therefore $\partial/\partial t \mathrm{grad}\varphi(0) = 0$.

For all $\psi \in H_0^1(\Omega)$ we get from (54)

$$(\varepsilon\tfrac{\partial}{\partial t}\mathrm{grad}\varphi, \mathrm{grad}\psi)_{2,\Omega} - \gamma(\sigma_0\boldsymbol{\mathcal{E}}, \mathrm{grad}\psi)_{2,\Omega} + \gamma(\sigma_0\mathrm{grad}\varphi, \mathrm{grad}\psi)_{2,\Omega}$$
$$= \gamma(\sigma_0\boldsymbol{E}^{\mathrm{ext}}, \mathrm{grad}\psi)_{2,\Omega}.$$

Taking the derivative of this equality in t and setting $\mathrm{grad}\psi = \partial/\partial t \mathrm{grad}\varphi$, we obtain

$$\varepsilon_1\|\tfrac{\partial}{\partial t}\mathrm{grad}\varphi\|_{2,\Omega}^2 + 2\gamma\sigma_{01}\int_0^t \|\tfrac{\partial}{\partial t}\mathrm{grad}\varphi\|_{2,\Omega}^2 dt$$
$$\leq 2\gamma\sigma_{02}\int_0^t \|\tfrac{\partial}{\partial t}(\boldsymbol{E}^{\mathrm{ext}} + \boldsymbol{\mathcal{E}})\|_{2,\Omega}\|\tfrac{\partial}{\partial t}\mathrm{grad}\varphi\|_{2,\Omega} dt.$$

Using (58), we have (59).

4.3   Non-relativistic Magnetic Approximation

The system of Maxwell's equations in the non-relativistic magnetic approximation is written, taking into account constitutive relations, as

$$\mathrm{curl}\boldsymbol{H} = \beta\gamma\sigma_0\boldsymbol{E} + \beta\gamma\sigma_0\boldsymbol{E}^{\mathrm{ext}}, \tag{60}$$

$$\mathrm{curl}\boldsymbol{E} = -\beta\frac{\partial}{\partial t}\mu\boldsymbol{H}. \tag{61}$$

System (60), (61) is considered under the boundary and initial conditions

$$\boldsymbol{E}(x, t) \times \boldsymbol{\nu}(x) = 0, \quad (x, t) \in \Gamma \times (0, T), \tag{62}$$

$$\boldsymbol{H}(x, 0) = \boldsymbol{h}(x), \quad x \in \Omega. \tag{63}$$

Problem (60)–(63) admits the following statement:
find $\boldsymbol{H} \in L_2(0, T, U_2(\mu; \Omega))$, that satisfies (63) and for all $\boldsymbol{v} \in U_2(\mu; \Omega)$

$$\beta^2\gamma\frac{d}{dt}(\mu\boldsymbol{H}, \boldsymbol{v})_{2,\Omega} + (\sigma_0^{-1}\mathrm{curl}\boldsymbol{H}, \mathrm{curl}\boldsymbol{v})_{2,\Omega} = \beta\gamma(\boldsymbol{E}^{\mathrm{ext}}, \mathrm{curl}\boldsymbol{v})_{2,\Omega}. \tag{64}$$

Theorem 6. *For all $\boldsymbol{h} \in K_0(\mathrm{div}\mu; \Omega)$, $\boldsymbol{E}^{\mathrm{ext}} \in L_2(0, T, \{L_2(\Omega)\}^3)$ there exists a unique solution $\boldsymbol{H} \in L_2(0, T, U_2(\mu; \Omega))$ to problem (64), (63). If $\boldsymbol{h} \in U_2(\mu; \Omega)$ and $\boldsymbol{E}^{\mathrm{ext}} \in L_2(0, T, H_0(\mathrm{curl}; \Omega))$, then $\partial/\partial t\boldsymbol{H} \in L_2(0, T, \{L_2(\Omega)\}^3)$.*

The theorem follows from the Lions theorem ([23], Ch. VIII).

Lemma 7. *Let $\boldsymbol{E}^{\mathrm{ext}} \in H^1(0, T, \{L_2(\Omega)\}^3)$, $\boldsymbol{h} \in U_2(\mu; \Omega)$ and*

$$\beta\gamma\boldsymbol{E}^{\mathrm{ext}}(0) - (\sigma_0)^{-1}\mathrm{curl}\boldsymbol{h} \in K_0(\mathrm{curl}; \Omega). \tag{65}$$

If $\boldsymbol{H}$ is a solution of problem (64), (63), then $\partial/\partial t\boldsymbol{H} \in L_\infty(0, T, \{L_2(\Omega)\}^3)$.

Lemma is proved in the same way as Theorem 5.1 of Chapter VII in [22].

Let $\boldsymbol{H} \in L_2(0, T, U_2(\mu; \Omega))$ be a solution of (64), (63). We may define function $\boldsymbol{E} \in L_2(0, T, \{L_2(\Omega)\}^3)$ from (60). Equation (61) is valid in the sense of distributions on $\Omega \times (0, T)$. If constrains for the initial data, formulated in Theorem 6 or Lemma 7, are met, then $\boldsymbol{E} \in L_2(0, T, H_0(\mathrm{curl}; \Omega))$, that is we have (62). Note also that (23) follows from (46).

5   Comparison of Solutions to Problems

Let $\Psi_0 \in V(\Omega)$, $\boldsymbol{E}^{\mathrm{ext}} \in H^1(0, T, \{L_2(\Omega)\}^3)$ and the constrains (46) on initial data are satisfied, $\{\boldsymbol{H}^n, \boldsymbol{E}^n\} \in L_2(0, T, V(\Omega))$ be a solution (42)–(45), $\boldsymbol{E}^n = \boldsymbol{\mathcal{E}}^n - \mathrm{grad}\varphi^n$, where $\boldsymbol{\mathcal{E}} \in L_2(0, T, U_1(\varepsilon; \Omega))$, $\mathrm{grad}\varphi^n \in L_2(0, T, K_0(\mathrm{curl}; \Omega))$. We denote by $\{\boldsymbol{H}^d, \mathrm{grad}\varphi^d\} \in L_2(0, T, V_0(\Omega))$, $\boldsymbol{\mathcal{E}}^d \in L_2(0, T, U_1(\varepsilon; \Omega))$ a solution of the problem (50)–(53), where $-\mathrm{grad}\varphi_0 = e$ and by $\boldsymbol{H}^m \in L_2(0, T, U_2(\mu; \Omega))$, $\boldsymbol{E}^m \in L_2(0, T, H_0(\mathrm{curl}; \Omega))$ a solution of the problem (60)–(63). Let $\kappa\rho^n = -\mathrm{div}\varepsilon\mathrm{grad}\varphi^n \in L_2(0, T, H^{-1}(\Omega))$, $\kappa\rho^d = -\mathrm{div}\varepsilon\mathrm{grad}\varphi^d \in L_2(0, T, H^{-1}(\Omega))$.

We denote $a = T\sigma_{01}/\varepsilon_1$ and define the following functions from γ:

$$f_\varphi(\gamma) = \frac{(1 - \exp(-a\gamma))^2}{\gamma}, \quad f_\varepsilon(\gamma) = \frac{1 - \exp(-a\gamma)}{\gamma}, \quad f_H(\gamma) = \frac{(1 - \exp(-a\gamma))}{\sqrt{\gamma}}.$$

The next estimates were established in [21].

Theorem 7

$$\|\mathrm{grad}\varphi^n - \mathrm{grad}\varphi^d\|_{2,Q} \le C_1 f_\varphi(\gamma)\|\partial/\partial t \boldsymbol{E}^{\mathrm{ext}}\|_{2,Q}, \tag{66}$$

$$\|\boldsymbol{\mathcal{E}}^n - \boldsymbol{\mathcal{E}}^d\|_{2,Q} \le C_2 f_\varepsilon(\gamma)\|\partial/\partial t \boldsymbol{E}^{\mathrm{ext}}\|_{2,Q}, \tag{67}$$

$$\|\boldsymbol{H}^n - \boldsymbol{H}^d\|_{L_\infty(0,T,\{L_2(\Omega)\}^3)} \le C_3 f_H(\gamma)\|\partial/\partial t \boldsymbol{E}^{\mathrm{ext}}\|_{2,Q}, \tag{68}$$

$$\|\boldsymbol{H}^n - \boldsymbol{H}^d\|_{L_2(0,T,H(\mathrm{curl};\Omega))} \le C_4 \beta(1 - \exp(-a\gamma))\|\partial/\partial t \boldsymbol{E}^{\mathrm{ext}}\|_{2,Q}, \tag{69}$$

$$\kappa\|\rho^n - \rho^d\|_{L_2(0,T,H^{-1}(\Omega))} \le C_5 f_\varphi(\gamma)\|\partial/\partial t \boldsymbol{E}^{\mathrm{ext}}\|_{2,Q}, \tag{70}$$

where positive constants C_i, $i = 1 - 5$, do not depend on β, γ.

We obtain proximity estimates for the potential components of the electric fields for the case of a weakly inhomogeneous medium.

Theorem 8. Let $\mathrm{grad}(\sigma_0\varepsilon^{-1}) \in \{L_\infty(\Omega)\}^3$. Then

$$\|\mathrm{grad}\varphi^n - \mathrm{grad}\varphi^d\|_{2,Q} \le C_6 f_\varphi(\gamma)\|\mathrm{grad}(\sigma_0\varepsilon^{-1})\|_{\infty,\Omega}\|\partial/\partial t \boldsymbol{E}^{\mathrm{ext}}\|_{2,Q}, \tag{71}$$

$$\kappa\|\rho^n - \rho^d\|_{L_2(0,T,H^{-1}(\Omega))} \le C_7 f_\varphi(\gamma)\|\mathrm{grad}(\sigma_0\varepsilon^{-1})\|_{\infty,\Omega}\|\partial/\partial t \boldsymbol{E}^{\mathrm{ext}}\|_{2,Q}, \tag{72}$$

where positive constants C_6 and C_7 do not depend on β, γ.

Proof. We denote $\boldsymbol{H} = \boldsymbol{H}^n - \boldsymbol{H}^d$, $\boldsymbol{E} = \boldsymbol{E}^n - \boldsymbol{E}^d = \boldsymbol{\mathcal{E}} - \mathrm{grad}\varphi$, $\rho = \rho^n - \rho^d$. Then $\mathrm{grad}\varphi(0) = 0$,

$$\mathrm{curl}\boldsymbol{H} = \beta\gamma\sigma_0\boldsymbol{E} + \beta\varepsilon\frac{\partial}{\partial t}(\boldsymbol{\mathcal{E}}^n - \mathrm{grad}\varphi). \tag{73}$$

Multiplying (73) by $\mathrm{grad}\varphi$, we obtain

$$(\varepsilon\mathrm{grad}\varphi, \mathrm{grad}\varphi)_{2,\Omega} + 2\gamma\int_0^t (\sigma_0\mathrm{grad}\varphi, \mathrm{grad}\varphi)_{2,\Omega}dt = 2\gamma\int_0^t (\sigma_0\boldsymbol{\mathcal{E}}, \mathrm{grad}\varphi)_{2,\Omega}dt.$$

Since

$$(\sigma_0\boldsymbol{\mathcal{E}}, \mathrm{grad}\varphi)_{2,\Omega} = (\sigma_0\varepsilon^{-1}\varepsilon\boldsymbol{\mathcal{E}}, \mathrm{grad}\varphi)_{2,\Omega} = -(\varphi\mathrm{grad}(\sigma_0\varepsilon^{-1}), \varepsilon\boldsymbol{\mathcal{E}})_{2,\Omega},$$

applying the Friedrichs inequality, we have

$$\varepsilon_1\|\mathrm{grad}\varphi\|_{2,\Omega}^2 + 2\gamma\sigma_{01}\int_0^t\|\mathrm{grad}\varphi\|_{2,\Omega}^2dt$$
$$\leq 2\gamma A(\Omega)\|\mathrm{grad}(\sigma_0\varepsilon^{-1})\|_{\infty,\Omega}\int_0^t\|\boldsymbol{\mathcal{E}}\|_{2,\Omega}\|\mathrm{grad}\varphi\|_{2,\Omega}dt,$$
$$\|\mathrm{grad}\varphi\|_{2,Q} \leq A(\Omega)^2\sigma_{01}^{-1}\|\mathrm{grad}(\sigma_0\varepsilon^{-1})\|_{\infty,\Omega}(1 - \exp(-\gamma\sigma_{01}T\varepsilon_1^{-1}))\|\boldsymbol{\mathcal{E}}\|_{2,Q}.$$

From (67) we obtain (71).

Because $\kappa\langle\rho, \psi\rangle = (\varepsilon\mathrm{grad}(\varphi, \mathrm{grad}\psi)_{2,\Omega}$ for all $\psi \in H_0^1(\Omega)$, (72) follows from (71).

Theorem 9. *The estimates*

$$\|\boldsymbol{E}^n - \boldsymbol{E}^m\|_{2,Q} \leq C_8 f_\varepsilon(\gamma)\|\partial/\partial t\boldsymbol{E}^{\mathrm{ext}}\|_{2,Q}, \tag{74}$$

$$\|\boldsymbol{H}^n - \boldsymbol{H}^m\|_{L_\infty(0,T,\{L_2(\Omega)\}^3)} \leq C_9 f_H(\gamma)\|\partial/\partial t\boldsymbol{E}^{\mathrm{ext}}\|_{2,Q}, \tag{75}$$

$$\|\boldsymbol{H}^n - \boldsymbol{H}^m\|_{L_2(0,T,H(\mathrm{curl};\Omega))} \leq \beta C_{10}(1 - \exp(-a\gamma))\|\partial/\partial t\boldsymbol{E}^{\mathrm{ext}}\|_{2,Q} \tag{76}$$

hold, where positive constants C_8–C_{10} *do not depend on* β, γ.

Proof. Denote $\boldsymbol{H} = \boldsymbol{H}^n - \boldsymbol{H}^m$, $\boldsymbol{E} = \boldsymbol{E}^n - \boldsymbol{E}^m = \boldsymbol{\mathcal{E}} - \mathrm{grad}\varphi$. Then $\boldsymbol{H}(0) = 0$,

$$\mathrm{curl}\boldsymbol{H} = \beta\gamma\sigma_0\boldsymbol{E} + \beta\frac{d}{dt}\varepsilon\boldsymbol{E}^n, \tag{77}$$

$$\mathrm{curl}\boldsymbol{E} = -\beta\frac{\partial}{\partial t}\mu\boldsymbol{H}. \tag{78}$$

Multiplying (77) by $\boldsymbol{E}$ and (78) by $\boldsymbol{H}$, we have

$$\frac{d}{dt}(\mu\boldsymbol{H}, \boldsymbol{H})_{2,\Omega} + 2\gamma(\sigma_0\boldsymbol{E}, \boldsymbol{E})_{2,\Omega} = -2(\varepsilon\frac{\partial}{\partial t}\boldsymbol{E}^n, \boldsymbol{E})_{2,\Omega}.$$

Integrating this equality and applying (49), we obtain (74), (75). From (77), (74) and (12) we get (76)

Theorem 10. *The next estimates are valid.*

$$\|\boldsymbol{E}^m - \boldsymbol{E}^d\|_{2,Q} \le D_1 f_{\mathcal{E}}(\gamma)\|\partial/\partial t \boldsymbol{E}^{\text{ext}}\|_{2,Q}, \tag{79}$$

$$\|\boldsymbol{H}^m - \boldsymbol{H}^d\|_{L_\infty(0,T,\{L_2(\Omega)\}^3)} \le D_2 f_H(\gamma)\|\partial/\partial t \boldsymbol{E}^{\text{ext}}\|_{2,Q}, \tag{80}$$

$$\|\boldsymbol{H}^m - \boldsymbol{H}^d\|_{L_2(0,T,H(\text{curl};\Omega))} \le \beta D_3(1 - \exp(-a\gamma))\|\partial/\partial t \boldsymbol{E}^{\text{ext}}\|_{2,Q}. \tag{81}$$

If $\text{grad}(\sigma_0\varepsilon^{-1}) \in \{L_\infty(\Omega)\}^3$ *then*

$$\|\boldsymbol{\mathcal{E}}^m - \boldsymbol{\mathcal{E}}^d\|_{2,Q} \le D_4 f_{\mathcal{E}}(\gamma)\|\text{grad}(\sigma_0\varepsilon^{-1})\|_{\infty,\Omega}\|\partial/\partial t \boldsymbol{E}^{\text{ext}}\|_{2,Q}, \tag{82}$$

$$\|\boldsymbol{H}^m - \boldsymbol{H}^d\|_{L_\infty(0,T,\{L_2(\Omega)\}^3)} \le D_5 f_H(\gamma)\|\text{grad}(\sigma_0\varepsilon^{-1})\|_{\infty,\Omega}\|\frac{\partial}{\partial t}\boldsymbol{E}^{\text{ext}}\|_{2,Q}. \tag{83}$$

Positive constants D_1–D_5 *do not depend on* β, γ.

The theorem is proved similarly to the Theorems 8, 9.

6   Discussion

The inequalities established in the paper allow us to evaluate the difference between quasi-stationary and non-stationary fields in different norms through norm of the time derivative of the external electric field ($\|\partial/\partial t \boldsymbol{E}^{\text{ext}}\|_{2,Q}$) with coefficients depending on the parameters $\beta = \Delta x/(c\Delta t)$ and $\gamma = 4\pi\sigma^*\Delta t$, which characterize the velocity of processes.

Note that $\beta \to 0$ and $\gamma \to \infty$ for slow processes with $\Delta t \to \infty$, and the smallness of β is sufficient for the proximity of magnetic fields for non-stationary Maxwell's equations, for the Darwin approximation and for the quasi-stationary magnetic approximation, although there are alternative estimates using γ. Accordingly, for the proximity of electric fields, as well as the charge densities, a smallness of $1/\gamma$ is sufficient. It can also be noted when $\gamma \to 0$ (fast processes) we have the proximity of the magnetic fields for all considered approximations, as well as the proximity of the potential components of the electric fields for the Darwin approximation and for the non-stationary Maxwell's equations.

The results obtained show that the quasi-stationary Darwin approximation for the totality of all fields (magnetic field strength, potential and solenoid components of the electric field) is closer to the non-stationary system of Maxwell's equations than the non-relativistic magnetic approximation. We can say that this approximation occupies an intermediate position between the non-stationary system of Maxwell's equations and the non-relativistic magnetic approximation.

The indicated hierarchy of quasi-stationary fields is most pronounced when considering problems in homogeneous media. In this case, the magnetic field and the divergence-free component of the electric field are the same for the Darwin approximation and for the quasi-stationary magnetic approximation, the potential component of the electric field is the same for the Darwin approximation, for quasi-stationary electric approximation and for the non-stationary Maxwell's equations. In general case we can estimate the difference between these fields depending on the degree of heterogeneity of the media characterized by the norm of $\text{grad}(\sigma\varepsilon^{-1})$.

References

1. Landau, L.D., Lifshitz, E.M.: Electrodynamics of Continuous Media. Course of Theoretical Physics, vol. 8. Pergamon Press, Oxford (1984)
2. Tolmachev, V.V., Golovin, A.M., Potapov, V.S.: Thermodynamics and electrodynamics of continuous media. Mosk. Gos. Univ, Moscow (1988)
3. Ammari, H., Buffa, A., Nedelec, J.-C.: A justification of eddy currents model for the Maxwell equations. SIAM J. Appl. Math. **60**(5), 1805–1823 (2000)
4. Rodriguez, A.A., Valli, A.: Eddy Current Approximation of Maxwell Equations: Theory, Algorithms and Applications. Springer, Milan (2010). https://doi.org/10.1007/978-88-470-1506-7
5. Galanin, M.P., Popov, Yu.P.: Quasi-stationary electromagnetic fields in inhomogeneous media: mathematical modeling. Fizmatlit, Moscow (1995)
6. Kolmbauer, M.: Existence and uniqueness of eddy current problems in bounded and unbounded domains. Numa-report 2011-03. Institute of Computational Mathematics, Linz (2011)
7. Kalinin, A.V., Sumin, M.I., Tyukhtina, A.A.: Stable sequential Lagrange principles in the inverse final observation problem for the system of Maxwell equations in the quasistationary magnetic approximation. Differ. Equ. **52**(5), 587–603 (2016). https://doi.org/10.1134/S0012266116050062
8. Kalinin, A.V., Tyukhtina, A.A.: Quasistationary electromagnetic fields in inhomogeneous media with non-conductive and low conductive inclusions. Zhurnal SVMO **4**(18), 119–133 (2016)
9. Kalinin, A.V., Sumin, M.I., Tyukhtina, A.A.: On the inverse problems of final observation for the system of Maxwell equations in the quasistationary magnetic approximation and stable sequential Lagrange principles of its solution. Comput. Math. Math. Phys. **2**(57), 189–210 (2017)
10. Kalinin, A.V., Tyukhtina, A.A.: L_p-estimates for scalar products of vector fields and their application to electromagnetic theory problems. Math. Methods Appl. Sci. **41**(18), 9283–9292 (2018)
11. Mareev, E.A.: Global electric circuit research: achievement and prospects. Phys. Usp. **5**(53), 504–511 (2010)
12. Kalinin, A.V., Slyunyaev, N.N., Mareev, E.A., Zhidkov, A.A.: Stationary and non-stationary models of the global electric circuit: well-posedness, analytical relations, and numerical implementation. Izv. Atmos. Ocean. Phys. **3**(50), 314–322 (2014). https://doi.org/10.1134/S0001433814030074
13. Kalinin, A.V., Slyunyaev, N.N.: Initial-boundary value problems for the equations of the global atmospheric electric circuit. J. Math. Anal. Appl. **450**(1), 112–136 (2017)
14. Boström, R., Fahleson, U.: Vertical propagation of time-dependent electric fields in the atmosphere and ionosphere. In: Dolezalek, H., Reiter, R. (eds.) Electrical Processes in Atmospheres, pp. 529–535. Steinkopff (1977)
15. Raviart, P.-A., Sonnendrücker, E.: A hierarchy of approximate models for the Maxwell equations. Numer. Math. **73**, 329–372 (1996). https://doi.org/10.1007/s002110050196
16. Larsson, J.: Electromagnetics from a quasistatic perspective. Am. J. Phys. **75**(3), 230–239 (2007)
17. Weitzner, H., Lawson, W.S.: Boundary conditions for the Darwin model. Phys. Fluids B **1**, 1953–1957 (1989)

18. Degond, P., Raviart, P.-A.: An analysis of the Darwin model of approximation to Maxwell's equations. Forum Math. **4**, 13–44 (1992)
19. Raviart, P.-A., Sonnendrücker, E.: Approximate models for the Maxwell equations. J. Comput. Appl. Math. **63**, 69–81 (1994)
20. Girault, V., Raviart, P.: Finite Element Methods for Navier-Stokes Equations. Springler, Heidelberg (1986). https://doi.org/10.1007/978-3-642-61623-5
21. Kalinin, A.V., Tyukhtina, A.A.: Darwin approximation for the system of Maxwell's equations in inhomogeneous conducting media. Comput. Math. Math. Phys. **8**(60), 1361–1374 (2020). https://doi.org/10.1134/S0965542520080102
22. Duvaut, G., Lions, J.L.: Inequalities in Mechanics and Physics. Springer, Heidelberg (1976). https://doi.org/10.1007/978-3-642-66165-5
23. Dautray, R., Lions, J.-L.: Mathematical Analysis and Numerical Methods for Science and Technology: Volume 5 Evolution Problems I. Springer, Heidelberg (2000). https://doi.org/10.1007/978-3-642-58090-1

Numerical Simulation of the Wind Resonance of the Bridge Based on Scale-Resolving Approaches

Nikita Ageev[✉], Olga Poddaeva, Pavel Churin, and Anastasia Fedosova

Moscow State University of Civil Engineering,
26, Yaroslavskoye Shosse, Moscow, Russia

Abstract. A series of computational and experimental studies of the wind resonance of the bridge span has been carried out. It is shown that for some wind directions the numerical solution of the Reynolds equations system (RANS) does not allow one to estimate the critical flow velocity at which wind resonance is observed. It has been established that modern CFD scale-resolving approaches (LES, DES) make it possible to obtain adequate estimates of the corresponding critical wind speed.

Keywords: Bridge · Wind resonance · Scale-resolving simulation · LES · DES · CFD · Numerical simulation

1 Introduction

The solution to the problem of the bridges sustainability against wind resonance is required due to the intensive road construction. The phenomenon of wind resonance occurs when the natural frequency of the bridge is close to the frequency of the exciting aerodynamic force (for example, the bridge across the Volga in the Volgograd region (2010) is an example of this phenomenon). It is necessary to develop a set of methods and approaches to simulate wind resonance. Exciting aerodynamic force is caused by vortex shedding around the bridge in the airflow. The Strouhal number characterizing this process is of the order of 0.05-0.2, which corresponds to relatively small vortex shedding frequency and process is quasi-stationary. In such processes, a fixed model is often used for numerical simulation. The vortex shedding frequency can be estimated by numerical modeling the bridge section aerodynamics in non-stationary formulation.

In this case, the system of Reynolds-Averaged Navier-Stokes (RANS) equations is often solved with the Shear Stress Transport (SST) turbulence model. The Reynolds averaging is a kind of time averaging, where the existence of the mean (time-average) component of the velocity is assumed. This mean velocity component should be significantly larger than velocity fluctuation. Usually it is correct for the well streamlined bodies. Strictly speaking it's not true for rough bulky bodies (like bridge) and flows with large turbulence level.

© Springer Nature Switzerland AG 2021
D. Balandin et al. (Eds.): MMST 2020, CCIS 1413, pp. 93–104, 2021.
https://doi.org/10.1007/978-3-030-78759-2_7

The RANS system of equations is non-closed. Different semi-empirical turbulence models are used for the closure of this equations system. These turbulent models are usually systems of differential (sometimes algebraic) equations. The most commonly used turbulence models are two-parametric $k - \epsilon$, $k - \omega$, their mix $k - \omega$ SST, one-parametric Spallart-Allmaras model, anisotropic Reynolds Stress Models. Most turbulence models were adjusted for the aerospace engineering applications, such as 2D airfoil or 3D wing and not adjusted for typical civil engineering objects. However, in some cases bulky bodies are producing large powerful essentially laminar vortices, so-called von Karman vortex street. In this mode the frequency of the vortex shedding still can be obtained by RANS approach.

The situation when the bulky body produces developed turbulent wake (without major laminar vortices) is more general case. In this situation turbulent vortex cascade is the main feature of the flow. There is no any "mean" velocity component for the vortices in the middle of the cascade. Obviously, the RANS approach assumptions are far from the real flow in this flow mode. Usually it is not possible to simulate the periodic process and to obtain the frequency of the exciting force in this case [1]. The numerical simulation based on scale-resolving approaches (such as Large Eddy Simulation, LES or Detached Eddy Simulation, DES) is more physically adequate. While the solution of the Reynolds equation system (RANS) simulates all scales of the turbulence by synthetic models, the approaches such as large eddy simulation (LES) or detached eddy simulation (DES) involve direct modeling of eddies whose characteristic size is larger than the mesh element. Small eddies are modeled using subgrid viscosity models (for example, Smagorinsky model). Than smaller the mesh element size is, the error caused by subgrid viscosity models is smaller. Thus, it is advisable to use sufficiently refined computational grids in numerical modeling using scale-resolving approaches.

2   Problem Statement

There is an example of modeling a triangular cylinder in cross flow using scale-resolving approaches in [2]. There are 26 elements across the edge of the cylinder there, and 81 elements along its length. The flow velocity is 17.3 m/s, and the Reynods number Re = 45,500. The time step is 10^{-5} s. This kind of simulation requires many computational resources. In addition, the aerodynamic force as function of time is nonharmonic. It complicates the interpretation of the results. Therefore, there is the basis for the development of a technique for assessment of the vibrations amplitude of the bridge based on scale-resolving approaches. The purpose of this work is to develop such technique. The technique includes practice description of the physical models, meshes, simulation techniques, data processing. The technique is confirmed by the comparison with the experimental data from the bridge section wind tunnel tests. Discussion, insights and ways of further technique improvement are also provided.

3   Technique of the Amplitude Estimation Based on Scale-Resolving Simulation

The technique is developed and applied for a real bridge. The main points of the technique and the results of its application are below. Comparison of the results of numerical simulation and tests in the large research gradient wind tunnel of Moscow State University of Civil Engineering is presented. The numerical simulation is performed in the scale of the wind tunnel model (1:70) for the correct comparison of the results.

The geometric model must be suitable for constructing a computational grid with prismatic elements in the boundary layer. When building a model, one should neglect the details that have size less than the characteristic mesh element size near the model. The characteristic cell size should allow approximately 20 cells along the main elements of the bridge (for a beam bridge - along the height of the beam). It is permissible to use a small length section in numerical modeling if the cross section of the natural size bridge is constant along the span. The section must include all major periodic structural elements. The geometric model used in the present work is shown in Fig. 1.

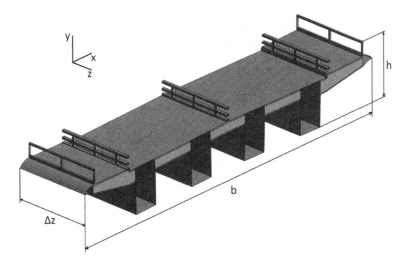

Fig. 1. Geometric model of the bridge element that is used for numerical simulation (CFD).

The computational mesh should be sufficiently detailed in the vicinity of the model and in the vortex shedding area. The boundaries of the computational domain should comply with the standard recommendations for aerodynamics of buildings [3]. The characteristic size maximum characteristic size of the object (usually the width of the bridge). Downstream of the bridge, the mesh should remain detailed at a distance $L_{refined}$. This distance should be not less than the distance corresponding to the 1 vortex period with a frequency equal to the natural frequency of the bridge model f_0 at the considered flow velocity V:

$$L_{refined} \geq \frac{V}{f_0} \tag{1}$$

An unstructured tetrahedral computational grid with prismatic layers is built for the considered bridge (Fig. 2). The characteristic cell size near the bridge model is 2 mm (scale of the wind tunnel model). The grid contains 10 layers of prismatic cells with a growth rate of 1.4 to simulate the boundary layer. The total number of grid elements is 6 820 520, the number of nodes is 1 520 966. Numerical modeling is performed using the detached eddy simulation method with the IDDES modification described in [4]. The pressure-velocity coupling scheme is SIMPLE. For angles of attack other than zero, it is recommended to use a corrected scheme (SIMPLEC) with a skewness correction (Skewness Correction is equal to 1). The calculation scheme is an implicit bounded second order of the bounded central difference type, the number of iterations per one time step is about 10. It is recommended to use a time step corresponding to the Courant number of about 1 (in this case, $dt = 0.0004$ seconds). The flow velocity V for simulation must correspond to the maximum considered flow velocity $V_{maxnatur}$. The transformation of the velocity from wind tunnel scale to natural scale is performed by assuming that the frequency of the vortex shedding is proportional to the flow velocity and inversely proportional to the size of the object (the Strouhal number is constant):

$$Sh = \frac{f_0 L}{V} = \frac{f_{0natur} L_{natur}}{V_{maxnatur}} \tag{2}$$

$$V = \frac{f_0 L}{f_{0natur} L_{natur}} V_{maxnatur} \tag{3}$$

In this case $V = 5$ m/s.

Total numerically simulated time should be of order 2–3 characteristic times τ, where

$$\tau = \frac{1}{\delta} \tag{4}$$

δ—damping coefficient. In this case the damping coefficient is 0.02 1/s. The equation of forcing oscillations with damping:

$$\ddot{y} + 2\delta\dot{y} + \omega_0^2 y = F_y \tag{5}$$

F_y—vertical component of the exciting force (aerodynamic in this case), y—vertical bridge coordinate, $\omega_0 = 2\pi f_0$—natural round frequency.

However, the required simulation time is about 100–150 s in this case. That needs quite a lot of calculation time. The simulation of 27 s of real time is performed. Recalculation to the required time is carried out. The recalculation is based on the copying of the part of the initial process with exclusion of the transition start process and the correct phase choice.

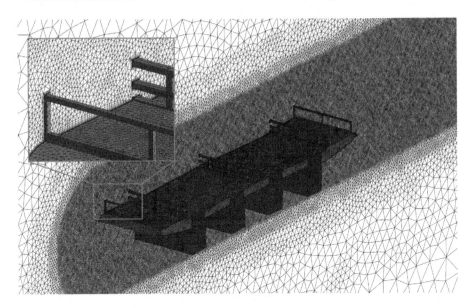

Fig. 2. Computational mesh.

Usually in aerodynamic research the C_y coefficient is invesigated:

$$F_y = C_y \frac{\rho V^2}{2} S \qquad (6)$$

where ρ—air density, S—characteristic area (in this case multiplication of the section length Δz by its width b: $S = b\Delta z$).

The computational fluid dynamics results for the coefficient $C_y(t)$ for fixed wing speed $V = 5$ m/s are presented in Fig. 3.

As one can see, before $t = 4$ s, there is some kind of start transition process. This part of data will be moved out the consideration. The spectrum of the Cy amplitude (at $t > 4$ s) is presented in Fig. 4.

The spectrum has complex pattern. It is impossible to isolate any fixed frequency. It is concerned with the difficulty of the physical flow pattern. Visualization of the flow as Q-criteria isosurface ($Q = 10001/s^2$) is presented in Fig. 5.

Instanteneous velocity field at one of the side calculation domain boundaries is presented in Fig. 6.

Numerically dynamics of the bridge as linear oscillator (that corresponds to the equation of the forced oscillations with damping) is simulated for the estimation of amplitude. Simplectic Euler method is applied for the simulation [5]:

$$V_{yi+1} = V_{yi} + a_{yi}dt \qquad (7)$$

$$y_{i+1} = y_i + V_{i+1}dt \qquad (8)$$

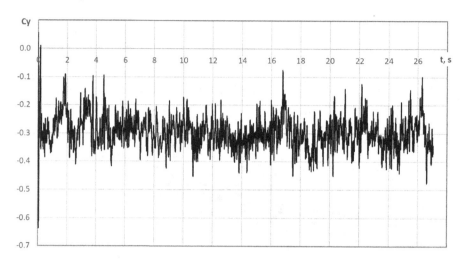

Fig. 3. Coefficient $C_y(t)$ at $V = 5\,\mathrm{m/s}$. IDDES, $dt = 0.0004\,\mathrm{s}$.

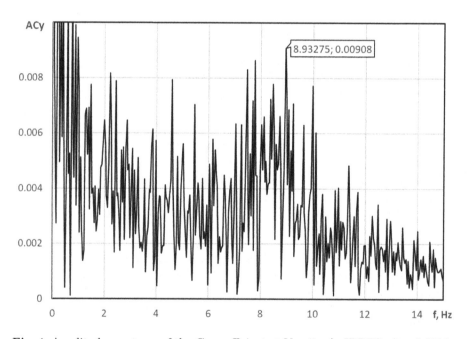

Fig. 4. Amplitude spectrum of the Cy coefficient at $V = 5\,\mathrm{m/s}$. IDDES, $dt = 0.0004\,\mathrm{s}$.

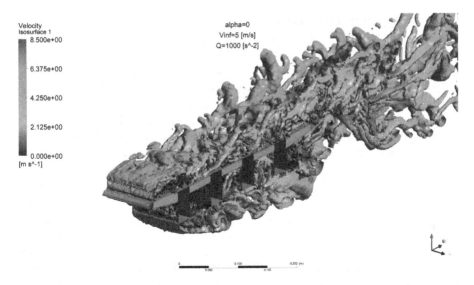

Fig. 5. Vortex system visualization. Isosurface of $Q = 1000 \text{ s}^{-2}$. IDDES, $dt = 0.0004 \text{ s}$.

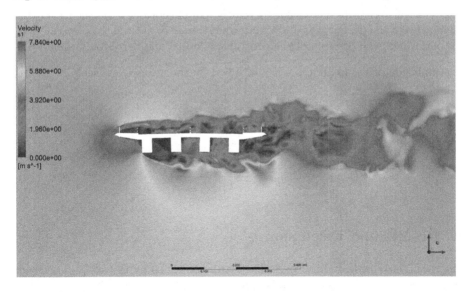

Fig. 6. Instanteneous velocity field at one of the side calculation domain boundaries. IDDES, $dt = 0.0004 \text{ s}$.

where

$$a_{yi} = \frac{F_y}{m} - \frac{k}{m} y_i - 2\delta V_{yi} \tag{9}$$

m—bridge section mass, k—hardness coefficient:

$$k = 4\pi^2 m f_0 \tag{10}$$

Simulation can be performed for any airflow velocity V_x by means of time step recalculation:

$$dt_x = \frac{V}{V_x} dt \tag{11}$$

Exciting aerodynamic force increase (that is proportional to squared flow velocity) should be taken into account.

Symplectic integrator conserves energy during any number of time steps.

The part of the stationary process (from 164080 to 323713 time step) is picked after simulation. The amplitude is obtained at this part of the process as

$$A = \frac{y_{max} - y_{min}}{2} \tag{12}$$

For example bridge vertical coordinate y as function of time at $V = 4.7\,\text{m/s}$ is presented in Fig. 7. Approximate oscillation limits $y = \bar{y} \pm A$ are also presented.

Simulation is carried out for wind tunnel flow velocities in the range of $V = 1.9\ldots5.4\,\text{m/s}$ with the step of $0.1\,\text{m/s}$ for further processing. Then velocity uncertainty of wind tunnel of about $0.1\,\text{m/s}$ (due to various factors such as turbulence and other unsteady processes) is taken into account. For it the averaging of every three adjacent in velocity amplitudes is performed (moving average is obtained). For transition from wind tunnel to natural scale two actions are performed (in approach of the constant Strouhal and Scrouton numbers):

1. Amplitude of the oscillations is multiplied by the geometrical scale

$$A_{natur} = \frac{L_{natur}}{L} A \tag{13}$$

2. Velocity is transformed by the formula:

$$V_{natur} = \frac{L_{natur}}{L} \frac{f_{natur}}{f_0} V \tag{14}$$

4 Results and Discussion

The results are presented in Fig. 8. The results obtained by the above-described technique are in satisfactory agreement with the experimental data. Nevertheless, at a flow velocity of about $22\,\text{m/s}$, a burst of vibration amplitudes is observed in numerical studies. This may be due to insufficiently accurate reproduction of the dependence of the direction of flow velocity on time. The oscillating system (bridge) has low damping. This leads to a very rapid change in the amplitude of the oscillatory process with a slight change in the parameters. It is necessary to take into account possible changes in the modulus and direction of the incoming flow velocity. Changes in the velocity modulus are primarily due to flow turbulence and are taken into account in the described technique by taking

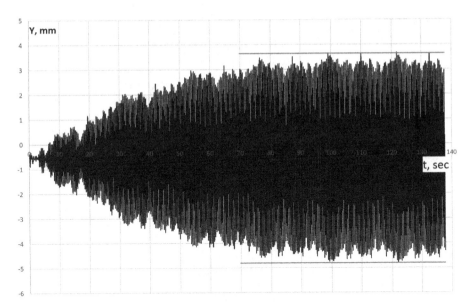

Fig. 7. Vertical bridge element shift as function of time. Numerical simulation of bridge dynamics, simplectic Euler method, $V = 4.7\,\mathrm{m/s}$

a moving average (the results obtained for 3 different velocities are averaged). The averaging is performed by the next way:

$$A_{natur}\left(V_{natur}\right)$$

$$= \frac{A_{natur}\left(V_{natur} - 0.1\,\mathrm{m/s}\right) + A_{natur}\left(V_{natur}\right) + A_{natur}\left(V_{natur} + 0.1\,\mathrm{m/s}\right)}{3}$$

(15)

The averaging step of 0.1 m/s is equal to estimated wind tunnel velocity scatter. Averaging has 2 major parameters: window size and step of the dynamics simulation. Basic 3 points window provide sufficiently good results, increase of the window size smoothens the chart, decrease of the window chart leads to the basic, very rapidly changing results, that is not physically valid (Fig. 9). Step of the dynamics simulation effects on the amplitude estimation. Smaller step is more valid. Greater step effects can be predicted from Fig. 9 (possible loss of amplitude peaks). Generally, this step should correspond to the uncertainty of the velocity.

The direction of speed can have the following sources of uncertainty:

- freestream turbulence;
- vertical vibration of the bridge (uncertainty of the order of 1°);
- model position inaccuracy.

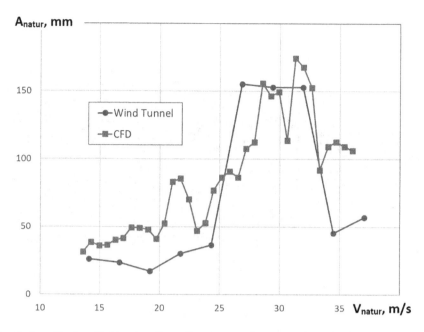

Fig. 8. Amplitude of bridge bending vibrations as function of wind velocity for numerical simulation estimation (CFD) and physical modeling (wind tunnel). Real scale.

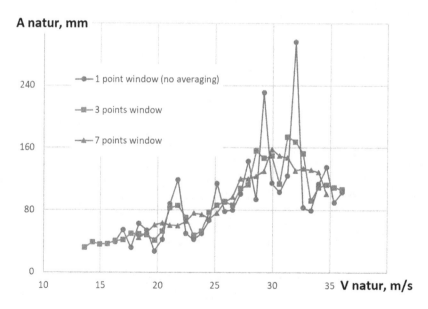

Fig. 9. Influence of the averaging window size.

It is necessary to investigate these parameters for a full-scale model and a physical experiment. This will also make it possible in the future to improve the numerical simulation technique and, probably, to achieve a better agreement with experimental data.

During last years there are attempts to use fluid-structure interaction (FSI) schemes for bridge resonance modeling (RANS or scale-resolving approaches are used) [6–8]. In this case the results of the aerodynamic problem solution are transferred to the dynamic (or structural) solver, the model of the bridge shifts, mesh deforms and the process is repeated (so-called two-way FSI). This approach let to obtain the natural velocity angle change, but is more computationally expensive. For some flow modes (with developed turbulent wake) it is necessary to use the scale-resolving approaches that makes problem quite complicated.

5   Conclusion

Scale-resolving approaches allow to obtain physically adequate flow patterns. A variable aerodynamic force is simulated, while in the case of using approaches based on the RANS equations, the aerodynamic force acting on the bridge is constant in time in some cases. A technique for estimating of the vibration amplitudes of a bridge is proposed. The technique is based on the use of DES methods in combination with dynamics modeling using the Euler symplectic method. Satisfactory agreement with the experimental results is obtained.

Acknowledgement. This work was financially supported by the Ministry of Science and Higher Education of the Russian Federation (Project: Theoretical and experimental design of new composite materials to ensure safety during the operation of buildings and structures under conditions of technogenic and biogenic threats FSWG-2020-0007).

References

1. Poddaeva, O., Ageev, N., Fedosova, A.: Investigation of the influence of various factors on the results of a calculation-experimental assessment of frequencies and amplitudes during vortex excitation of bending vibrations of building structures. In: Journal of Physics: Conference Series, vol. 1425, no. 1, p. 012135. IOP Publishing (2019)
2. Menter, F.R., Schutze, J., Gritskevich, M.: Global vs. zonal approaches in hybrid RANS-LES turbulence modelling. In: Fu, S., Haase, W., Peng, S.H., Schwamborn, D. (eds.) Progress in Hybrid RANS-LES Modelling, pp. 15–28. Springer, Heidelberg (2012). https://doi.org/10.1007/978-3-642-31818-4_2
3. Franke, J., et al.: Recommendations on the use of CFD in predicting pedestrian wind environment. Cost action C, vol. 14, May 2004
4. Gritskevich, M.S., Garbaruk, A.V., Schutze, J., Menter, F.R.: Development of DDES and IDDES formulations for the $k - \omega$ shear stress transport model. Flow Turbul. Combust. **88**(3), 431–449 (2012)
5. Donnelly, D., Rogers, E.: Symplectic integrators: an introduction. Am. J. Phys. **73**(10), 938–945 (2005)

6. Bai, Y., Sun, D., Lin, J.: Three dimensional numerical simulations of long-span bridge aerodynamics, using block-iterative coupling and DES. Comput. Fluids **39**(9), 1549–1561 (2010)
7. Farsani, H.Y., Valentine, D.T., Arena, A., Lacarbonara, W., Marzocca, P.: Indicial functions in the aeroelasticity of bridge decks. J. Fluids Struct. **48**, 203–215 (2014)
8. Szabo, G., Gyorgyi, J., Kristof, G.: Three-dimensional FSI simulation by using a novel hybrid scaling-application to the Tacoma Narrows Bridge. Periodica Polytechnica Civ. Eng. **64**(4), 975–988 (2020)

Influence of Material Damage on the Parameters of a Nonlinear Flexible Wave Which Spread in a Beam

Dmitry M. Brikkel[1]([✉]) [ID] and Vladimir I. Erofeev[1,2] [ID]

[1] National Research Lobachevsky State University, Gagarin Avenue 23,
603950 Nizhny Novgorod, Russia
[2] Mechanical Engineering Research Institute of RAS, Belinsky Street 85,
603024 Nizhny Novgorod, Russia

Abstract. In this paper, a mathematical model is formulated (in linear and nonlinear formulations) and investigated, which makes it possible to describe the propagation of flexural waves in a beam taking into account the damage of its material. An approach is proposed that determines new dependences of the flexural waves parameters on the material damage degree. This approach makes it possible to formulate a self-consistent problem that includes the equations of material dynamics and the conditions for its destruction. In the framework of a geometrically nonlinear model of a damaged rod, the problem of the intense bending waves formation of a stationary profile is considered. It is shown that such essentially non-sinusoidal waves can be either periodic or solitary (localized in space). The dependencies connecting the parameters of the waves (amplitude, width, wavelength) with the damage to the material are determined. It is shown that the periodic waves amplitude and the solitary waves amplitude increase with increasing material damage parameter, in turn, the periodic waves length and the solitary waves width decrease with increasing this parameter.

Keywords: Material damage · Bending wave · Longitudinal wave · Material damage degree · Nonlinearity

1 Introduction

Today, the mechanics of a damaged continuum is intensively developed by many authors. The first works in this field were fundamental studies by L. M. Kachanov, which are summarized in his monograph [1], and the detailed investigations and analysis by Yu. N. Rabotnov that are generalized in [2]. The significance of these pioneer works, which presently are recognized as classical,

This work was financially supported by RFBR grant 19-38-90282.

D. Balandin et al. (Eds.): MMST 2020, CCIS 1413, pp. 105–116, 2021.
https://doi.org/10.1007/978-3-030-78759-2_8

consists in the possibility of using a unified approach for description of the damage of elastic and elastoplastic bodies. The damage is usually understood as a reduction of an elastic response of the body due to decreasing of the effective area, through which the internal forces are transmitting from one part of the body to another. This phenomenon is caused by the appearance and spreading of the scattered field of microdefects (the microcracks in the case of elasticity, the dislocations in the case of plasticity, the micropores in the case of creep, and the surface microcracks in the case of fatigue) [3].

The damage, i.e. the degradation of the mechanical properties of a solid material, cannot be measured directly in the same manner as, for example, velocity, force, or temperature. The damage can be detected indirectly only by analyzing the response of the elastic structure on the various external impacts. According to experimental knowledge, the presence of a damages field inside a solid material can be observed also by changing of physical features of the structure. For example, it may be the decreasing of velocity of ultrasonic signal propagation [4–6], a decrease in the Young's modulus (the modulus defect) [7], a decrease in material density (loosening) [8], a hardness change [9], a decrease in the stress amplitude under the cyclic testing [10,11], and an acceleration of the tertiary creep [12].

The purpose of the present study is the modeling of the process of acoustic wave propagation through the damaged material, and estimation of influence of damage on the phase velocity and attenuation of that wave.

As a rule, in the mechanics of solids, problems of dynamics are considered separately from problems of damage accumulation. When developing such methods, it is customary to postulate in advance that the elastic wave velocity is a given damage function, and then experimentally determine the proportionality coefficients. The phase wave velocity and wave attenuation are usually considered to be power functions of frequency and linear functions of damage [13]. With its undoubted advantages (simplicity), this approach has a number of disadvantages, like any approach that is not based on mathematical models of processes and systems.

The authors of [14–16] consider the problem to be self-consistent, including, in addition to the equation of damage development, the dynamic equation of the theory of elasticity. This approach made it possible to consider a number of applied problems of wave dynamics of damaged materials and structural elements [17–23].

2   Mathematical Model

A beam performing bending vibrations is considered. We take into account the geometric nonlinearity of the beam (i.e., the nonlinear relationship between deformation and displacement), assuming that the middle line of the beam remains inextensible. We suppose that the beam was under the static or cyclic loads, and damage might accumulate in its material. To describe the measure of damage, we introduce the Kachanov – Rabotnov function $\Psi(x, t)$ which equals to zero when there are no damages and is close to 1 when the material is damaged [1,2].

We designate the displacements of the middle axis during bending as $W(x,t)$.

The dynamics of the beam is described by a set of equations with consideration of the damage of its material:

$$\begin{cases} \dfrac{\partial^2 W}{\partial t^2} + c_s^2 r_y^2 \dfrac{\partial^4 W}{\partial x^4} - \dfrac{c_s^2}{2}\dfrac{\partial}{\partial x}\left[\left(\dfrac{\partial W}{\partial x}\right)^3\right] = \beta_1 \dfrac{\partial \Psi}{\partial x} \\ \dfrac{\partial \Psi}{\partial t} + \alpha\Psi = \beta_2 E \dfrac{\partial W}{\partial x}. \end{cases} \quad (1)$$

Here $c_s = \sqrt{E\rho^{-1}}$, $r_y = \sqrt{J_y F^{-1}}$ where E is the Young's modulus, ρ is the material density, J_y is the axial moment of inertia, F is the beam's cross-sectional area, α, β_1, β_2 are constants which characterize the material damage ($\alpha = T^{-1}$, where T is the relaxation time [18], while the physical meaning of other coefficients is not so obvious $\beta_1\beta_2 < 0$).

System (1) is reduced to one equation with respect to the transverse displacement $W(x,t)$, which has the following form:

$$\begin{aligned} &\dfrac{\partial^2 W}{\partial t^2} - \dfrac{\beta_1\beta_2 E}{\alpha}\dfrac{\partial^2 W}{\partial x^2} + c_s^2 r_y^2 \dfrac{\partial^4 W}{\partial x^4} + \dfrac{1}{\alpha}\dfrac{\partial^3 W}{\partial t^3} \\ &+ \dfrac{c_s^2 r_y^2}{\alpha}\dfrac{\partial^5 W}{\partial x^4 \partial t} - \dfrac{c_s^2}{2}\dfrac{\partial}{\partial x}\left[\left(\dfrac{\partial W}{\partial x}\right)^3\right] - \dfrac{c_s^2}{2\alpha}\dfrac{\partial^2}{\partial x \partial t}\left[\left(\dfrac{\partial W}{\partial x}\right)^3\right] = 0 \end{aligned} \quad (2)$$

This equation in dimensionless variables $U = W(W_0^{-1})$; $z = x(r_y^{-1})$; $\tau = c_s(r_y^{-1})t$ will be written as:

$$\begin{aligned} &\dfrac{\partial^2 U}{\partial \tau^2} + a_1 \dfrac{\partial^2 U}{\partial z^2} + a_2 \dfrac{\partial^3 U}{\partial \tau^3} + \dfrac{\partial^4 U}{\partial z^4} + a_2 \dfrac{\partial^5 U}{\partial z^4 \partial \tau} \\ &- \dfrac{a_3}{2}\dfrac{\partial}{\partial z}\left[\left(\dfrac{\partial U}{\partial z}\right)^3\right] - \dfrac{a_2 a_3}{2}\dfrac{\partial^2}{\partial z \partial \tau}\left[\left(\dfrac{\partial U}{\partial z}\right)^3\right] = 0 \end{aligned} \quad (3)$$

Where $a_1 = -\beta_1\beta_2 E(\alpha c_s^2)^{-1}$, $a_2 = c_s(r_y\alpha)^{-1}$, $a_3 = W_0^2(r_y^2)^{-1}$.

3   Nonlinear Stationary Waves

We will seek a solution to Eq. (3) in the class of stationary waves:

$$U = U(\xi) \quad (4)$$

where $\xi = z - V\tau$ is the "running" coordinate, $V = const$ is the velocity of wave.

For small relaxation times $T \to 0$ (i.e. for $\alpha \to \infty$) with the substitution of (4), Eq. (3) reduces to the Duffing equation [24]:

$$\frac{d^2\Theta}{d\xi^2} + m_1\Theta + m_2\Theta^3 = 0, \tag{5}$$

$$\Theta = dW(d\xi)^{-1}, \tag{6}$$

$$m_1 = a_1 + V^2, \tag{7}$$

$$m_2 = -0,5a_3. \tag{8}$$

Here Θ is angle of rotation of the beam's cross-section.

The signs of the coefficients m_1 and m_2 indicate the possibility of the existence of nonlinear stationary flexural waves. In this case, the first coefficient is always positive ($m_1 > 0$), and the second is always negative ($m_2 < 0$).

Equation (5) has the first integral:

$$\frac{1}{2}\left(\frac{d\Theta}{d\xi}\right)^2 = E - \frac{m_1}{2}\Theta^2 - \frac{m_1}{4}\Theta^4 \tag{9}$$

which can be interpreted as the energy conservation law for an anharmonic oscillator. Here E – the constant of integration, which has the meaning of the initial energy of the system, and the function (10) has the sense of potential energy.

$$f(\Theta) = \frac{m_1}{2}\Theta^2 - \frac{m_2}{4}\Theta^4 \tag{10}$$

Equation (9) presumes the separation of variables:

$$\sqrt{2}d\xi = \frac{d\Theta}{\sqrt{E - f(\Theta)}}, \tag{11}$$

and has limited solutions in the region between any real roots of the polynomial $E - f(\Theta)$, where $E - f(\Theta) > 0$.

The potential energy function (10) has local maximum $f_{max} = -m_1^2/4m_2$ when $\Theta = \pm\sqrt{-m_1/m_2}$ and local minimum $f_{min} = 0$ when $\Theta = 0$ (Fig. 1). Because of that fact on the phase plane $(T; dT/d)$ the point $(0; 0)$ is a stable equilibrium position of the "center" type, and the points $(\pm\sqrt{(m_1/m_2)}; 0)$ - unstable equilibrium positions of the "knot" type. The phase portrait of the system is shown in Fig. 2.

In this case, bounded solutions of Eq. (9) exist only for $0 \le E \le f_{max}$. Moreover, the polynomial $E - f(\Theta)$ has four real roots $\Theta_{1,2} = \pm\alpha_0, \Theta_{3,4} = \pm\beta_0$, where

$$\alpha_0^2 = \frac{m_1 - \sqrt{m_1^2 + 4m_2E}}{-m_2}, \tag{12}$$

$$\beta_0^2 = \frac{m_1 + \sqrt{m_1^2 + 4m_2E}}{-m_2}, \tag{13}$$

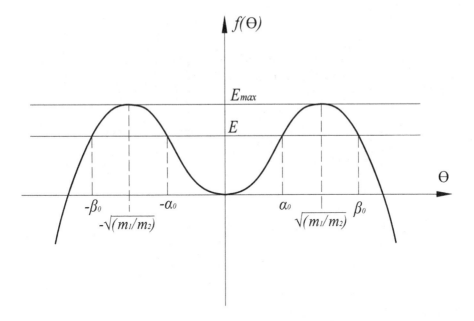

Fig. 1. Potential energy of a nonlinear oscillator

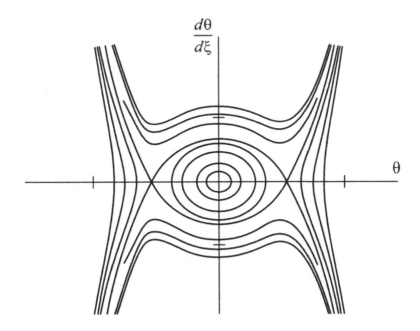

Fig. 2. Phase portrait of a nonlinear oscillator.

$$\alpha_0^2 > \beta_0^2, \tag{14}$$

and takes positive values when $-\alpha_0 < \Theta < \alpha_0$. In case when $E = f_{max} = -m_1/4m_2$ the roots match in pairs $\Theta_1 = \Theta_3, \Theta_2 = \Theta 4$, which corresponds to the motion along the separatrix on the phase plane.

The Eq. (11) takes the following form:

$$\alpha_0 \beta_0 \sqrt{-\frac{m_2}{2}} d\xi = \frac{d\Theta}{\sqrt{(1 - (\Theta^2/\alpha_0^2))(1 - (\Theta^2/\beta_0^2))}}, \tag{15}$$

And with the help of substitution

$$\frac{\Theta}{\alpha_0} = Z, \tag{16}$$

can be reduced to an elliptic integral of the first kind

$$\sqrt{\frac{m_2}{2}} (\xi - \xi_0) = \frac{1}{\beta_0} \int_0^Z \frac{dZ}{\sqrt{(1 - Z^2)(1 - S^2 Z^2)}}, \tag{17}$$

where $S^2 = \alpha_0^2/\beta_0^2$.

Inverting the elliptic integral on the right-hand side of (17) for $Z = 0, \xi_0 = 0$, we obtain a solution describing nonlinear periodic oscillations in the form:

$$\Theta(\xi) = \alpha_0 sn \left(-\sqrt{\frac{1}{2} m_2 \beta_0^2} \xi, S \right). \tag{18}$$

In expression (18), we introduce the notations:

$$A = \alpha_0 = \sqrt{\frac{(m_1 - \sqrt{m_1^2 + 4m_2 E})}{-m_2}}, \tag{19}$$

$$k = \sqrt{-\frac{1}{2} m_2 \beta_0^2} = \sqrt{\frac{1}{2} \left(m_1 + \sqrt{m_1^2 + 4m_2 E} \right)}, \tag{20}$$

$$S^2 = \frac{\alpha_0^2}{\beta_0^2} = \frac{m_1 - \sqrt{m_1^2 + 4m_2 E}}{m_1 + \sqrt{m_1^2 + 4m_2 E}}, \tag{21}$$

where A – amplitude of a stationary wave, k – nonlinear analog of wavenumber, S – the modulus of the elliptic function, and the wavelength Λ is $\Lambda = 4K(S)/k$. From relations (19, 20, 21) it follows that when E varies from 0 to $E_{max} = -m_1^2/4m_2$ the vibration frequency decreases from $k = \sqrt{m_1}$ to $k = \sqrt{m_1/2}$, and the amplitude of the oscillations changes within $0 \leq A \leq A^{(c)} = \sqrt{m_1/m_2}$, where $A^{(c)}$ – the amplitude of the oscillation corresponding to the motion along the separatrix on the phase plane. The modulus of the elliptic function (coefficient of linear distortion) at the same time changes in the interval $0 \leq S^2 \leq 1$.

Similarly, excluding E from expressions (19, 20, 21), we obtain relations between the parameters A, k, S in the solution (18):

$$S^2 = -\frac{m_1 A^2}{2m_1 + m_2 A^2}, \tag{22}$$

$$k = \sqrt{\frac{2m_1 + m_2 A^2}{2}}, \tag{23}$$

$$\Lambda = \frac{4 K(S)}{\sqrt{m_1 + \frac{m_2 A^2}{2}}}. \tag{24}$$

Another form of these expressions:

$$A = \pm\sqrt{-\frac{2m_1}{m_2}\frac{S^2}{1 + S^2}}, \tag{25}$$

$$k = \sqrt{\frac{m_1}{1 + S^2}}. \tag{26}$$

Taking into account the introduced designations, solution (18), which describes nonlinear periodic oscillations along closed phase trajectories near the separatrix, can be represented as an elliptic sine (Fig. 3):

$$\Theta(\xi) = A sn(k\xi, S) \tag{27}$$

The parameters of a torsional stationary wave are related by relation (27). When substituting expressions m_1, m_2 in (27), we obtain:

$$A = \sqrt{\frac{4\left(\frac{-\gamma}{c_s^2} + V^2\right) r_y^2 S^2}{W_0^2(1 + S^2)}}, \tag{28}$$

$$k = \sqrt{\frac{\left(\frac{-\gamma}{c_s^2} + V^2\right)}{(1 + S^2)}}, \tag{29}$$

A solitary stationary wave has the form of a drop (kink) (Fig. 4) and is described by the hyperbolic tangent

$$\Theta(\xi) = A^{(c)} th\left(\frac{\xi}{\Delta}\right), \tag{30}$$

where

$$A^{(c)} = \pm\sqrt{\frac{2\left(\frac{-\gamma}{c_s^2} + V^2\right) r_y^2}{W_0^2}} \tag{31}$$

– wave amplitude,

$$\Delta = \sqrt{\frac{2}{\left(\frac{-\gamma}{c_s^2} + V^2\right)}}$$ (32)

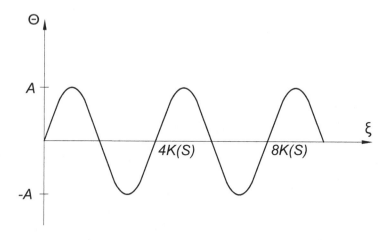

Fig. 3. Nonlinear periodic stationary wave profile.

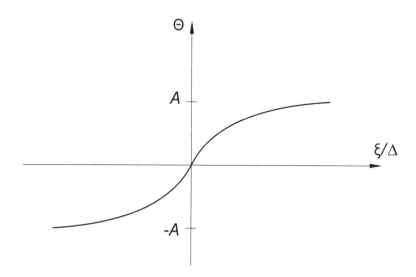

Fig. 4. Profile of a nonlinear solitary stationary wave.

– wave width.
Where $\gamma = \frac{\beta_1\beta_2 E}{\alpha}$ - is coefficient characterizing the damage to the material.

It follows from (28) and (31) that the amplitude of the periodic wave A and the amplitude of the solitary wave $A(c)$ increase with increasing material damage parameter. In turn, the length of a periodic wave (see (29)) and the width of a solitary wave (see (32)) decrease with increasing damage parameter.

It is noteworthy that the ratio of the stationary wave amplitude to the wavenumber is a constant which is determined only by the inertia radius of the beam's cross section.

$$\frac{A}{k} = \frac{2r_y S}{W_0} = const, \tag{33}$$

Notice, that the product of the wave amplitude by its width is also a constant.

$$A^{(c)} \Delta = \frac{2r_y}{W_0} = const. \tag{34}$$

For $S = 1$, expressions (33) and (34) are identical. This is obvious, since at this value the elliptic sine transforms into a hyperbolic tangent. For a beam of circular cross-section, the axial inertia radius is half the radius. Relation (33) in this case can be rewritten as, where d is the beam diameter.

4   Alternative Mathematical Model

When formulating the mathematical model (1), it was assumed that the main source of material damage is the angle of rotation of the beam cross-section during bending. It is no less natural to assume that damage mainly depends on volumetric deformation and this will lead instead of (1) to the following system of equations:

$$\begin{cases} \dfrac{\partial^2 W}{\partial t^2} + c_s^2 r_y^2 \dfrac{\partial^4 W}{\partial x^4} - \dfrac{c_s^2}{2} \dfrac{\partial}{\partial x} \left[\left(\dfrac{\partial W}{\partial x} \right)^3 \right] = \beta_1 \dfrac{\partial \Psi}{\partial x} \\[4mm] \dfrac{\partial \Psi}{\partial t} + \alpha \Psi = \beta_3 E \dfrac{\partial^2 W}{\partial x^2}. \end{cases} \tag{35}$$

where $\beta_3 \neq \beta_2$

This system is reduced to one equation about lateral displacement:

$$\begin{aligned} &\frac{\partial^2 W}{\partial t^2} + c_s^2 r_y^2 \frac{\partial^4 W}{\partial x^4} + \frac{1}{\alpha} \frac{\partial^3 W}{\partial t^3} - \frac{\beta_1 \beta_3 E}{\alpha} \frac{\partial^3 W}{\partial x^3} \\ &+ \frac{c_s^2 r_y^2}{\alpha} \frac{\partial^5 W}{\partial x^4 \partial t} - \frac{c_s^2}{2} \frac{\partial}{\partial x} \left[\left(\frac{\partial W}{\partial x} \right)^3 \right] - \frac{c_s^2}{2\alpha} \frac{\partial^2}{\partial x \partial t} \left[\left(\frac{\partial W}{\partial x} \right)^3 \right] = 0 \end{aligned} \tag{36}$$

Note that Eq. (36) contains a term $\sim \frac{\partial^3 W}{\partial x^3}$, in contrast to Eq. (2), which contains a term $\sim \frac{\partial^2 W}{\partial x^2}$.

If we look for a solution to Eq. (36) in the form of a traveling stationary wave:

$$W(x, t) = W(\eta), \tag{37}$$

where $\eta = x - Vt$.

Then it is easy to see that for a small relaxation time, i.e. at $\alpha \to \infty$, this equation will not be reduced to the equation of the classical Duffing oscillator, but will have the form:

$$\frac{\partial^2 \Theta}{\partial \eta^2} + m_1 \Theta + l_1 \frac{\partial \Theta}{\partial \eta} + m_2 \Theta^3 = 0 \tag{38}$$

The notation is introduced here:

$$\Theta = \frac{\partial W}{\partial \eta}, \tag{39}$$

$$m_1 = \frac{V^2}{c_s^2 r_y^2}, \tag{40}$$

$$m_2 = \frac{c_s^2}{2 r_y^2}, \tag{41}$$

$$l_1 = -\frac{\beta_1 \beta_3}{\alpha c_s^2 r_y^2}, \tag{42}$$

In the phase portrait of the oscillator (38), a singular point of the "center" type will go to a singular point of the "focus" type, and a singular point of the "saddle" type to a singular point of the "node" type.

According to the mathematical model (35), it is impossible to form nonlinear stationary flexural waves in a beam, in the material of which damage accumulates.

5   Conclusion

Analytical solutions are obtained for the dependences of the nonlinear bending waveparameters, such as amplitude, width and wavelength on the parameter characterizing the accumulation of damage in the material.

Notice, that the developed approach, which makes it possible to formulate and solve a self-consistent problem, including the equation of beam dynamics and the kinetic equation of damage to its material, can be used in the development of methods for acoustic control of materials and structural elements.

References

1. Kachanov, L.: Introduction to Continuum Damage Mechanics. Springer, Dordrecht (1986). https://doi.org/10.1007/978-94-017-1957-5
2. Rabotnov, Yu.: Creep Problems in Structural Members. North-Holland Series in Applied Mathematics and Mechanics. North-Holland Publishing Company, Amsterdam/London (1969)
3. Maugin, G.: The Thermomechanics of Plasticity and Fracture. Cambridge University Press, Cambridge (1992)
4. Zuev, L., Murav'ev, V., Danilova, Yu.: Criterion for fatigue failure in steels. Tech. Phys. Lett. **25**(5), 352–353 (1999). https://doi.org/10.1134/1.1262478
5. Hirao, O., Ogi, H., Suzuki, N., Ohtani, T.: Ultrasonic attenuation peak during fatigue of polycrystalline copper. Acta Mater. **48**(2), 517–524 (2000)
6. Wang, J., Fang, Q., Zhu, Z.: Sensitivity of ultrasonic attenuation and velocity change to cyclic deformation in pure aluminum. Phys. Status Solidi (a) **169**(1), 43–48 (1998)
7. Klepko, V., Lebedev, E., Kolupaev, B.B., Kolupaev, B.S.: Energy dissipation and modulus defect in heterogeneous systems based on flexible-chain linear polymers. Polym. Sci., Ser. B **49**(1–2), 18–21 (2007). https://doi.org/10.1134/S1560090407010058
8. Volkov, I., Korotkikh, Yu.: Equations of State of Viscoelastoplastic Media with Damages. Fizmatlit, Moscow (2008)
9. Collins, J.: Failure of Materials in Mechanical Design: Analysis, Prediction, Prevention, 2nd edn. Wiley, New York (1993)
10. Makhutov, N.: Deformation Criteria of Fracture and Calculation of Construction Elements for Strength. Mashinostroenie, Moscow (1981)
11. Romanov, A.: Fracture Under Small-Cycle Loading. Nauka, Moscow (1988)
12. Berezina, T., Mints, I.: Heat-Strength and Heat-Resistant of Metallic Materials. Nauka, Moscow (1976)
13. Uglov, A., Erofeev, V., Smirnov, A.: Acoustic control of equipment during manufacture and operation. In: Mitenkov, F. Nauka, Moscow (2009)
14. Erofeev, V., Nikitina, E.: Self-consistent dynamic problem of the evaluation of damage by acoustic method. Acoust. J. **1**(1), 554–557 (2010)
15. Erofeev, V., Nikitina, E.: Localization of the deformation wave propagating in the damaged material. Probl. Mech. Eng. Reliab. Mach. **39**(6), 559–561 (2010)
16. Erofeyev, V.I., Nikitina, E.A., Sharabanova, A.V.: Wave propagation in damaged materials using a new generalized continuum model. In: Maugin, G., Metrikine, A. (eds.) Mechanics of Generalized Continua. AMMA, vol. 21, pp. 143–148. Springer, New York (2010). https://doi.org/10.1007/978-1-4419-5695-8_15
17. Erofeev, V., Nikitina, E., Smirnov, S.: Acoustoelasticity of damaged materials. Control Diagn. **1**(3), 24–26 (2010)
18. Stulov, A., Erofeev, V.I.: Frequency-dependent attenuation and phase velocity dispersion of an acoustic wave propagating in the media with damages. In: Altenbach, H., Forest, S. (eds.) Generalized Continua as Models for Classical and Advanced Materials. ASM, vol. 42, pp. 413–423. Springer, Cham (2016). https://doi.org/10.1007/978-3-319-31721-2_19
19. Dar'enkov, A., Plekhov, A., Erofeev, V.: Effect of material damage on parameters of a torsional wave propagated in a deformed rotor. Procedia Eng. **150**, 86–90 (2016)

20. Erofeev, V., Lisenkova, E.: Excitation of waves by a load moving along a damaged flexible one-dimensional guide lying on an elastic foundation. Probl. Mech. Eng. Reliab. Mach. **45**(6), 495–499 (2016)
21. Erofeev, V., Leontyeva, A., Malkhanov, A.: Influence of material damage on the propagation of a longitudinal magnetoelastic wave in a rod. Comput. Mech. Contin. Media **11**(4), 397–408 (2018)
22. Antonov, A., Erofeev, V., Leontyeva, A.: Influence of material damage on Rayleigh wave propagation along the half-space boundary. Comput. Mech. Contin. Media **12**(3), 293–300 (2019)
23. Brikkel, D., Erofeev, V., Nikitina, E.: Influence of material damage on the parameters of a nonlinear longitudinal wave which spread in a rod. In: IOP Conference Series: Materials Science and Engineering, vol. 747, pp. 1–5 (2020)
24. Moiseev, N.: Asymptotic Methods of Nonlinear Mechanics. Nauka, Moscow (1981)

Chaotic Change of Extracellular Matrix Molecules Concentration in the Presence of Periodically Varying Neuronal Firing Rate

Maiya A. Rozhnova[1](✉)📖, Daniil V. Bandenkov[1], Victor B. Kazantsev[2,3,4]📖, and Evgeniya V. Pankratova[1]📖

[1] Department of Applied Mathematics, Institute of Information Technologies, Mathemaics and Mechanics, Lobachevsky State University of Nizhni Novgorod, Nizhny Novgorod, Russia
rozhnova@itmm.unn.ru
[2] Neurotechnology Department, Lobachevsky State University of Nizhni Novgorod, Nizhny Novgorod, Russia
[3] Neuroscience and Cognitive Technology Laboratory, Center for Technologies in Robotics and Mechatronics Components, Innopolis University, Innopolis, Russia
[4] Center of Neurotechnologies and Machine Learning, Immanuel Kant Baltic Federal University, Kaliningrad, Russia

Abstract. Transmission and processing of information in the brain are highly complicated processes that are defined by a lot of non-trivial interconnections of structural elements of neural networks. To shed light on peculiarities of such communication, a lot of mathematical models were introduced and computer simulations were carried out. In this work, a model describing the impact of neural activity on changes in brain extracellular matrix (ECM) molecules concentration was considered. It was assumed that the rate of neural activity is periodically changed. For this case, various regular and chaotic modes in dynamics of ECM-molecules concentration were observed. The role of the amplitude and frequency of the periodically varying neuronal firing rate in transitions between various dynamical modes in the ECM-model were examined. Bifurcational mechanisms for chaotic oscillations appearance were demonstrated.

Keywords: Brain extracellular matrix · Complicated dynamics · Bifurcations · Emergence of chaos

1 Introduction

Rhythmic behavior of neuronal electrical activity is believed to be responsible for an information coding and its further processing in the brain. Indeed,

This work is supported by grant of the President of the Russian Federation for state support of leading scientific schools No. NSh-2653.2020.2.

© Springer Nature Switzerland AG 2021
D. Balandin et al. (Eds.): MMST 2020, CCIS 1413, pp. 117–128, 2021.
https://doi.org/10.1007/978-3-030-78759-2_9

both carefully controlled experiments and rigorous mathematical modeling have demonstrated neuronal activity dependence on various external driving [1–6]. At the same time, recent studies shown that type of the neuronal response also depends on non-trivial interconnections of the brain structural elements. Concentration of neurotransmitters, gliotransmitters, proteases and other components of the brain extracellular matrix (ECM) can significantly modify the neuronal response [7–10]. It is now known that neuronal activity dependent concentration of ECM molecules affect synaptic plasticity and such feedback can have both positive and negative outcomes.

In this work, we focus on dynamical regimes observed in a model proposed recently for description of changes in ECM molecules concentration [11]. In contrast to [11], we assume that level of neuronal activity is a slowly periodically varying function. Due to existence of the natural periodic environmentally-induced oscillations known as circadian rhythms or the sleep-wake cycle, such assumption seems to be matter of course. To study the role of such driving in dynamical change of ECM molecules concentration, we consider three types of monostable regimes observed in autonomous system [11].

2   Mathematical Model

The mathematical mechanism behind the neural activity dependent dynamical changes of ECM molecules concentration was recently described in [8]. In this study, following by the work [11], we focus on simplified version of the model that is described by the following nonlinear differential equations:

$$
\begin{aligned}
\frac{dZ}{dt} &= -(\alpha_Z + \gamma_P P)Z + \beta_Z \left[Z_0 - \frac{Z_0 - Z_1}{1 + \exp\left(-\frac{Q + \alpha_Q Z - \theta_Z}{k_Z}\right)} \right], \\
\frac{dP}{dt} &= -\alpha_P P + \beta_P \left[P_0 - \frac{P_0 - P_1}{1 + \exp\left(-\frac{Q + \alpha_Q Z - \theta_P}{k_P}\right)} \right],
\end{aligned}
\tag{1}
$$

where the variable Z corresponds to the concentration of the ECM molecules, P is the concentration of proteases. The term $Q = Q_0 + Q_s \sin \omega_s t$ reflects the effect of neural activity on changes in Z and P. Here, similarly to previous works [8,11,12], we considered $Q_0 = 5$ and studied the role of the parameters Q_s and ω_s in dynamical change of ECM-molecules concentration. Another parameters of the system (1) were the following: impact of excitatory ECM-neuron interaction is $\alpha_Q = 0.23$, the rates of concentration degradation for ECM-molecules and proteases are $\alpha_Z = 0.0001 \text{ ms}^{-1}$ and $\alpha_P = 0.001 \text{ ms}^{-1}$, respectively, the corresponding activation rates are $\beta_Z = 0.01 \text{ ms}^{-1}$ and $\beta_P = 0.01 \text{ ms}^{-1}$, gain parameter is $\gamma_P = 0.001$, the asymptotic levels with $Q_s \to \pm\infty$ for correspondent activation functions are $Z_0 = 0$, $Z_1 = 1$ and $P_0 = 0$, $P_1 = 1$, the inverse slopes of the activation curves are $k_Z = 0.15$ and $k_P = 0.05$, activation midpoint for proteases-curve is $\theta_P = 6$. In order to study the role of periodically changed neuronal activity we considered various value of the parameter θ_Z.

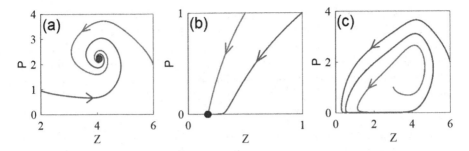

Fig. 1. Phase portraits for three different monostable regimes of the system (1): (a) stable focus for $\theta_Z = 5.5$; (b) stable node $\theta_Z = 6$; (c) stable limit cycle observed for $\theta_Z = 5.77$. In all cases, transients are shown by gray curves.

3   ECM Dynamics for Various Types of Neuronal Activity

Dynamical regimes of the system (1) for $Q_s = 0$ were studied in details in [11]. Various types of monostable and multistable behavior were revealed. In this work, we focus on the first type of regimes for ECM molecules concentration, and study the role of a weak periodical change in neuronal firing rate in dynamics of the system (1). With this aim, we consider three qualitatively different monostable modes observed for various constant levels of neuronal activity, particularly, we focus on three values of the activation midpoint θ_Z of the ECM-molecules activation curve.

3.1   Stable ECM-modes for Constant Levels of Neuronal Firing Rate

It is known from [11], that three types of asymtotically stable regimes, namely, stable focus, stable node and stable limit cycle can be observed in system (1) for $Q_s = 0$. In Fig. 1, phase portraits for these modes are presented. Particularly, in Fig. 1(a), the stable focus obtained for $\theta_Z = 5.5$ is shown by green point. This state has the coordinates $(Z^*, P^*) = (4.08, 2.23)$ and corresponds to a high level of ECM molecules concentration. In Fig. 1(b), the stable node obtained for $\theta_Z = 6$ is shown by blue point. This state has coordinates $(Z^*, P^*) = (0.163, 0)$ and corresponds to a low level of ECM molecules concentration. In Fig. 1(c) the stable limit cycle obtained for $\theta_Z = 5.77$ is shown by blue closed curve. Gray curves in Fig. 1 correspond to transients from various initial points and illustrate the behavior of phase trajectories near vicinity of stable states.

3.2   ECM Dynamics for Periodically Varying Neuronal Firing Rate

To demonstrate changes in the dynamics of the ECM molecules concentration in the presence of small periodic deviations of Q, bifurcations diagrams were calculated. Namely, the distributions of the maximal values of Z were numerically obtained versus the frequency ω_s for different values of the amplitude Q_s. Integration of system was carried out by the fourth-order Runge-Kutta method with a step of 0.01. To obtain a sufficient amount of data for bifurcation diagrams, the time interval from 0 to $2 \cdot 10^5$ was considered, while the points obtained for $t < 7 \cdot 10^4$ were not taken into account. The parameter ω_s was assumed to be changed in the range from 0 to 0.012 with a step of $h_\omega = 5 \cdot 10^{-5}$.

Chaos Emergence from Stable Focus. Let us start with the set of parameters providing the stable focus shown in Fig. 1(a). One-parameter bifurcation diagrams obtained for seven values of Q_s and for $\omega_s \in (0, 0.012)$ are presented in Fig. 2 and Fig. 3. For small driving amplitudes Q_s, Fig. 2(a), for each value of the frequency ω_s one value of Z_{max} was obtained. This means that all the attractors here are stable limit cycles. It should be noted that for considered parameters, the stable focus of autonomous system has the following complex eigenvalues: $\lambda_{1,2} = -0.00129 \pm 0.00569i$. Its imaginary part defines the frequency where Z_{max} (as well as the amplitude of oscillations) is maximal. Thus, for small driving amplitudes, typical resonance is observed. With the increase of Q_s the

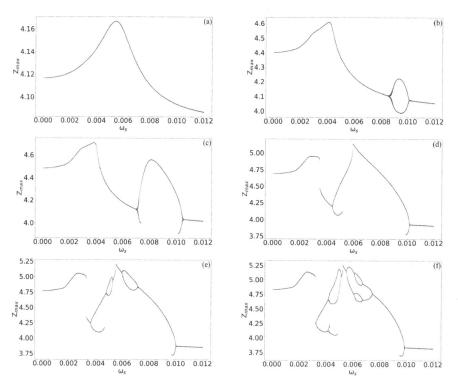

Fig. 2. One-parameter bifurcation diagram obtained for $\theta_Z = 5.5$ and different values of the parameter Q_s: (a) $Q_s = 0.01$; (b) $Q_s = 0.08$; (c) $Q_s = 0.1$; (d) $Q_s = 0.15$; (e) $Q_s = 0.17$; (f) $Q_s = 0.185$.

maximum of the curve moves to lower frequencies: for $Q_s = 0.01$ the maximal value of Z_{max} is observed for $\omega_s^{max} \approx 0.00545$, if $Q_s = 0.07$ then $\omega_s^{max} \approx 0.00425$ (not shown). Additionally, due to nonlinearity of the system, for larger driving amplitudes the shape of the bifurcation curve becomes complicated. The increase of ω_s leads to appearance of the second maximum in dependence $Z_{max}(\omega_s)$ as, for example, for $Q_s = 0.08$, Fig. 2(b). Note, that for this case, a frequency range exists where period-2 oscillations are observed. The increase of Q_s leads to the increase of the second maximum value of the dependence $Z_{max}(\omega_s)$. The diagram obtained for $Q_s = 0.1$ is shown in Fig. 2(c). In this case, when ω_s approaches 0.0071, the period doubling bifurcation occurs. Further increase of ω_s leads to sharp growth of Z_{max} values, but within the range $\omega_s \in (0.0075, 0.0101)$ simple period-1 cycle is observed, i.e. Z_{max} has only one large value. For $Q_s = 0.15$, Fig. 2(d), bifurcation diagram illustrates non-trivial ω_s-dependent changes in the amplitude of the concentration of ECM molecules with larger values of Z_{max} for larger driving frequencies. With further increase of Q_s, additional period doubling bifurcations occur, Fig. 2(e) and Fig. 2(f) that finally leads to emergence of irregular dynamics.

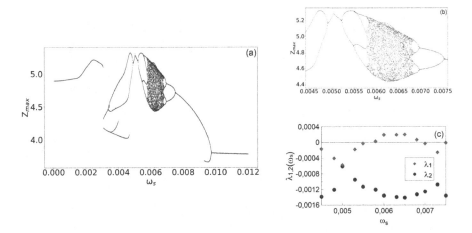

Fig. 3. (a) One-parameter bifurcation diagram obtained for $\theta_Z = 5.5$ and $Q_s = 0.2$; (b) enlarged part with chaotic dynamics of the diagram shown in (a); (c) Lyapunov exponents.

For example, in Fig. 3(a), for $Q_s = 0.2$, driving frequency range Ω_{ch} exists where a large number of Z_{max} for each value of ω_s can be obtained. Note that, the increase of integration time allows obtaining more unrepeatable values of Z_{max} demonstrating dense filling of the entire area with the points. Magnified part with this complicated distribution of the calculated Z_{max} values is presented in Fig. 3(b).

Complicated dynamics, and, particularly, chaotic dynamics is ubiquitous in nonlinear dynamical systems used for simulation of biological systems behavior [13–22]. In order to show that dynamics is chaotic, Lyapunov exponents are usually calculated. Therefore, in Fig. 3(c) the change of Lyapunov exponents with the increase of the driving frequency is presented. Since the system (1) is non-autonomous, the spectrum does not contain zero values. As seen from the data, within the range Ω_{ch} the maximal exponent becomes positive confirming the chaotic character of behavior.

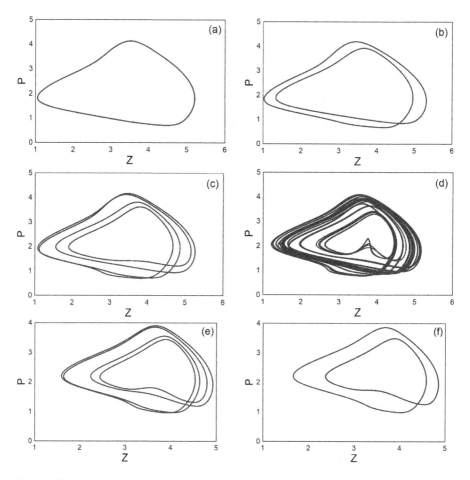

Fig. 4. Phase portraits obtained for $\theta_Z = 5.5, Q_s = 0.2$: (a) $\omega_s = 0.0052$; (b) $\omega_s = 0.0054$; (c) $\omega_s = 0.0057$; (d) $\omega_s = 0.0062$; (e) $\omega_s = 0.0069$; (f) $\omega_s = 0.007$

The mechanism of transition to chaos observed within the range Ω_{ch} for $Q_s = 0.2$ is demonstrated in Fig. 4, where phase portraits for six values of ω_s are presented. Limit cycle for $\omega = 0.0052$ is shown in Fig. 4(a). Such type of attractor exists within the range $\omega_s \in (0.005, 0.00525)$: within this range, one value of Z_{max} is observed in Fig. 2(e). For $\omega_s > 0.00525$ bifurcation diagram shows that for each w_s several values of Z_{max} exist. Phase portrait after the first period doubling bifurcation is shown in Fig. 4(b) where period-2 cycle is presented. Then the period doubles again and for driving frequencies within the range $w_s \in (0.00525, 0.0058)$ a period-4 stable limit cycle is observed, Fig. 4(c). Such cascade of doublings of the cycle period (known as Feigenbaum's scenario of transition to chaos [23–26]) leads to emergence of a chaotic attractor shown in Fig. 4(d). Further increase of the driving frequency provokes the backward period-doubling cascade leading to transition from chaos to various periodic solutions. Namely, a period-4 stable limit cycle shown in Fig. 4(e) is observed within the range $w_s \in (0.00685, 0.00698)$, whereas for $w_s \in (0.00698, 0.00754)$ a period-2 cycle is observed, Fig. 4(f).

Increase of ECM Molecules Concentration Level for the Stable Node Regime. For the case shown in Fig. 1(b), the bifurcation diagrams were calculated for two values of Q_s. Note that, in the presence of periodic driving, there is no any complicated behavior here. In Fig. 5, blue points are the Z^*-coordinates of the stable node at $w_s = 0$. For each frequency within the considered range, Z_{max} has a unique value, in particular, $Z_{max} > Z^*$ for all $w_s \neq 0$. It should be noted that, for low values of the frequency w_s, the values of Z_{max} are much higher than for $w_s = 0$. This can be explained by the analysis of numerator in the degree of exponent in the system (1). Obviously, that in the course of time the term

$$Q_0 + Q_s \sin(w_s t) - \theta_Z \qquad (2)$$

periodically approaches its minimal

$$Q_0 - Q_s - \theta_Z = Q_0 - (\theta_Z + Q_s) \qquad (3)$$

and maximal value

$$Q_0 + Q_s - \theta_Z = Q_0 - (\theta_Z - Q_s) = Q_0 - \theta_Z^*. \qquad (4)$$

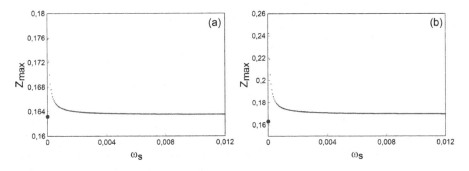

Fig. 5. One-parameter bifurcation diagram obtained for $\theta_Z = 6$ and different values of the parameter Q_s: (a) $Q_s = 0.01$; (b) $Q_s = 0.05$

From the bifurcation diagram obtained in [11], it is known that within the considered range, Z decreases with the increase of θ_Z^*, i.e. has got maximal value for $\theta_Z^* = \theta_Z - Q_s$. When ω_s is close to zero, signal changes slowly. For this range, to obtain Z_{max} we can consider θ_Z^* as bifurcation parameter of autonomous system. For example, for $\theta_Z = 6$ and slow periodic driving $Q_s \sin(\omega_s t)$ with $Q_s = 0.01$ we have $\theta_Z^* = 6 - 0.01 = 5.99$, for which autonomous system has a stable node at $(Z^*, P^*)=(0.179, 0)$ (for details of the bifurcation diagram obtained for autonomous system see [11]). As seen from in Fig. 5(a), for non-autonomous system (1) the closest to zero value of ω_s, e.g. $\omega_s = 2 \cdot 10^{-5}$, gives $Z_{max} \approx 0.178$. Similarly, for slow periodic driving $Q_s \sin(\omega_s t)$ with $Q_s = 0.05$ we have $\theta_Z = 6 - 0.05 = 5.95$, for which autonomous system has a stable node at $(Z^*, P^*)=(0.267, 0)$. Namely this value is reached by Z_{max} for small frequencies, e.g. for $\omega_s = 5 \cdot 10^{-5}$ the maximal concentration of ECM molecules is $Z_{max} \approx 0.24$, while for $\omega_s = 2 \cdot 10^{-5}$ is $Z_{max} \approx 0.259$, Fig. 5(b).

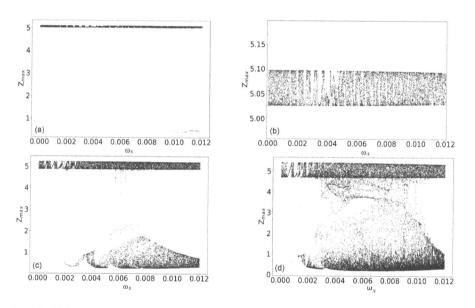

Fig. 6. Bifurcation trees obtained for $\theta_Z = 5.77$ and different values of the parameter Q_s: (a) $Q_s = 0.01$; (b) enlarged part with complicated dynamics of the diagram obtained for $Q_s = 0.01$ for $Z_{max} > 5$; (c) $Q_s = 0.05$; (d) $Q_s = 0.1$

Chaos Emergence from Oscillatory Mode. For the case shown in Fig. 1(c), the bifurcation diagrams were obtained for three values of Q_s. Particularly, in Fig. 6(a), the diagram for $Q_s = 0.01$ is presented. Figure 6(b) is an enlarged part of Fig. 6(a). As seen from these diagrams, most of the points are placed near $Z_{max} \approx 5$ that is defined by the maximal Z for the stable cycle of autonomous system $Z_{max} = 5.11$, Fig. 1(c). For large frequencies, several point are located near $Z_{max} \approx 0.5$ that corresponds to the minimal Z of the stable cycle in autonomous system, $Z_{min} = 0.51$. This can be explained as follows: due to the bifurcation mechanism of the cycle emergence when $Q_s = 0$, movement of the phase point is significantly slower near its minimum (see [11] for details). Therefore, for high frequencies, driving oscillations begin to reveal itself. This becomes more evident for larger values of the driving amplitudes: in Figs. 6(c) and (d) a lot of points are located in the lower part of the diagrams. Particularly, for $Q_s = 0.1$, the phase portraits with small oscillations are presented in Fig. 7(d, e, f).

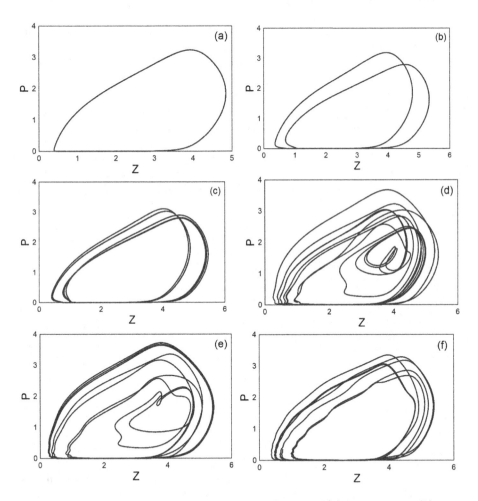

Fig. 7. Phase portraits obtained for $\theta_Z = 5.77, Q_s = 0.1$: (a) $\omega_s = 0.0005$; (b) $\omega_s = 0.0012$; (c) $\omega_s = 0.00125$; (d) $\omega_s = 0.004$; (e) $\omega_s = 0.006$; (f) $\omega_s = 0.012$

In Fig. 7(a), limit cycle for $\omega_s = 0.0005$ is shown. Such type of dynamics is observed for small values of ω_s where several periodicity windows exist. Within these parts of the diagrams only one value of Z_{max} was obtained. Note that, the frequency range of the first periodicity window is near the frequency of the cycle shown in Fig. 1(c), i.e. is near ≈ 0.0005, is observed. Small increase of the driving frequency near the ranges of periodicity windows leads to emergence of more complicated periodic attractors. Namely, for $\omega_s = 0.0012$, period-2 solution is observed, Fig. 7(b), whereas for $\omega_s = 0.00125$, period-4 cycle shown in Fig. 7(c) can be revealed.

4   Discussion

Finally, comparing the chaotic attractors observed in our study, we should take some notes on their similarity and difference. In both cases, we revealed the transition to chaos through the well-known Feigenbaum's scenario. At the same time, the emerged chaotic attractors differ significantly in time series representation, see Figs. 8(a) and (b) obtained for the same parameters of the driving. The point is, that for the last case, the origin of the driving-induced chaos is a limit cycle emerged from the loop of the saddle node separatrix. Therefore, the dynamics near the smallest values of variables is significantly slower than for another part of the attractor, Fig. 8(b). This particularly means that during the same time the change of ECM-molecules concentration is much faster for the case with the smaller activation midpoint for ECM concentration curve.

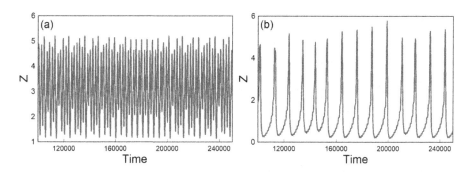

Fig. 8. Time series for ECM-molecules concentration $Z(t)$ obtained for the neuronal firing rate periodically varying with $Q_s = 0.2$ and $\omega_s = 0.006$ when (a) $\theta_Z = 5.5$ and (b) $\theta_Z = 5.77$.

Thus, the revealed peculiarities of chaotic change of ECM-molecules concentration in the presence of periodically varying neuronal firing rate shed light on the role of the natural periodic environmentally-induced oscillations known as circadian rhythms or the sleep-wake cycle in dynamics of the brain extracellular matrix and can be helpful for the specialists interested in neurodynamical modeling.

5 Conclusions

In this work, the dynamics of ECM molecules concentration in the presence of an external slowly varying periodic driving has been examined. This driving is assumed to be dependent on some level of neural network activity that can naturally changed by well-known circadian rhythms. To study the impact of these rhythms, three various monostable regimes have been considered. For the regime with stable node in autonomous case, the increase of ECM molecules concentration level has been revealed. For the regimes with stable focus and stable limit cycle in autonomous case, emergence of chaos has been observed. Summarizing the results, for all the considered regimes the role of external slowly varying periodic driving is significant. Even weak change of its frequency can move the considered system to disordered/diseased output mode or, contrariwise, facilitate the stabilization of its behavior.

References

1. Matsumoto, G., Aihara, K., Hanyu, Y., Takahashi, N., Yoshizawa, S.: Nagumo: Chaos and phase locking in normal squid axons. J. Phys. Lett. A **123**, 162–166 (1987)
2. Swadlow, H.: Monitoring the excitability of neocortical efferent neurons to direct activation by extracellular current pulses. J. Neurophysiol. **68**, 605–619 (1992)
3. Lee, S.-G., Kim, S.: Parameter dependence of stochastic resonance in the stochastic Hodgkin-Huxley neuron. Phys. Rev. E. **60**, 826–830 (1999)
4. Pikovsky, A., Rosenblum, M., Kurths, J.: Synchronization: A Universal Concept in Nonlinear Science. Cambridge University Press, Cambridge (2001)
5. Pankratova, E., Belykh, V., Mosekilde, E.: Role of the driving frequency in a randomly perturbed Hodgkin-Huxley neuron with suprathreshold forcing. Eur. Phys. J. B **53**, 529–536 (2006)
6. Belykh, V., Pankratova, E., Mosekilde, E.: Dynamics and synchronization of noise perturbed ensembles of periodically activated neuron cells. Int. J. Bifurcat. Chaos **18**(09), 2807–2815 (2008)
7. Dityatev, A., Schachner, M., Sonderegger, P.: The dual role of the extracellular matrix in synaptic plasticity and homeostasis. Nat. Rev. Neurosci. **11**, 735–746 (2010). https://doi.org/10.1038/nrn2898
8. Kazantsev, V., Gordleeva, S., Stasenko, S., Dityatev, A.: A homeostatic model of neuronal firing governed by feedback signals from the extracellular matrix. PLoS ONE **7**(7) (2012). https://doi.org/10.1371/journal.pone.0041646
9. Pankratova, E., Kalyakulina, A., Stasenko, S., Gordleeva, S., Lazarevich, I., Kazantsev, V.: Neuronal synchronization enhanced by neuron-astrocyte interaction. Nonlinear Dyn. **97**(1), 647–662 (2019)
10. Rozhnova, M.A., Kazantsev, V.B., Pankratova, E.V.: Brain extracellular matrix impact on neuronal firing reliability and spike-timing jitter. In: Kryzhanovsky, B., Dunin-Barkowski, W., Redko, V., Tiumentsev, Y. (eds.) NEUROINFORMATICS 2019. SCI, vol. 856, pp. 190–196. Springer, Cham (2020). https://doi.org/10.1007/978-3-030-30425-6_22

11. Lazarevich, I., Stasenko, S., Rozhnova, M., Pankratova, E., Dityatev, A., Kazantsev, V.: Activity-dependent switches between dynamic regimes of extracellular matrix expression. PLoS ONE **15**(1) (2020). https://doi.org/10.1371/journal.pone.0227917

12. Rozhnova, M., Pankratova, E., Stasenko, S., Kazantsev, V.: Bifurcation analysis of multistability and oscillation emergence in a model of brain extracellular matrix. Chaos, Soliton & Fractals (in Press)

13. Degn, H., Holden, A., Olsen, L. (eds.): Chaos in Biological Systems. Plenum Press, New York (1987)

14. Elbert, T., Ray, W., Kowalik, Z., Skinner, J., Graf, K., Birbaumer, N.: Chaos and physiology. Physiol. Rev. **74**, 1–47 (1994)

15. Chay, T., Fan, Y., Lee, Y.: Bursting, spiking, chaos, fractals, and universality in biological rhythms. Int. J. Bifurcat. Chaos **5**(3), 595–635 (1995)

16. Korn, H., Faure, P.: Is there chaos in the brain? II. Experimental evidence and related models. C. R. Biol. **326**(9), 787–840 (2003)

17. Rulkov, N.: Regularization of synchronized chaotic bursts. Phys. Rev. Lett. **86**, 183–186 (2001)

18. Belykh, V., Pankratova, E.: Chaotic synchronization in ensembles of coupled neurons modeled by the FitzHugh-Rinzel system. Radiophys. Quantum Electron. **49**(11), 910–921 (2006)

19. Papasavvas, C., Wang, Y., Trevelyan, A., Kaiser, M: Gain control through divisive inhibition prevents abrupt transition to chaos in a neural mass model. Phys. Rev. E Stat. Nonlinear Soft Matter Phys. **92**(3), 032723 (2015). https://doi.org/10.1103/PhysRevE.92.032723

20. Pankratova, E., Kalyakulina, A.: Environmentally induced amplitude death and firing provocation in large-scale networks of neuronal systems. Regular Chaotic Dyn. **21**, 840–848 (2016)

21. Lavrentovich, M., Hemkin, S.: A mathematical model of spontaneous calcium (II) oscillations in astrocytes. J. Theor. Biol. **251**, 553–560 (2008)

22. Sinitsina, M., Gordleeva, S., Kazantsev, V., Pankratova, E.: Calcium concentration in astrocytes: emergence of complicated spontaneous oscillations and their cessation. Izvestiya VUZ, Applied Nonlinear Dynamics (2021, in Press)

23. Feigenbaum, M.: Quantitative universality for a class of nonlinear transformations. J. Stat. Phys. **19**(1), 25–52 (1978)

24. Canavier, C.C., Clark, J.W., Byrne, J.H.: Routes to chaos in a model of a bursting neuron. Biophys. J. **57**(6), 1245–1251 (1990)

25. Sinitsina, M., Gordleeva, S., Kazantsev, V., Pankratova, E.: Emergence of complicated regular and irregular spontaneous Ca2+ oscillations in astrocytes. In: 4th Scientific School on Dynamics of Complex Networks and their Application in Intellectual Robotics, DCNAIR, Innopolis, Russia, pp. 217–220 (2020)

26. Li, Y., Xiao, L., Wei, Z., Zhang, W.: Zero-Hopf bifurcation analysis in an inertial two-neural system with delayed Crespi function. Eur. Phys. J. Spec. Top. **229**(6–7), 953–962 (2020)

Quiescence-to-Oscillations Transition Features in Dynamics of Spontaneous Astrocytic Calcium Concentration

Maria S. Sinitsina[1]([✉]) [iD], Susanna Yu. Gordleeva[3] [iD], Victor B. Kazantsev[2,3,4] [iD], and Evgeniya V. Pankratova[1] [iD]

[1] Department of Applied Mathematics, Institute of Information Technologies, Mathematics and Mechanics, Lobachevsky State University of Nizhni Novgorod, 23, Gagarin Avenue, Nizhny Novgorod 603950, Russia
sinitsina@itmm.unn.ru
[2] Neurotechnology Department, Lobachevsky State University of Nizhni Novgorod, 23, Gagarin Avenue, Nizhny Novgorod 603950, Russia
[3] Neuroscience and Cognitive Technology Laboratory, Center for Technologies in Robotics and Mechatronics Components, Innopolis University, 1, Universitetskaya Street, Innopolis 420500, Russia
[4] Neuroscience Research Institute, Samara State Medical University, 89, Chapaevskaya Street, Samara 443099, Russia

Abstract. Recent experimental data show that neural network dynamics significantly depends on the properties of an active nonlinear environment in the brain. Chemical activity of astrocytes is one of the main factor modifying the excitability of the neuronal membrane and regulating the efficiency of signal transmission between neurons. It is known that astrocytes can demonstrate both spontaneous changes in calcium concentration and calcium signals caused by neuronal activity. We focus on spontaneous calcium concentration dynamics in astrocytes. Nonlinear differential equations describing molecular t between neuronstransport in the astrocytes are investigated. Particularly, within the frame of Lavrentovich-Hemkin mathematical model for calcium dynamics, the bifurcation mechanisms of spontaneous calcium concentration change are determined. We show that both soft (emergence of small-amplitude oscillations via supercritical Andronov-Hopf bifurcation) and hard (instantaneous emergence of large-amplitude oscillations via fold limit cycle bifurcation) ways for oscillations emergence can be realized in the astrocyte dynamics.

Keywords: Mathematical modeling · Calcium concentration in astrocytes · Oscillatory and stationary modes

This work was supported by grant of the President of the Russian Federation for state support of leading scientific schools No. NSh-2653.2020.2. SG work was supported by the RFBR grants No. 20-32-70081, 18-29-10068 and the grant of the Ministry of Science and Higher Education of the Russian Federation project No. 0729-2020-0061.

D. Balandin et al. (Eds.): MMST 2020, CCIS 1413, pp. 129–137, 2021.
https://doi.org/10.1007/978-3-030-78759-2_10

1   Introduction

Brain is a complex multicomponent structure which includes neurons, glial cells and other elements placed in intercellular space. Wherein, the neurons are the main signaling cells of the nerve system. In terms of nonlinear dynamics the neurons are generators of electrical pulse signals regulated by an excitation threshold [1–4]. When the threshold is reached, an electrical pulse is generated and transmitted to another element of a neural network [5–10]. It is believed that the processes of generation, transmission and transformation of such pulses and their sequences in neural systems are the basis for information processing in the brain. Despite the great variety of experimental observations, the main principles of such processing are still not completely clear, and their search is one of the most exciting interdisciplinary task of the modern nonlinear science.

For a long time, it was believed that glial cells perform a number of functions that support the vital activity of neurons. Recent studies, however, show that astrocytes representing a type of glial cell can generate pulses of chemical activity in response to the passage of electrical pulse signals through the neural network. Such pulses represent a short-term increase in intracellular calcium concentration. An important aspect of the dynamics of a neural network is the influence of an active nonlinear environment on the generation and transmission of signals between neurons. Astrocytes play the role of such an active nonlinear environment for neural networks in the brain. Research results show that astrocytes are characterized by both spontaneous changes in calcium concentration [11] and calcium signals caused by the activity of neurons [12,13]. During the generation of calcium signals, the astrocytes are able to affect signaling functions of neurons, regulating the excitability of the neuronal membrane and the efficiency of signal transmission between neurons [14–18,26]. The kinetic equations of biochemical transformations in astrocytes are known and can be formalized in the form of systems of nonlinear differential equations. Particularly, in this work, a detailed study of the calcium concentration dynamics in astrocytes, namely, bifurcation mechanisms of oscillation emergence are presented.

2   Description of the Mathematical Model

In accordance with Lavrentovich-Hemkin model [19], calcium concentration changes in the cytosol of astrocytes and in its endoplasmic reticulum, and Ca^{2+}-dependent dynamics of inositol-1, 4, 5-triphosphate (IP_3) concentration are governed by the following equations:

$$\frac{d[Ca^{2+}]_{cyt}}{dt} = J_{in} - k_{out}[Ca^{2+}]_{cyt} + J_{CICR} - J_{serca} \\ + k_f([Ca^{2+}]_{ER} - [Ca^{2+}]_{cyt}), \tag{1}$$

$$\frac{d[Ca^{2+}]_{ER}}{dt} = J_{serca} - J_{CICR} + k_f([Ca^{2+}]_{cyt} - [Ca^{2+}]_{ER}), \tag{2}$$

$$\frac{[IP_3]_{cyt}}{dt} = J_{PLC} - k_{deg}([IP_3]_{cyt}), \tag{3}$$

where the expressions for J_{serca}, J_{CICR} and J_{PLC} are:

$$J_{serca} = v_{M2} \left(\frac{[Ca^{2+}]^2_{cyt}}{[Ca^{2+}]^2_{cyt} + k_2^2} \right), \tag{4}$$

$$J_{CICR} = 4v_{M3} \left(\frac{k_{CaA}^2 [Ca^{2+}]^n_{cyt}}{([Ca^{2+}]^n_{cyt} + k_{CaA}^n)([Ca^{2+}]^n_{cyt} + k_{CaI}^n)} \right)$$
$$\left(\frac{[IP_3]^m_{cyt}}{[IP_3]^m_{cyt} + k_{ip3}^m} \right) \left([Ca^{2+}]_{ER} - [Ca^{2+}]_{cyt} \right), \tag{5}$$

$$J_{PLC} = v_p \left(\frac{[Ca^{2+}]^2_{cyt}}{([Ca^{2+}]^2_{cyt} + k_p^2)} \right). \tag{6}$$

The variables of the considered system are the intracellular calcium concentration $[Ca^{2+}]_{cyt}$, the calcium concentration in the internal storage - the endoplasmic reticulum (ER) $[Ca^{2+}]_{ER}$, and the concentration of the secondary messenger inositol 1, 4, 5-triphosphate $[IP_3]_{cyt}$, which helps to open the channels and remove calcium into the cytosol. The parameters of the system were chosen according to the data given in [19], namely: $k_2 = 0.1\,\mu M$, $k_{CaA} = 0.15\,\mu M$, $k_{CaI} = 0.15\,\mu M$, $k_{ip3} = 0.1\,\mu M$, $k_p = 0.3\,\mu M$, $k_{deg} = 0.08\,s^{-1}$, $k_{out} = 0.5\,s^{-1}$, $k_f = 0.5\,s^{-1}$, $n = 2.02$, $m = 2.2$. For this set of parameters, and when $v_{M3} = 40\,s^{-1}$, $v_p = 0.05\,\mu M/s$ $v_{M2} = 15\,\mu M/s$, the dynamics was studied in [19]. The role of extracellular calcium flow in emergence of chaotic astrocytic spontaneous calcium oscillations was examined in [20]. In the present study, the rate of calcium flow through SERCA to the ER from the cytosol, v_{M2}, the rate of calcium flow through IP$_3$ from the ER to the cytosol, v_{M3}, and the amount of feedback between calcium in the cytosol and IP$_3$, v_p, are assumed to be varied. Role of these parameters in emergence of regular spontaneous calcium oscillations in the presence of various level of extracellular calcium J_{in} is examined.

3   Results

It was shown in [21] that Lavrentovich-Hemkin system has a unique equilibrium state. We start with the study of its local stability within the framework of well-known linearization method. Figure 1(a) shows the bifurcation diagram obtained by changing the parameters v_p and v_{M2}. Different colored domains show areas with different types of phase trajectory behavior in the vicinity of the equilibrium state. The areas are also numbered for ease of description. Thus, in regions 1 and 4, the equilibrium state is a stable focus and a stable node, respectively. In regions 2 and 3, the equilibrium state is not stable: in region 3 there is a saddle with two-dimensional unstable and one-dimensional stable manifolds, whereas in region

2 there is a saddle-focus with two-dimensional unstable and one-dimensional stable manifolds. Within these areas of parameters, due to the dissipativity of the considered system, oscillatory modes of spontaneous calcium concentration in the astrocyte are observed. Moreover, detailed analysis of the behavior of the system near the boundary between regions 1 and 2 that is depicted by red dots, showed the difference in emergence of oscillations. In particular, it was found that the lower part of this curve (when $v_{M2} < v_{M2}^*$) corresponds to a transition with a so-called *soft emergence* of a stable limit cycle (supercritical Andronov-Hopf bifurcation), and the upper parts of this curve (when $v_{M2} > v_{M2}^*$) correspond to a transition with the birth of an unstable limit cycle - a subcritical Andronov-Hopf bifurcation occurs here. Note that, the value v_{M2}^* defines the point corresponding to Bautin bifurcation where the branches of sub- and supercritical Andronov-Hopf bifurcations meet with the fold limit cycle bifurcation (not shown).

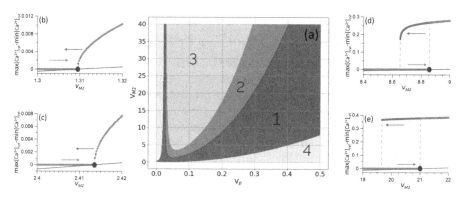

Fig. 1. (a) Two-parameter bifurcation diagram on the plane (v_p, v_{M2}). Numbered colored domains indicate areas with different types of the system behavior in vicinity of the equilibrium state: 1 - stable focus, 2 - saddle-focus, 3 - saddle, 4 - stable node. Red boundary between regions 1 and 2 corresponds to the transition through the zero value of the real part of the characteristic exponents (Andronov-Hopf bifurcation); $J_{in} = 0.03\,\mu M/s$; Peculiarities of transition via Andronov-Hopf bifurcation is presented in (b) for $v_p = 0.05\,\mu M/s$; (c) for $v_p = 0.1\,\mu M/s$; (d) for $v_p = 0.2\,\mu M/s$; (e) for $v_p = 0.3\,\mu M/s$. In (b)–(e), Andronov-Hopf bifurcation is depicted by black circle, arrows show the direction of v_{M2} parameter change (blue color is for the increase, green is for the decrease of v_{M2}), red curve shows the change of the real part of complex roots of characteristic equation. (Color figure online)

Thus, due to the dissipativity of the system, passing through the upper branches of the red curve, there must already exist a stable limit cycle of a large amplitude (varying v_p from the region 1 to the domain 2). Its birth occurs for larger values of the parameter v_p (for the right branch) and smaller v_p (for the left branch) as a result of fold limit cycle bifurcation. This means that some bistability range exists, where two types of attracting sets coexist: a stable equilibrium state and a stable limit cycle.

To demonstrate different scenarios of oscillations emergence, four values of the parameter v_p were chosen, and the difference between the maximal and the minimal values of calcium concentration in cytosol were calculated. Zero values of this difference means that the equilibrium point is stable.

For $v_p = 0.05\,\mu M/s$ and $v_p = 0.1\,\mu M/s$, in Figs. 1(b) and (c), respectively, *soft emergence* of a stable limit cycle via supercritical Andronov-Hopf bifurcation is presented. Blue symbols show the difference between the maximal and the minimal values of calcium concentration in cytosol obtained with the increase of the parameter v_{M2}, whereas the green symbols depict the data obtained with the decrease v_{M2}. In both cases, emergence of oscillations is observed for the value shown by black circle. This is the bifurcation value where the real part of the complex roots of characteristic equation becomes equal to zero. The change of the real part of the complex roots with the change of the parameter v_{M2} is shown by red curve in Figs. 1(b)–(e).

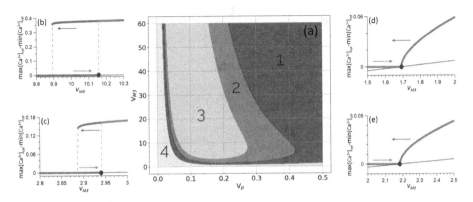

Fig. 2. (a) Two-parameter bifurcation diagram on the plane (v_p, v_{M3}). As in Fig. 1, numbered colored domains indicate areas with different types of the system behavior in vicinity of the equilibrium state: 1 - stable focus, 2 - saddle-focus, 3 - saddle, 4 - stable node. Red boundary between regions 1 and 2 corresponds to the transition through the zero value of the real part of the characteristic exponents (Andronov-Hopf bifurcation); $J_{in} = 0.03\,\mu M/s$; Peculiarities of transition via Andronov-Hopf bifurcation is presented in (b) for $v_p = 0.05\,\mu M/s$; (c) for $v_p = 0.1\,\mu M/s$; (d) for $v_p = 0.2\,\mu M/s$; (e) for $v_p = 0.3\,\mu M/s$. In (b)–(e), Andronov-Hopf bifurcation is depicted by black circle, arrows show the direction of v_{M3} parameter change (blue color is for the increase, green is for the decrease of v_{M3}), red curve shows the change of the real part of complex roots of characteristic equation. (Color figure online)

For $v_p = 0.2\,\mu M/s$ and $v_p = 0.3\,\mu M/s$, in Figs. 1(d) and (e), respectively, *hard emergence* of a large-amplitude stable limit cycle occurs with the increase of the parameter v_{M2} after the passing through the supercritical Andronov-Hopf bifurcation shown by black circle. Here, in contrast to the previous case, various data for the increase (blue points) and decrease (green points) of the parameter v_{M2} are obtained: with the increase of v_{M2} the stability of equilibrium point changes via supercritical Andronov-Hopf bifurcation, whereas the oscillation death with the decrease of v_{M2} occurs via fold limit cycle.

Similar partition into regions with different types of local stability of the equilibrium state was obtained for (v_p, v_{M3})-parameter plane, Fig. 2(a). Figures 2(b)–(e) present similar to observed before scenarios of regular spontaneous calcium oscillations emergence.

Comparative analysis of the obtained diagrams allows us to conclude that, for considered range of the parameter v_p, at small values of v_{M2} and v_{M3}, transitions to the oscillatory regime do not occur. In this case, an increase in the v_{M2} parameter leads to an increase of the range with oscillatory mode, whereas an increase in v_{M3} leads to decrease of this range.

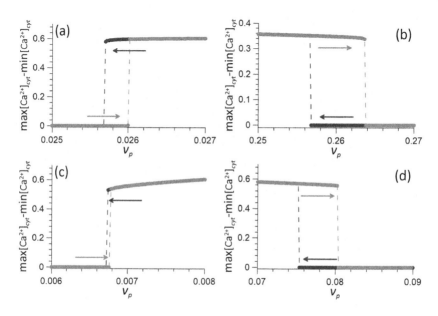

Fig. 3. Difference between the maximal and minimal values of calcium oscillations in cytosol with the change of v_p near the Andronov-Hopf bifurcations for left and right boundaries obtained when (a)–(b) $J_{in} = 0.03\,\mu\text{M/s}$ (near the red boundaries for $v_{M2} = 15\,\mu\text{M/s}$ and $v_{M3} = 40\,\text{s}^{-1}$ in the diagrams of Fig. 1 and Fig. 2, respectively), and (c)–(d) $J_{in} = 0.05\,\mu\text{M/s}$. Colored arrows show the direction of v_p parameter change (blue color is for the increase, green is for the decrease of v_p). (Color figure online)

It should be noted that the diagrams shown in Fig. 1(a) and Fig. 2(a) were obtained when the flux of calcium ions from the extracellular space to the cytosol was $J_{in} = 0.03\,\mu\text{M/s}$. An increase in this parameter leads to compression of all areas of the diagrams to the direction of lower v_p values and to decrease of the range with an oscillatory mode. To demonstrate this, Fig. 3 presents the difference between the maximal and the minimal values of calcium oscillations in cytosol for $J_{in} = 0.03\,\mu\text{M/s}$, Figs. 3(a), (b) and for $J_{in} = 0.05\,\mu\text{M/s}$, Figs. 3(c), (d). Figures 3(a) and (b), as well as Figs. 3(c) and (d) present the system behavior near the left and right boundary of oscillations emergence. As seen from the figures, both the values of v_p where the boundaries are observed and the values of the bistability range width, become less for larger flux of calcium ions from the extracellular space to the cytosol of astrocyte.

4 Conclusions

In this work, within the framework of Lavrentovich-Hemkin model the bifurcation mechanisms of regular spontaneous calcium oscillations emergence had been studied. Influence of the rate of calcium flow into the endoplasmic reticulum from the cytosol, v_{M2}, the rate of calcium flow in the opposite direction, v_{M3}, and the magnitude of the feedback between calcium in the cytosol and IP_3, v_p, on the change in areas with monostable and bistable behavior of the system had been examined. The biophysical mechanisms of the complex repertoire of intracellular Ca^{2+} signalling obtained in this work underlie the physiological and pathophysiological properties of the astrocytic activity [22,23]. Taking into account the current findings about the possible oscillatory modes of astrocytes, decoding mechanisms of astrocytic signalling in various brain circuits will be helpful to fully understanding how neuron-astrocytic interaction originate and become dysregulated in disease [24,25].

References

1. Anderson, J., Binzegger, T., Kahana, O., Martin, K., Segev, I.: Dendritic asymmetry cannot account for directional responses of neurons in visual cortex. Nat. Neurosci. **2**, 820–824 (1999)
2. Gasparini, S., Migliore, M., Magee, J.C.: On the initiation and propagation of dendritic spikes in CA1 pyramidal neurons. J. Neurosci. **24**, 11046–11056 (2004)
3. Pankratova, E.V., Belykh, V.N., Mosekilde, E.: Role of the driving frequency in a randomly perturbed Hodgkin-Huxley neuron with suprathreshold forcing. Eur. Phys. J. B **53**, 529–536 (2006)
4. Sardi, S., Vardi, R., Sheinin, A., Goldental, A., Kanter, I.: New types of experiments reveal that a neuron functions as multiple independent threshold units. Sci. Rep. **7**, 18036 (2017)
5. Belykh, V.N., Pankratova, E.V.: Chaotic synchronization in ensembles of coupled neurons modeled by the FitzHugh-Rinzel system. Radiophys. Quantum Electron. **49**(11), 910–921 (2006). https://doi.org/10.1007/s11141-006-0124-z
6. Eytan, D., Marom, S.: Dynamics and effective topology underlying synchronization in networks of cortical neurons. J. Neurosci. **26**(33), 8465–8476 (2006)
7. Belykh, V.N., Pankratova, E.V.: Synchronization and control in ensembles of periodic and chaotic neuronal elements with time dependent coupling. IFAC Proc. Vol. **40**(14), 120–125 (2007)
8. Larkum, M.E., Nevian, T.: Synaptic clustering by dendritic signalling mechanisms. Curr. Opin. Neurobiol. **18**, 321–331 (2008)

9. Belykh, V.N., Pankratova, E.V., Mosekilde, E.: Dynamics and synchronization of noise perturbed ensembles of periodically activated neuron cells. Int. J. Bifurcat. Chaos **18**(9), 2807–2815 (2008)

10. Pankratova, E.V., Kalyakulina, A.I.: Environmentally induced amplitude death and firing provocation in large-scale networks of neuronal systems. Regul. Chaot. Dyn. **21**(7–8), 840–848 (2016). https://doi.org/10.1134/S1560354716070078

11. Wu, Y.W., Gordleeva, S., Tang, X., Shih, P.Y., Dembitskaya, Y., Semyanov, A.: Morphological profile determines the frequency of spontaneous calcium events in astrocytic processes. Glia **67**(2), 246–262 (2018)

12. Gordleeva, S.Y., Ermolaeva, A.V., Kastalskiy, I.A., Kazantsev, V.B.: Astrocyte as spatiotemporal integrating detector of neuronal activity. Front. Physiol. **10**, 294 (2019)

13. Gordleeva, S.Y., Lebedev, S.A., Rumyantseva, M.A., Kazantsev, V.B.: Astrocyte as a detector of synchronous events of a neural network. JETP Lett. **107**, 440–445 (2018). https://doi.org/10.1134/S0021364018070032

14. Pankratova, E.V., Kalyakulina, A.I., Stasenko, S.V., Gordleeva, S.Y., Lazarevich, I.A., Kazantsev, V.B.: Neuronal synchronization enhanced by neuron-astrocyte interaction. Nonlinear Dyn. **97**(1), 647–662 (2019). https://doi.org/10.1007/s11071-019-05004-7

15. Makovkin, S.Y., Shkerin, I.V., Gordleeva, S.Y., Ivanchenko, M.V.: Astrocyte-induced intermittent synchronization of neurons in a minimal network. Chaos, Solitons Fractals **138**, 109951 (2020)

16. Kanakov, O., Gordleeva, S., Ermolaeva, A., Jalan, S., Zaikin, A.: Astrocyte-induced positive integrated information in neuron-astrocyte ensembles. Phys. Rev. E **99**, 012418 (2019)

17. Abrego, L., Gordleeva, S., Kanakov, O.I., Krivonosov, M., Zaikin, A.A.: Estimating integrated information in bidirectional neuron-astrocyte communication. Phys. Rev. E **103**, 002400 (2021)

18. Gordleeva, S.Y., Stasenko, S.V., Semyanov, A.V., Dityatev, A.E., Kazantsev, V.B.: Bi-directional astrocytic regulation of neuronal activity within a network. Front. Comput. Neurosci. **6**(92), 1–11 (2012)

19. Lavrentovich, M., Hemkin, S.: A mathematical model of spontaneous calcium (II) oscillations in astrocytes. J. Theor. Biol. **251**, 553–560 (2008)

20. Sinitsina, M.S., Gordleeva, S.Y., Kazantsev, V.B., Pankratova, E.V.: Calcium concentration in astrocytes: emergence of complicated spontaneous oscillations and their cessation. Izvestiya VUZ, Appl. Nonlinear Dyn. **29**(3), 440–448 (2021). https://doi.org/10.18500/0869-6632-2021-29-3-440-448

21. Sinitsina, M.S., Gordleeva, S.Yu., Kazantsev, V.B., Pankratova, E.V.: Emergence of complicated regular and irregular spontaneous Ca^{2+} oscillations in astrocytes. In: 4th Scientific School on Dynamics of Complex Networks and their Application in Intellectual Robotics (DCNAIR), Innopolis, Russia, pp. 217–220 (2020)

22. Santello, M., Toni, N., Volterra, A.: Astrocyte function from information processing to cognition and cognitive impairment. Nat. Neurosci. **22**, 154–166 (2019)

23. Whitwell, H.J., et al.: The human body as a super network: digital methods to analyze the propagation of aging. Front. Aging Neurosci. **12**, 136 (2020)

24. Gordleeva, S., Kanakov, O., Ivanchenko, M., Zaikin, A., Franceschi, C.: Brain aging and garbage cleaning. Semin. Immunopathol. **42**(5), 647–665 (2020). https://doi.org/10.1007/s00281-020-00816-x

25. Gordleeva, S., et al.: Modelling working memory in spiking neuron network accompanied by astrocytes. Front. Cell Neurosci. **15**, 631485 (2021). https://doi.org/10.3389/fncel.2021.631485

26. Matrosov, Valeri., Gordleeva, Susan., Boldyreva, Natalia., Ben-Jacob, Eshel., Kazantsev, Victor, De Pittà, Maurizio: Emergence of Regular and Complex Calcium Oscillations by Inositol 1,4,5-Trisphosphate Signaling in Astrocytes. In: De Pittà, Maurizio, Berry, Hugues (eds.) Computational Glioscience. SSCN, pp. 151–176. Springer, Cham (2019). https://doi.org/10.1007/978-3-030-00817-8_6

An Architecture for Real-Time Massive Data Extraction from Social Media

Aigerim B. Mussina[1]($\boxtimes$) ⓘ, Sanzhar S. Aubakirov[1] ⓘ, and Paulo Trigo[2] ⓘ

[1] Al-Farabi Kazakh National University, Almaty, Kazakhstan
[2] GulAA, ISEL - Instituto Superior de Engenharia de Lisboa, LASIGE,
Faculdade de Ciências, Universidade de Lisboa, Lisbon, Portugal
paulo.trigo@isel.pt

Abstract. Social media already plays a significant role in human's daily life. Therefore, social media and associated network connections have become an arena of enormous opportunities to perform data analysis. Its impact on daily life covers areas as diverse as digital marketing, social opinion analysis, political situation monitoring and natural disaster notification. Event detection is a "building-block" that sustains goal-oriented analytics, such as the "real-time evaluation of peoples' reaction to certain event(s)". We propose to develop social models aimed at the extraction and prediction of event patterns based on online social media event detection in real-time. At its current stage the research is focused on the development of highly loaded, fault-tolerant, scalable system for social media data extraction and real-time analysis.

Keywords: Event detection · Highly loaded · Fault-tolerant · Scalable architecture · Telegram

1 Introduction

Social media is mostly free and thus a powerful and highly-spreaded infrastructure for communication with a large audience. For example, a company may get quick feedback on a certain brand and make influence analysis of its posts [11]. Different Online Social Networks (OSN) establish and strengthen their own relationships between consumers and producers [22].

Our research goal is to analyse social behavior in the context of messaging and posts reaction within Telegram OSN messenger. At the beginning of 2019 there were approximately 200 000 users from Kazakhstan [18]. As of April 2020 the Kazakhstan's audience reached 400 000 users. In January 2021 its world-wide monthly active users reached 500 million users. Despite its fast growth, Telegram is a relatively young OSN and has not yet become a subject of intense research. The authors of Telegram OSN provide many public tools for developers to build their own client application [19].

Event detection process has been applied for various OSNs and purposes. Also, an "event" has different definitions according to context and application.

© Springer Nature Switzerland AG 2021
D. Balandin et al. (Eds.): MMST 2020, CCIS 1413, pp. 138–145, 2021.
https://doi.org/10.1007/978-3-030-78759-2_11

First we consider an event as "an occurrence causing change in the volume of text data that discusses the associated topic at a specific time. This occurrence is characterized by topic and time, and often associated with entities such as people and location" [3].

Event detection in OSN is important for social analysis, because it allows to estimate public interest in an occurred event. Moreover, events analysis lead to the detection of substantial sub-events [20]. For example, at the beginning of the 2020 year the whole world faced new virus COVID-19. Such high-level event subsumes many significant low-level sub-events such as illegal reselling essential items, bullying activity, misinformation spreading, growth of free online services and infection of celebrities.

Our goal is focused on the intellectual social modelling and event patterns prediction based on online social media in real-time. To achieve that goal we have implemented a highly loaded, fault-tolerant, scalable architecture for massive data extraction from social media.

2   Related Work

Event detection is a topic with high adherence from researchers. A comprehensive survey on event detection [3] describes four challenges on this topic:

1. New Event Detection (NED)
2. Event Tracking
3. Event Summarization
4. Event Associations

Another survey includes research on event detection about disasters, news, outbreaks and traffic [15]. The work was focused on analysing only the Twitter.

The natural disasters are causes of common interest that motivates various researchers. People post messages with essential information during and after a disaster according to their feelings and real-time situation. Some research approaches concentrate on the evolution of a rare event, like a storm, in the real world by analysing activities in a virtual world and constructing temporal patterns [14].

Event detection technique depends on event characteristics and its category. A survey on event detection techniques covered events about natural disaster, trending topics and public opinion in newswire, web forums, emails, blogs and microblogs [7]. The survey includes researches on OSNs like Twitter, Facebook, Instagram, Youtube and Pinterest. Domain dependence is a huge challenge for researchers because techniques are extremely situational dependent. Time constraint is another characteristic which is also varying according to event category. Authors discussed techniques grouped by information flow between users: thematic, temporal, spatial and network structure.

A huge survey proposes definitions and categorizations in event detection process [2]. The authors classified 34 works by event types, pivot techniques, detection method, detection task and application. The majority of researches were for detecting general interest events. Only 20% of examined works were oriented on real-time event detection.

Researches on event detection valuable for social science. Journalistic's event descriptors 5W1H (who? what? when? where? why? how?) were applied for event detection in news articles [8,21,24]. Such approach needs deep knowledge of NLP (Natural Language Processing) and depends on language and well-labeled dataset.

Researchers highlight relevance of event detection in contextual decision making, emergence notification tools implementation, better understanding of social interest [7,15,20].

3   Requirements and Design Approach

Social media generates massive amount of continuously increasing data therefore imposing stringent requirements for any crawler-based data extraction architecture. Some major requirements include:

1. Crawl over different Online Social Networks (OSN). Since social media crawler should work with different OSNs then external APIs will be different. However internal data processing will be identical for all.
2. Plug into new OSN. This process should be gentle and simple.
3. Scalable regarding data pre-processing and analysing.

In order for our architecture to meet the requirements above, it must have the following properties.

Elasticity. It is the degree to which the system is able to adapt to changes in workload by providing and removing resources in an autonomous approach, such that at any given time, the available resources are as close as possible to current demand [9].

Scalability. It can be differentiated into "structural scalability and load scalability. Structural scalability is the ability of a system to expand in a chosen dimension without significant changes in its architecture. Load scalability is the ability of the system to work correctly when the offered traffic increases [1].

Self-service deployment. It is understood as part of the application deployment topology to implement a specific technical unit [10]. More often, a unit of deployment is understood as a "standard container". The goal of a standard container is to encapsulate a software component and all of its dependencies in a self-describing and portable format so that any compatible runtime can run it without additional dependencies, regardless of the underlying machine and the contents of the container. This is a definition from the Open Container Initiative (OCI), which is explained in 5 principles of standard containers [17].

In this section we describe the technology that fits our architecture requirements. Nowadays there is a paradigm of software development, which lays in business processes and architecture solutions to the above issues. The paradigm is called Cloud-Native Application Development (CNA Development) [6]. This concept refers to a set of technologies and design patterns that have become the standard for building large-scale cloud applications. Software development in this paradigm provides the properties of successful cloud applications, including dynamic scalability, ultimate resiliency, non-disruptive upgrades, and security. To enable the creation of applications that meet these requirements, we describe a microservices architecture that is central to cloud design.

In a huge survey more than 50 works related to the development of cloud applications were analysed, collected and summarized approaches, methods and terms [12]. As a result, authors defined the term Cloud-Native Application as follows: "A cloud application (CNA) is a distributed, flexible and scale-out system of (micro) services that isolates state in a minimum of stateful components. The application and each individual deployment module of this application are designed according to cloud-centric design patterns and run on a flexible self-service platform."

It was proposed to call such applications IDEAL, so that the application was [Isolated state] isolated, [Distributed] had a distributed architecture, was [Elastic] flexible in the sense of horizontal scaling, was controlled using [Automated] automated systems, and its components must be loosely coupled [4]. Creation of cloud applications in this paradigm leads to the following results [23]:

1. faster provision of software solutions to the customer
2. fault tolerance
3. automation of recovery
4. easy and fast horizontal scaling of applications
5. the ability to process a huge amount of data

Our design approach was leaded by our first case study OSN - Telegram. Nowadays social networks developers usually provide public tools or libraries to interact with their system and its data. It was mentioned above that Telegram becomes more and more popular. We decided to construct our social media crawler firstly based on Telegram OSN. According to official Telegram Database Library (TDLib) provided by Telegram, we have constructed our Client API application [4].

4   Cloud-Based Architecture and Data Model Design

In this section we describe our architecture for social media data extraction and processing. The database structure also presented in Sect. 4.2.

4.1   Cloud-Based Architecture

During this stage of the research we have developed a cloud architecture based on microservices. Microservices are the decomposition of monolithic business systems into independently deployable services that perform a single task. The main way to communicate between services in a cloud application architecture is through published and versioned APIs (API-based collaboration). In our architecture microservices communicate via message queue.

The individual architecture deployment units are designed and interconnected according to a set of cloud-oriented patterns such as a twelve-factor app [25], a Circuit Breaker [5]. The Twelve-Factor App methodology is a methodology for building software-as-a-service applications. A Circuit Breaker used in microservices architecture to prevent cascade fails during services communication. These best practices are designed to enable applications to be built with portability and resilience when deployed to the web.

We use the flexible OpenStack platform, which is used to deploy and operate these microservices through autonomous deployment units (containers). This platform provides additional operational capabilities on top of IaaS infrastructures, such as auto-scaling application instances and scaling on demand, application health management, dynamic routing, load balancing, and log and metrics aggregation.

In Fig. 1 we depicted the parsing process flow through our cloud architecture and general view of architecture. Each New message created in OSN is detected and parsed by a crawler. Firstly, the crawler will save message in database. Secondly, it will put the textual content of the message in queue. Queue service has an exchange area and defined queues. Tokenization consumer receive its data through queue. Consumers are located in a data processing microservice. This microservice will be exposed later with other processing tools. After data pre-processing and processing, all extracted information go to database.

4.2   Data-Model Design

The Telegram open-API allows to collect a lot of information about users, chats and messages. The Fig. 2 shows the conceptual data-model of our relational database.

The first implemented pre-processing task was message text tokenization. The token in our case is unigram. Before tokenization, text is cleared from Russian stop-words taken from RussianAnalyzer in Apache Lucene library [16]. Each token has its count within users' messages and chats' messages. For example, in Kazakhstan during COVID-19 pandemic and state of emergency, government pays 42500 tenge of compensation for people who lost their job or income. This '42500' token has total count greater than 11000 among 34 chats, but it was mainly due to conversations in one news channel.

The second pre-processing tool was topic extraction including lemmatization of Russian and English language messages. For Russian messages we used Lucene based library [13]. The topic extraction simply made by tf-idf calculation. We consider each message as a "document" and combine all messages from chats during one day and calculate the tf-idf for each token. The top-K tokens with highest tf-idf are taken as the K topics of the day (we are exploring $K = 5$).

We are collecting messages, since 19.02.2020, from public groups and channels from Kazakhstan. At the beginning of April we included chats from Russia, Belarussia, Ukraine and Uzbekistan. On the 21.01.2021 database consists of 2621 chats, 285 000 users and 1 093 298 messages. The data growth is presented in Fig. 3.

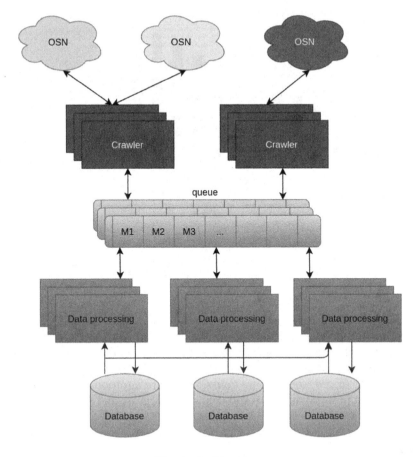

Fig. 1. Architecture

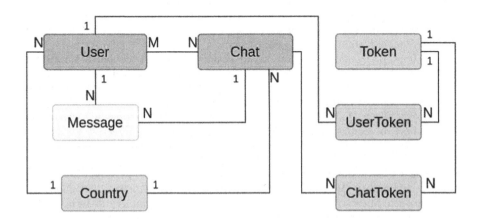

Fig. 2. Data conceptual model

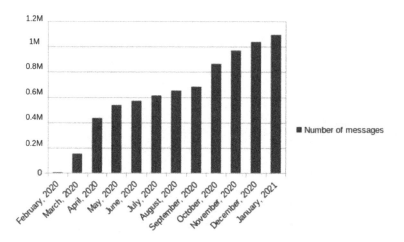

Fig. 3. Extracting data growth

5   Conclusion

We have developed highly loaded, fault-tolerant, scalable architecture for social media data extraction and storage using Cloud Native Application Development paradigm. Crawler collects messages from 2141 Telegram chats. Data processing microservice designed as scalable unit which could be easily exposed with additional tools. In future work we are going to explore additional related work on event detection in social networks and define the accurate and effective algorithm for event detection and event embedding.

References

1. Bondi, A.B.: Characteristics of scalability and their impact on performance. In: Proceedings of the 2nd International Workshop on Software and Performance, WOSP 2000, pp. 195–203. Association for Computing Machinery, New York (2000). https://doi.org/10.1145/350391.350432
2. Cordeiro, M., Gama, J.: Online social networks event detection: a survey. In: Michaelis, S., Piatkowski, N., Stolpe, M. (eds.) Solving Large Scale Learning Tasks. Challenges and Algorithms. LNCS (LNAI), vol. 9580, pp. 1–41. Springer, Cham (2016). https://doi.org/10.1007/978-3-319-41706-6_1
3. Dou, W., Wang, X., Ribarsky, W., Zhou, M.: Event detection in social media data. In: Proceedings of the IEEE VisWeek Workshop on Interactive Visual Text Analytics-Task Driven Analytics of Social Media Content, pp. 971–980, January 2012
4. Fehling, C., Leymann, F., Retter, R., Schupeck, W., Arbitter, P.: Cloud Computing Patterns. Springer, Vienna (2014). https://doi.org/10.1007/978-3-7091-1568-8
5. Fowler, M.: Microservices - a definition of this new architectural term (2014). http://martinfowler.com/articles/microservices.html. Accessed 20 May 2020
6. Gannon, D., Barga, R., Sundaresan, N.: Cloud-native applications. IEEE Cloud Comput. **4**, 16–21 (2017). https://doi.org/10.1109/MCC.2017.4250939

7. Goswami, A., Kumar, A.: A survey of event detection techniques in online social networks. Soc. Netw. Anal. Min. **6** (2016). https://doi.org/10.1007/s13278-016-0414-1
8. Hamborg, F., Breitinger, C., Gipp, B.: Giveme5W1H: a universal system for extracting main events from news articles. In: Proceedings of the 13th ACM Conference on Recommender Systems, 7th International Workshop on News Recommendation and Analytics (INRA 2019), September 2019
9. Herbst, N., Kounev, S., Reussner, R.: Elasticity in cloud computing: what it is, and what it is not. In: International Conference on Autonomic Computing, pp. 23–27, January 2013
10. Inzinger, C., Nastic, S., Sehic, S., Vögler, M., Li, F., Dustdar, S.: MADCAT - a methodology for architecture and deployment of cloud application topologies, April 2014. https://doi.org/10.1109/SOSE.2014.9
11. Klepek, M., Starzyczná, H.: Marketing communication model for social networks. J. Bus. Econ. Manag. **19**, 500–520 (2018). https://doi.org/10.3846/jbem.2018.6582
12. Kratzke, N., Quint, P.C.: Understanding cloud-native applications after 10 years of cloud computing - a systematic mapping study. J. Syst. Softw. **126**, 1–16 (2017). https://doi.org/10.1016/j.jss.2017.01.001
13. Kuznetsov, A.: Russian morphology for Apache Lucene. https://github.com/AKuznetsov/russianmorphology. Accessed 20 Nov 2020
14. Lu, S., Zhou, M., Qi, L., Liu, H.: Clustering-algorithm-based rare-event evolution analysis via social media data. IEEE Trans. Comput. Soc. Syst. **PP**, 1–10 (2019). https://doi.org/10.1109/TCSS.2019.2898774
15. Nurwidyantoro, A.: Event detection in social media: a survey, pp. 1–5, June 2013. https://doi.org/10.1109/ICTSS.2013.6588106
16. Apache Software Foundation: Lucene 4.0.0 analyzers-common API. https://lucene.apache.org/core/4_0_0/analyzers-common/. Accessed 15 Sept 2020
17. Open Container Initiative: Open container runtime specification (2020). https://opencontainers.org/. Accessed 27 May 2016
18. Telegram Analytics: Telegram channels Kazakhstan (2020). https://kaz.tgstat.com/. Accessed 18 Apr 2020
19. Telegram Library: Telegram database library, version 1.6.0, 31 January 2020. https://core.telegram.org/tdlib. Accessed 10 Feb 2020
20. Saravanou, A., Katakis, I., Valkanas, G., Gunopulos, D.: Detection and delineation of events and sub-events in social networks, pp. 1348–1351, April 2018. https://doi.org/10.1109/ICDE.2018.00147
21. Sharma, S., Kumar, R., Bhadana, P., Gupta, S.: News event extraction using 5W1H approach & its analysis. Int. J. Sci. Eng. Res. (IJSER) **4**(5), 2064–2068 (2013)
22. Sharma, S., Verma, H.V.: Social media marketing: evolution and change. In: Hegde, G., Shainesh, G. (eds.) Social Media Marketing, pp. 19–36. Springer, Singapore (2018). https://doi.org/10.1007/978-981-10-5323-8_2
23. Stine, M.: Migrating to Cloud-Native Application Architectures. O'Reilly Media Inc., Sebastopol (2015)
24. Wang, W., Zhao, D., Wang, D.: Chinese news event 5W1H elements extraction using semantic role labeling. In: 2010 Third International Symposium on Information Processing, pp. 484–489 (2010). https://doi.org/10.1109/ISIP.2010.112
25. Wiggins, A.: The twelve-factor app (2014). http://12factor.net. Accessed 14 Feb 2020

Node Degree Dynamics in Complex Networks Generated in Accordance with a Modification of the Triadic Closure Model

Sergei Sidorov$^{(\boxtimes)}$ ⓘ, Sergei Mironov ⓘ, Alexey Faizliev ⓘ,
and Alexey Grigoriev ⓘ

Saratov State University, Saratov 410012, Russian Federation
sidorovsp@sgu.ru
http://www.sgu.ru

Abstract. The triadic closure model attempts to reflect the well-known fact that many real-world networks have a much higher likelihood of a link between a pair of nodes with a common neighbor than in the random null model. The effect is especially evident in social networks and is also widely present in cooperation networks in the field of knowledge, citation and research, and many others. It should be noted that in the triadic closure model proposed by P. Holme and B.J. Kim, at the triad formation step, the new link is attached to any random neighbor (which is chosen evenly) of the node selected in the previous step (which is chosen using the preferential attachment mechanism). However, in many real-world complex networks, the second link is also usually selected using the preferential attachment mechanism. In this paper, we propose to make adjustments to the triadic closure model for the case when the nodes are selected using the preferential attachment mechanism at the triad formation step. We empirically investigate the dynamics of the mean value of node degree in the networks generated by this model and show that it follows a power law.

Keywords: Complex networks · Social networks · Preferential attachment · Triadic closure · High clustering · Growth model

1 Introduction

One of the most common tool for developing communities in different complex networks is the so-called triadic closure [12]. It is well-known that many real networks have a much higher probability of having a link between a pair of

This work was supported by the Ministry of science and education of the Russian Federation in the framework of the basic part of the scientific research state task, project FSRR-2020-0006.

D. Balandin et al. (Eds.): MMST 2020, CCIS 1413, pp. 146–153, 2021.
https://doi.org/10.1007/978-3-030-78759-2_12

nodes with a common neighbor with comparison to a random null model. The effect is especially evident in social networks [2–5,7–11,14–16].

The so-called triadic closure model was developed in the work [6] and is an extension of the Barabási-Albert model which uses the preferential attachment mechanism [1]. Both models generate networks with scale-free structure and the power law degree distribution. However, the triadic closure model induces networks with a higher clustering than the BA model, and therefore the networks obtained with use of the triadic closure model are more similar to real social networks in that sense. In the model, when a new node is added to network, it forms a link with an existing node in the network, elected with a probability proportional to its degree (*preferential attachment*). The remaining $m - 1$ links bringing forth by the new node are tied with a probability p to a randomly chosen neighbor of the node which received the most recent preferentially attached link (*triad formation*), while with a probability $1 - p$ the new node links with a node chosen with use of preferential attachment mechanism. The triadic closure model produces networks with varying levels of clustering by differing p. On the other hand, the degree distribution is the same as in the BA model and follows a power law with $\gamma = -3$ for any p.

Note that in the basic model proposed by P. Holme and B. J. Kim, during the triad formation step a new link attaches to any random neighbor (*which is uniformly chosen*) of the node picked at the previous step (*which is chosen using preferential attachment mechanism*). However, empirical evidence shows that in many complex networks the second link is chosen also with use of preferential attachment mechanism. Therefore, in this paper we propose to amend the triadic closure model for the case when nodes are chosen *preferentially* at the triadic formation step (*not uniformly* as it is in the basic model). We empirically investigate the properties of the networks generated by this model and show that the dynamics of the expected degree of any node follows the power law.

2 The Modification of the Triadic Closure Model and Its Analysis

2.1 The Model Description

We propose a modification of the triadic closure model by P. Holme and B. J. Kim. Let us first describe the rules, by which the network grows. Let t denote the current iteration. Then at each iteration t:

1. a new node t is added to the network;
2. it would be connected with m other nodes in the network. These nodes are chosen as follows:
   (a) one of the existing nodes i in the network is chosen with a probability proportional to its degree k_i;
   (b) the remaining $m - 1$ links connect the newborn node t with other nodes in the network, which are selected as follows:

- *(b1)* (*Triad formation*)
  with probability p, the new node is connected with one of the neighbors j of the node i, with a probability proportional to k_j, i.e. the degree of vertex j.
- *(b2)* with probability $1 - p$, the link is attached to one of the vertices s of the network (not necessarily adjacent to the node i) with a probability proportional to the degree of this node k_s.

To better replicate a great deal of complex networks, the changes are made to its core triad formation step (b1). Instead of selecting the arbitrary node, the selection is carried out with preferential attachment mechanism involved. Another modification of the triadic closure model has been recently introduced in [13].

2.2   The Model Analysis in Case $m = 2$ and $p = 1$

In this subsection we examine the time evolution of networks generated in accordance with the partial case of the model ($m = 2$ and $p = 1$). The first parameter $m = 2$ leads to only two links being added at each iteration. The second parameter $p = 1$ means that after the first link is attached to a node selected with preferential attachment, it is also connected with a neighbor of that node, which is selected with preferential attachment as well.

We start with the analysis of the time-dependent degree for an arbitrary node. Denote $\bar{k}_i(t)$ the (expected) degree of node i at iteration $t \geq i$.

Let $N(t)$ be the number of vertices in the graph at iteration t, $M(t)$ be the number of links at the iteration. It should be noted that $N(t) = N_0 + (t - 1)$, $M(t) = M_0 + 2(t - 1)$, where N_0 and M_0 are the initial number of nodes and links in the model, respectively.

There are two cases, in which an existing node i can increase its degree each time a new node enters the network. The first opportunity occurs if node i is selected at step 2(a), and the second chance takes place when node i is selected at step 2(b).

For convenience we will approximate the degree k_i by means of a continuous real variable, representing its expectation value over many realizations of the growth process. Therefore, we calculate the rate at which an existing node i acquires links at iteration t is

$$\frac{dk_i(t)}{dt} = p_a(i) + p_b(i), \tag{1}$$

where $p_a(i)$ and $p_b(i)$ are the probabilities of choosing node i at steps 2(a) and 2(b) respectively.

The probability of the node i to be chosen from $N(t)$ nodes in the network at the first preferential attachment step is proportional to its degree, i.e.

$$p_a(i) = \frac{k_i(t)}{2M(t)}. \tag{2}$$

We have $2M(t) \sim 4t$. Let $p(j, i) = 1$, if the link (j, i) exists, and $p(j, i) = 0$, otherwise. By definition of $p(j, i)$ we have $\sum_{j=1}^{N(t)} p(j, i) = k_i(t)$. Let $s_i(t)$

denote the sum of the degrees for all neighbors of node j at time t, i.e. $s_i(t) = \sum_{j=1}^{t} p(i,j)k_j(t)$.

The probability of choosing node i at step 2(b) is equal to the sum over all nodes adjacent to node i of products of

- the probability of choosing node j at step 2(a), i.e. $\frac{k_j(t)}{2M(t)}$,
- the probability of choosing node i among $k_j(t)$ neighbors of node j, i.e. $\frac{k_i(t)}{s_j(t)}$,

so we get

$$p_b(i) = \sum_{j=1}^{N(t)} p(j,i) \frac{k_j(t)}{2M(t)} \frac{k_i(t)}{s_j(t)} = \frac{k_i(t)}{2M(t)} \sum_{j=1}^{N(t)} p(j,i) \frac{k_j(t)}{s_j(t)}. \tag{3}$$

Let γ_i be defined as follows:

$$\gamma_i(t) := \sum_{j=1}^{N(t)} p(j,i) \frac{k_j(t)}{s_j(t)}. \tag{4}$$

It follows from (1), (2) and (3) (and since for large t the $M(t) \sim 2t$) that

$$\frac{dk_i(t)}{dt} = (1 + \gamma_i(t)) \frac{k_i(t)}{4t}. \tag{5}$$

If we integrate Eq. (5), we have the following dynamics for large t:

$$k_i(t) = c_i t^{\frac{1}{4}} \exp \left(\int \frac{\gamma_i(t)}{4t} dt \right),$$

where c_i can be found from the initial condition $k_i(i) = 2$, since each node i has degree 2 at the initial moment $t_i = i$.

The following proposition can be proved.

Lemma 1. *For each $i = 1, 2, \ldots$, there is a finite limit $\gamma_i := \lim_{t \to \infty} \gamma_i(t)$ and $0 \leq \ldots \leq \gamma_{i+1} \leq \gamma_i \leq \ldots \leq \gamma_1 \leq 3$.*

Equation (7) describes the dynamical behavior of the degree in the network of single node i. The dynamical exponents have different values for different nodes, in contrast with the Barabási-Albert model in which the dynamic exponents are the same for all nodes.

Equations (6) and (7) predict that the degree of a node in the network increases following a power-law with the dynamical exponent $\beta_i \leq 1$. However, different nodes follow different dynamical law with different value of the dynamical exponent.

The random graph model described above can easily be implemented and applied to generate graphs. Now we will show empirically that degree evolution of each node i follows the power law with β_i depending on i:

$$k_i(t) \sim c \left(\frac{t}{i} \right)^{\beta_i}. \tag{6}$$

Figure 1(a) shows the dynamics of degree for different nodes $i = 1, 10, 50, 100$ for $t = 1, 2, \ldots, 10^6$. All degree values are non decreasing with maximum values $k_2^{\max} = 33223.4$, $k_5^{\max} = 6033$, $k_{10}^{\max} = 394.4$, $k_{50}^{\max} = 84.4$, $k_{100}^{\max} = 26.6$, achieved at iteration $t = 100,000$. Therefore, to present the plots on the figure we normalized each of them by dividing on the maximum degree for corresponding node. The results were averaged for 10 independent runs (all networks are of size $N = 100,000$). The figure shows that all nodes have different dynamics. Figure 1(b) presents the log-log plot of degree dynamics for the same nodes $i = 1, 10, 50, 100$ on $t \in [1, 10^6]$. All linear regressions are with $R^2 = 0.999$, and obtained slopes are 1, 0.71, 0.55 and 0.46, respectively.

The empirical results show that the dynamics of the degree growth of each node follows a power law. However, all nodes have different exponent values β_i (which is 1 for the node $i = 1$, and β_i rapidly decreases to $\frac{1}{4}$ with the increase of i). This is a significant difference from the BA model or the triadic closure model, for which the exponents are the same and equal to 0.5 for all nodes, regardless of when a particular node appeared.

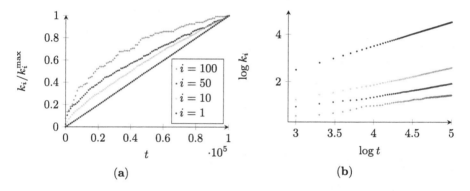

Fig. 1. (a) Degree dynamics of nodes $i = 1, 10, 50, 100$. We normalized each of them by dividing on the maximum degree for corresponding node. (b) Log-log plots of degree dynamics of nodes $i = 1, 10, 50, 100$. The slopes (which are 1, 0.71, 0.55 and 0.46, respectively) are obtained using the linear regression model. The results were averaged for 10 independent runs (all networks are of size $N = 100,000$).

The empirical experiments also confirm that there is a limit of $\gamma_i(t)$ as $t \to \infty$, $\gamma_i := \lim_{t\to\infty} \gamma_i(t)$. Figure 2 presents the evolution of normalized γ_i, i.e. the values $\frac{\gamma_i}{\gamma_{i,max}}$, for nodes $i = 2, 5, 10, 100, 1000$ over time. The results were averaged for 10 independent runs (all networks are of the same size $N = 100,000$).

Thus, we have approximation for large t

$$k_i(t) \sim 2 \left(\frac{t}{i}\right)^{\frac{1}{4}(1+\gamma_i)}. \tag{7}$$

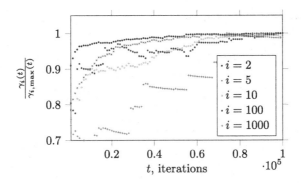

Fig. 2. The dynamics of normalized γ_i for nodes $i = 2, 5, 10, 100, 1000$ over time. The results were averaged for 10 independent runs (all networks are of size $N = 100,000$).

2.3   Node Degree Dynamics (General Case)

In this section we find the node degree dynamics for the model described in Sect. 2.1.

This model is different from the previous as $b2$ step is included, while 2b algorithm repeats $m - 1$ times. This means that the node i can increase its degree each time a new node enters the network in three cases: at steps 2(a), 2(b1) and 2(b2).

Similarly to previous section, we approximate the degree k_i by means of a continuous real variable through representing its expectation value. The existing node i increases its degree after new nodes are connected to it on iteration t (which consists of steps 2(a), 2(b1) and 2(b2)) at a rate described by

$$\frac{dk_i(t)}{dt} = p_a(i) + p_{b1}(i) + p_{b2}(i), \tag{8}$$

where $p_a(i)$, $p_{b1}(i)$ and $p_{b2}(i)$ are the probabilities of choosing node i at steps 2(a), 2(b1) and 2(b2) respectively. We have

$$p_a(i) = \frac{k_i}{2M(t)} \sim \frac{k_i}{4t}. \tag{9}$$

The probability of choosing node i at step 2(b1) is equal to

$$p_{b1}(i) = p(m - 1)p_b(i) = p(m - 1)\frac{k_i}{2M(t)} \sum_{j=1}^{N(t)} p(j, i)\frac{k_j}{s_j} \sim p(m - 1)\frac{k_i}{4t}\gamma_i, \tag{10}$$

where $p_b(i)$ is obtained in (3) and γ_i is defined in (4).

We have

$$p_{b2}(i) = (1 - p)(m - 1)\frac{k_i}{2M(t)} \sim (1 - p)(m - 1)\frac{k_i}{4t}. \tag{11}$$

It follows from (8), (9), (10) and (11) that

$$\frac{d\bar{k}_i}{dt} = \left(1 + p(m-1)\gamma_i + (1-p)(m-1)\right)\frac{\bar{k}_i}{4t}. \tag{12}$$

After we integrate the Eq. (12), we acquire the following dynamics for large t:

$$\bar{k}_i \sim c\left(\frac{t}{i}\right)^{\beta_i}, \tag{13}$$

where

$$\beta_i = \frac{1}{4}\left(m(1-p) + p\big(1 + (m-1)\gamma_i\big)\right).$$

Equation (13) describes the dynamical behavior of the degree in the network of single node i. In case $m = 2$ and $p = 1$, the dynamics are equal to (7), which has been acquired for the simple variant of model as described in Sect. 2.2. The dynamical exponents are different for nodes i and j, $i \neq j$, unlike the Barabási-Albert model in which the exponent of the dynamics is the same for every node.

Equation (13) offers a number of predictions:

- The degree of a node in the network increases while following a power-law with the dynamical exponent β_i. However, it should be noted that each node follows different dynamical law with varying value of the exponent.
- The earlier node i was added, the higher its degree $k_i(t)$ and the higher its exponent β_i is. Hence, hubs are larger in this model in comparison to BA or the triadic closure models.

3   Conclusion

This paper considers a modification of the triadic closure model, in which at the triad closure step, the mechanism of preferential attachment with respect to the neighbors of the node selected at the previous step is used. The goal of the paper is to obtain the trajectories describing the expected values of the degrees for nodes in the networks that are generated using this model. The results show that the dynamics of the degree of each node in the network follows a power law. However, unlike the triadic closure or the Barabási-Albert models, for which the exponent of the power law is the same for all vertices and is equal to $\frac{1}{2}$, in the modified model the exponents are different for different nodes. For example, for the nodes that appeared earlier, it is close to 1, while for the nodes that appeared in the last iterations, it approaches $\frac{1}{4}$. Thus, in this model, the nodes which were added at the early stages of network growth, get an even greater advantage in the sense of degree growth rate than in classical models.

References

1. Barabási, A.L., Albert, R.: Emergence of scaling in random networks. Science **286**(5439), 509–512 (1999). https://doi.org/10.1126/science.286.5439.509
2. Bianconi, G., Darst, R., Iacovacci, J., Fortunato, S.: Triadic closure as a basic generating mechanism of communities in complex networks. Phys. Rev. E Stat. Nonlinear Soft Matter Phys. **90**(4) (2014). https://doi.org/10.1103/PhysRevE.90.042806
3. Chen, B., Poquet, O.: Socio-temporal dynamics in peer interaction events, pp. 203–208 (2020). https://doi.org/10.1145/3375462.3375535
4. Fang, Z., Tang, J.: Uncovering the formation of triadic closure in social networks, January 2015, pp. 2062–2068 (2015)
5. Fang, Z., Tang, J.: Uncovering the formation of triadic closure in social networks. In: Proceedings of the 24th International Conference on Artificial Intelligence, IJCAI 2015, pp. 2062–2068. AAAI Press (2015)
6. Holme, P., Kim, B.J.: Growing scale-free networks with tunable clustering. Phys. Rev. E **65** (2002). https://doi.org/10.1103/PhysRevE.65.026107
7. Huang, H., Tang, J., Liu, L., Luo, J., Fu, X.: Triadic closure pattern analysis and prediction in social networks. IEEE Trans. Knowl. Data Eng. **27**(12), 3374–3389 (2015). https://doi.org/10.1109/TKDE.2015.2453956
8. Huang, H., Tang, J., Liu, L., Luo, J., Fu, X.: Triadic closure pattern analysis and prediction in social networks. IEEE Trans. Knowl. Data Eng. **27**(12), 3374–3389 (2015)
9. Huang, H., Dong, Y., Tang, J., Yang, H., Chawla, N.V., Fu, X.: Will triadic closure strengthen ties in social networks? ACM Trans. Knowl. Discov. Data **12**(3) (2018). https://doi.org/10.1145/3154399
10. Li, M., et al.: A coevolving model based on preferential triadic closure for social media networks. Sci. Rep. **3** (2013). https://doi.org/10.1038/srep02512
11. Linyi, Z., Shugang, L.: The node influence for link prediction based on triadic closure structure. In: 2017 IEEE 2nd Information Technology, Networking, Electronic and Automation Control Conference (ITNEC), pp. 761–766 (2017)
12. Rapoport, A.: Spread of information through a population with socio-structural bias: I. The assumption of transitivity. Bull. Math. Biophys. **15**, 523–533 (1953). https://doi.org/10.1007/BF02476440
13. Sidorov, S., Mironov, S.: Growth network models with random number of attached links. Physica A **576** (2021). https://doi.org/10.1016/j.physa.2021.126041
14. Song, T., Tang, Q., Huang, J.: Triadic closure, homophily, and reciprocation: an empirical investigation of social ties between content providers. Inf. Syst. Res. **30**(3), 912–926 (2019). https://doi.org/10.1287/isre.2019.0838
15. Wharrie, S., Azizi, L., Altmann, E.: Micro-, meso-, macroscales: the effect of triangles on communities in networks. Phys. Rev. E **100**(2) (2019). https://doi.org/10.1103/PhysRevE.100.022315
16. Yin, H., Benson, A.R., Leskovec, J.: The local closure coefficient: a new perspective on network clustering. In: Proceedings of the Twelfth ACM International Conference on Web Search and Data Mining, WSDM 2019, pp. 303–311. Association for Computing Machinery, New York (2019). https://doi.org/10.1145/3289600.3290991

On a General Approach for Numerical Solving Singular Integral Equations in the Scalar Problems of Diffraction by Curvilinear Smooth Screens

Aleksei Tsupak[(✉)] [iD]

Penza State University, 40 Krasnaya, Penza, Russia

Abstract. The problems of diffraction of monochromatic acoustic waves by smooth infinitely thin curvilinear sound-soft and sound-hard screens are considered. The singular integral equations of the diffraction problems are numerically solved using Galerkin method. A general approach for definition of basis functions with compact support is proposed in the case of arbitrary smooth (or piecewise smooth) parameterized screens. Several examples of such basis functions on non-planar screens are presented. It is shown that the basis functions possess the denseness property. Convergence of the Galerkin method is established in appropriate Sobolev spaces on manifolds with boundary. The parallel implementation of the Galerkin method is used. Several numerical tests are carried out; in particular, the approximate solutions of a test problem are compared with the known analytical solution. The proposed technique can be used for solving more complicated problems, i.e., problems of diffraction by systems of solids and screens, or by partially shielded solids without requiring consistency of grids on volumetric scatterers and surfaces.

Keywords: Acoustic diffraction · Curvilinear parameterized screens · Basis functions · Convergence of Galerkin method

1 Introduction

We study the problem of diffraction of an acoustic wave by smooth open curvilinear non-planar (sound-soft or sound-hard) oriented screens, which can be reduced [8] to singular integral equations over the surfaces of the screens.

The integral operators are considered as elliptic pseudo-differential operators [9] in appropriate Sobolev spaces. Ellipticity and invertibility of the operators result in the convergence of the Galerkin method with any basis functions v_k that satisfy the approximation condition [3].

In the case of a plane screen, the piecewise-linear functions with a hexagonal support can be chosen as basis functions [4,6].

In this work, we consider smooth parameterized curvilinear screens. In spite of piecewise flat approximation [2] of a screen's surface we properly determine

© Springer Nature Switzerland AG 2021
D. Balandin et al. (Eds.): MMST 2020, CCIS 1413, pp. 154–160, 2021.
https://doi.org/10.1007/978-3-030-78759-2_13

basis functions on the curvilinear screen and show that such basis functions represent complete systems of functions in solutions' spaces.

In order to illustrate the proposed approximation technique we show some numerical results of solving the problem by Galerkin method (the parallel version of the method was implemented using the MPI [10] interface). In particular, we represent the comparison of obtained approximate solutions and the well-known analytical solution on the sound-hard sphere of the diffraction problem on the unit sphere.

2 Integro-Differential Equation of the Diffraction Problem

Scattering of an acoustic wave of the form $U(x,t) = u_0(x)e^{-i\omega t}$ by an acoustically hard screen Ω is described [8] by the hypersingular integral equation

$$A_h\varphi = \int_\Omega \frac{\partial}{\partial \mathbf{n}_x} \frac{\partial}{\partial \mathbf{n}_y} G(x,y)\varphi(y)ds_y = \frac{\partial}{\partial \mathbf{n}_x} u_0(x), \quad x \in \Omega, \tag{1}$$

whereas scattering from by a sound-soft screen is reduced to a weakly singular equation

$$A_s\varphi = -\int_\Omega G(x,y)\varphi(y)ds_y = u_0(x), \quad x \in \Omega. \tag{2}$$

Here is $G = \frac{e^{ik_0|x-y|}}{4\pi|x-y|}$ is the Green's function of the Helmholtz equation in R^3.

We consider C^∞-smooth oriented bounded surfaces parameterized by a given vector function

$$x(t) = (x_1(t_1,t_2), x_2(t_1,t_2), x_3(t_1,t_2)), \quad t \in T \subset R^2, \tag{3}$$

whose Jacobi matrix's rank is 2 in $\overline{T}$. The parameters domain T is bounded, with piecewise smooth boundary ∂T.

The operators A_h and A_s are ΨD operators of order -1 and 1, respectively, in the following Sobolev spaces:

$$A_h : \tilde{H}^{1/2}(\overline{\Omega}) \to H^{-1/2}(\Omega), \quad A_s : \tilde{H}^{-1/2}(\overline{\Omega}) \to H^{1/2}(\Omega),$$

which follows from the Gårding inequalities [8].

To define the Sobolev spaces we consider Ω as an open submanifold of a C^∞ two-dimensional smooth closed oriented manifold Ω_0. Let $\{U_\alpha\}$ be a finite cover of Ω_0 with coordinate diffeomorphisms $\kappa_\alpha : U_\alpha \to V_\alpha$ onto $V_\alpha \subset R^2$, and $\{\rho_\alpha\}$ be a smooth unity partition subordinating to the cover $\{U_\alpha\}$. Define the norm of the Sobolev space $H^s(\Omega_0)$ [1],

$$\|u\|_{H^s(\Omega_0)} = \sum_\alpha \left\|u(\kappa_\alpha^{-1}(t))\rho_\alpha(\kappa_\alpha^{-1}(t)))\right\|_{H^s(R^2)}. \tag{4}$$

Then, $\tilde{H}^s(\overline{\Omega})$ is the space of functions from $H^s(\Omega_0)$ with compact support in $\overline{\Omega}$, whereas $H^s(\Omega)$ is the space of restrictions of functions from $H^s(\Omega_0)$ to Ω.

3   Basis Functions and Galerkin Method

We consider the following formulation of Galerkin method

$$\langle A\varphi_n, v_i \rangle = \langle f, v_i \rangle, \quad i = 1, ..., n, \tag{5}$$

using the notation $\langle \cdot, \cdot \rangle$ for the antidual pairing of spaces $\tilde{H}^{1/2}(\overline{\Omega})$ and $H^{-1/2}(\Omega)$ (or $\tilde{H}^{-1/2}(\overline{\Omega})$ and $H^{1/2}(\Omega)$). The approximate solution to (5) is denoted by $\varphi_n := \sum_{i=1}^{n} c^i v_i$.

Let us first describe the way to determine basis functions $v_i(x)$, $x \in \Omega$ so as to satisfy the approximation condition. To this end, we consider the parameter's domain T and define some basis functions $v_i^{(0)}(t)$, $t = (t_1, t_2) \in T$, $i = 1, ..., n$, for any natural n. The basis functions $v_i(x)$ on Ω are defined as follows

$$v_i(x) = v^{(0)}(\kappa(x)), \quad x \in \Omega, \quad \text{or} \quad v_i(x(t)) = v^{(0)}(t), \quad t \in T. \tag{6}$$

where $\kappa(x) : \Omega \to T$ is the inverse mapping with respect to $x(t) : T \to \Omega$. Let $X_n^{(0)} = span\{v_1^{(0)}, ..., v_n^{(0)}\}$ and $X_n = span\{v_1, ..., v_n\}$.

Theorem 1. *Let Ω be a smooth oriented surface in R^3 parameterized by a smooth vector function $x(t)$, $t \in T \subset R^2$. If subspaces $X_n^{(0)}$ possess the denseness property in $\tilde{H}^s(\overline{T})$ then X_n possess the denseness property in $\tilde{H}^s(\overline{\Omega})$.*

Proof. We will show that $v_i(x)$ form a complete system of functions in $\tilde{H}^s(\overline{\Omega})$, i.e., that for any element $u \in \tilde{H}^s(\overline{\Omega})$ and for any $\varepsilon > 0$ there exist n and $u_n \in X_n$ such that

$$\|u - u_n\|_{\tilde{H}^s(\overline{\Omega})} < \varepsilon.$$

Introduce an open circle $B_1 = B_r(t_0)$ of radius $r > 0$ centered at $t_0 \in R^2$ such $\overline{T} \subset B$ and $dist(\partial T, \partial B) > 0$. Let Ω_1 be the image of the ball B_1 under the same parameterizing mapping $x(t)$ that defines Ω. C^∞-smoothness of the mapping $x(t)$ implies that Ω_1 is a smooth surface such that $\Omega \subset \Omega_1$ and $dist(\partial \Omega, \partial \Omega_1) > 0$.

Let u be an arbitrary function in $\tilde{H}^s(\overline{\Omega})$. From definition of the latter space it follows that the continuation by zero $\mathcal{E}_0 u$ of the function u from the screen Ω to Ω_1 belongs to $\tilde{H}^s(\overline{\Omega}_1)$. In addition, $\mathcal{E}_0 u \in H^s(\Sigma)$ for any smooth closed oriented surface $\Sigma \supset \Omega_1$.

Let $(U_\alpha, \kappa_\alpha)$, $\alpha = 1, ..., m$, be an atlas on Σ, and $\{\rho_\alpha\}$ represent a smooth partition of unity subordinate to $\{U_\alpha\}$ such that for some index α_0

$$U_{\alpha_0} = \Omega_1, \quad \kappa_{\alpha_0}(x) = \kappa(x).$$

Since $u \in \tilde{H}^s(\overline{\Omega})$ and $x \in C^\infty(T, \Omega)$ then [1] $u_T(t) := u(x(t)) \in \tilde{H}^s(\overline{T})$. As the denseness property for $X_n^{(0)}$ holds by assumption, then there exist constant $c_j (j = 1, ..., n)$ such that

$$\|u_T - u_{n,T}\|_{\tilde{H}^s(\overline{T})} = \|\mathcal{E}_0 u_T - \mathcal{E}_0 u_{n,T}\|_{\tilde{H}^s(B_1)} = \|\mathcal{E}_0 u_T - \mathcal{E}_0 u_{n,T}\|_{H^s(R^2)} < \varepsilon. \tag{7}$$

Further, we obtain

$$\|u - u_n\|_{\tilde{H}^s(\Omega)} = \|\mathcal{E}_0 u - \mathcal{E}_0 u_n\|_{\tilde{H}^s(\Omega_1)} = \|\mathcal{E}_0 u - \mathcal{E}_0 u_n\|_{H^s(\Sigma)}$$
$$= \sum_{\alpha=1}^{m} \|\mathcal{E}_0 u(\kappa_\alpha^{-1}(t))\varphi_\alpha(\kappa_\alpha^{-1}(t)) - \mathcal{E}_0 u_n(\kappa_\alpha^{-1}(t))\varphi_\alpha(\kappa_\alpha^{-1}(t))\|_{H^s(R^2)} \quad (8)$$
$$= \|\mathcal{E}_0 u(\kappa^{-1}(t))\varphi(\kappa^{-1}(t)) - \mathcal{E}_0 u_n(\kappa^{-1}(t))\varphi(\kappa^{-1}(t))\|_{H^s(R^2)}$$
$$= \|\mathcal{E}_0 u_T - \mathcal{E}_0 u_{n,T}\|_{H^s(R^2)} < \varepsilon.$$

Thus, the denseness property of X_n in $\tilde{H}^s(\Omega)$ is satisfied.

Let us now describe the sets of most simple basis functions suitable for solving scalar problems of diffraction.

Consider the rectangle $T = [a_1, b_1] \times [a_2, b_2]$ and define for any natural m the uniform grid

$$t_{i,k} = a_i + k h_i, \ i = 1, 2, \ h_i = (b_i - a_i)/m; \quad k = 0, ..., m.$$

We will define $v_k^{h,(0)}(k)$ with hexagonal support,

$$v_{k_1,k_2}^{h,(0)}(t) = \tilde{v}^{(0)}((t_1 - a_1)/h_1 - k_1, (t_2 - a_2)/h_2 - k_2), \quad (9)$$

where

$$\tilde{v}^{(0)}(t_1, t_2) = \begin{cases} 1 - t_1, \ t_1 \in [0, 1], & t_2 \in [0, t_1], \\ 1 - t_2, \ t_1 \in [0, 1], & t_2 \in [t_1, 1], \\ 1 + t_1 - t_2, \ t_1 \in [-1, 0], \ t_2 \in [0, t_1 + 1], \\ 1 + t_1, \ t_1 \in [-1, 0], \ t_2 \in [t_1, 0], \\ 1 + t_2, \ t_1 \in [-1, 0], \ t_2 \in [-1, t_1], \\ 1 - t_1 + t_2, \ t_1 \in [0, 1], & t_2 \in [t_1 - 1, 0]. \end{cases} \quad (10)$$

We will also use functions piecewise constant functions defined in T,

$$v_k^{s,(0)}(t) = v_{k_1,k_2}^{s,(0)}(t) = \chi_{T_k}(t), \ t \in T. \quad (11)$$

where χ_{T_k} is the indicator of the subset $T_k = (t_{1,k_1}, t_{1,k_1+1}) \times (t_{2,k_2}, t_{2,k_2+1})$, $k_i = 0, ..., m - 1$.

Thus, definition of the basis functions $v_k^h(x)$ on a sound-hard or functions $v_k^s(x)$ on a sound-soft screen is reduced to application of formula (6) to functions (10) and (11), respectively.

Let us prove the theorem on convergence of the Galerkin method.

Theorem 2. *Let $\Omega \subset R^3$ be a C^∞-smooth oriented open surface parameterized by a given function $x(t)$, $t \in T \subset R^2$. Let the basis functions v_i be defined by formulas (6), (10) or (6), (11). Then the Galerkin method converges for the operators A_h and A_s in $\tilde{H}^{1/2}(\overline{\Omega})$ and $\tilde{H}^{-1/2}(\overline{\Omega})$, respectively.*

Proof. As indicated above, A_s and A_h are elliptic operators in spaces chosen. Moreover the operators are injective. Indeed, the homogeneous integral equations (2) and (1) are equivalent [6] to Dirichlet and Neumann boundary value problems for the Helmholtz equation. These homogeneous problems have only the trivial solution.

Consequently, convergence of the Galerkin method follows from the denseness property of basis functions' subspaces X_n in the solutions' space X. It is known that the functions (10) and (11) possess the denseness property in $\tilde{H}^{1/2}(\overline{T})$ and $\tilde{H}^{-1/2}(\overline{T})$, respectively, for any rectangle T. The smoothness of parametrization function $x(t) : T \to \Omega$ now implies [1] that $v_k^h \in \tilde{H}^{1/2}(\overline{\Omega})$ and $v_k^s \in \tilde{H}^{1/2}(\overline{\Omega})$. Then, the basis functions on a curvilinear screen satisfy the approximation condition, which follows from Theorem 1.

Thus, Galerkin method converges and there holds an error estimate [3]

$$\|u - u_n\|_{\tilde{H}^{\pm 1/2}(\overline{\Omega})} \le \inf_{\psi \in X_n} \|u - \psi\|_{\tilde{H}^{\pm 1/2}(\overline{\Omega})}. \tag{12}$$

3.1 Numerical Tests

In this subsection several results of solving the diffraction problem are shown.

First, we consider a test problem of scattering form a sound-hard unit sphere $S_1(0)$ centered at the origin.

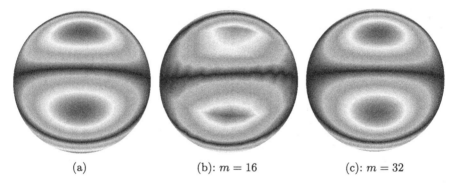

(a)   (b): $m = 16$ (c): $m = 32$

Fig. 1. Modulus of the analytical (a) and approximate (b, c) solutions φ_m of the model scattering problem on the centered sound-hard unit sphere.

The integro-differential equation (1) is known [2] to have eigenfunctions defined by formulas

$$s_{n,m}(x) = P_n^m(\cos\theta)\cos(m\varphi), \quad P_n^m(u) = (1 - u^2)^{1/2}\frac{d^m}{du^n}P_n(u), \tag{13}$$

where P_n are the Legendre polynomials, and φ, θ are the spherical on $S_1(0)$.

Figure 1(a) shows the modulus of the analytical solution $\varphi(x)$ to (1) with the function $s_{3,3}(x)$ representing the right-hand side of the equation, whereas Fig. 1(b), and (c) represent the approximate solutions $\varphi_n(x)$. The total number of basis functions is $n = (m-1)^2$, and m denotes the grid's dimensions in parameters t_1, t_2. Note, that we consider only uniform grids γ_m in the domain T. If T is a rectangle $[a_1, b_1] \times [a_2, b_2]$, then γ_m is defined as below,

$$\gamma_m = \{(t_{1,i_1}, t_{2,i_2}) : t_{k,i_k} = a_k + (b_k - a_k) \cdot i_k/m, \ i_k = 0, ..., m\}.$$

Further, the diffraction of an incident plane wave by the sound-hard unit sphere and a sound-soft torus were studied (see Fig. 1) (Fig. 2).

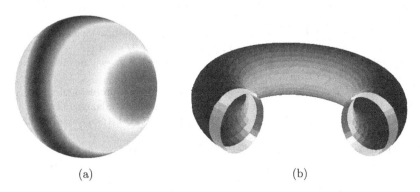

(a)   (b)

Fig. 2. Modulus of the potential density φ. (a) : diffraction of a plane wave $u_0 = e^{ik_0 x_2}$ by the sound-hard sphere $S_1(0)$, (b): diffraction of a plane wave $u_0 = e^{ik_0 x_2}$ by the sound-soft 3/4 torus ($R = 4$, $r = 1$, $\phi \in [0; 3\pi/2]$).

The described technique of definition of basis function exempts from the need for matching the computational grids on solids and screens that arises when solving problems of scattering from partially shielded bodies [7]. It is, thus, possible to construct grids and determine the basis functions independently on scatterers of various dimensions. The quantitative parameters of the grids (e.g., the number and sizes of finite elements) can be selected independently of each other. The latter implies that the order of the main matrix and the block sizes (corresponding to scatterers of different dimensions) can be specified in an almost arbitrary way, which is especially convenient for parallel implementation of the numerical method.

Conclusion
Scalar problems of scattering of monochromatic acoustic waves form non-planar hard and soft screens are considered. The Galerkin method for solving the corresponding integral equations on the screens' surfaces is formulated. The basis functions are defined on arbitrary parameterized curvilinear screens; the denseness property is proved for the basis functions in appropriate Sobolev spaces on manifolds with boundary. Several tests were carried out to check the convergence of Galerkin method. In particular, the exact solution of the problem on a sound-hard sphere was compared with approximate solutions.

References

1. Agranovich, M.S.: More general spaces and their applications. Sobolev Spaces, Their Generalizations and Elliptic Problems in Smooth and Lipschitz Domains. SMM, pp. 193–311. Springer, Cham (2015). https://doi.org/10.1007/978-3-319-14648-5_4

2. Daeva, S.G., Setukha, A.V.: Numerical simulation of scattering of acoustic waves by inelastic bodies using hypersingular boundary equation. In: AIP Conference Proceedings 2015, vol. 1648, pp. 39004-1-390004-4 (2015). https://doi.org/10.1063/1.4912614

3. Kress, R.: Linear Integral Equations. Applied Mathematical Sciences. Springer, New York (1989). https://doi.org/10.1007/978-3-642-97146-4

4. Marchuk, G.I.: Vvyedyeniye v proyektsionno-syetochnyye myetody. Nauka, Moscow (1981). (in Russian)

5. Medvedik, M.Y., Smirnov, Y.G., Tsupak, A.A.: Scalar problem of plane wave diffraction by a system of nonintersecting screens and inhomogeneous bodies. Comput. Math. Math. Phys. **54**(8), 1280–1292 (2014). https://doi.org/10.1134/S0965542514080089

6. Smirnov, Y.G., Tsupak, A.A.: Method of integral equations in the scalar problem of diffraction on a system consisting of a "soft" and a "hard" screen and an inhomogeneous body. Diff. Eqn. **50**, 1150–1160 (2014). https://doi.org/10.1134/S0012266114090031

7. Smirnov, Y.G., Tsupak, A.A.: Investigation of electromagnetic wave diffraction from an inhomogeneous partially shielded solid. Appl. Anal. **97**(11), 1881–1895 (2018). https://doi.org/10.1080/00036811.2017.1343467

8. Stephan, E.P.: Boundary integral equations for screen problems in R^3. Integr. Eqn. Potential Theory **10**(10), 236–257 (1987). https://doi.org/10.1007/BF01199079

9. Taylor, M.E.: Pseudodifferential Operators. Princeton University Press, Princeton (1981)

10. Voevodin, V.l., et al.: Supercomputer Lomonosov-2: large scale, deep monitoring and fine analytics for the user community. Supercomputing Front. Innov. **6**(2), 4–11 (2019). https://doi.org/10.14529/jsfi190201

Noise Influence on the Estimation of Characteristics of Intermittent Generalized Synchronization Using Local Lyapunov Exponents

Evgeniy V. Evstifeev[1,2(✉)] and Olga I. Moskalenko[1,2]

[1] Saratov State University, Saratov 410012, Russia
[2] Regional Scientific and Educational Mathematical Center
"Mathematics of Future Technologies", Saratov 410012, Russia

Abstract. Using the calculation of local Lyapunov exponents the influence of the stationary noise on statistical characteristics of intermittent generalized synchronization and critical coupling parameter value corresponding to the generalized synchronization regime onset has been studied. Two unidirectionally and mutually coupled chaotic Lorenz oscillators with a complex (two-sheeted) topology of attractors, characterized by the jump-intermittency, have been chosen as the systems under study. The dependence of the critical value of the coupling parameter on the noise intensity has been estimated. We calculated such general characteristics of the intermittency as the distributions of the laminar phase lengths for fixed values of the coupling parameter and the dependence of the mean length of the laminar phases on the criticality parameter. We have shown that the numerically obtained statistical characteristics greatly correspond to theoretical exponential laws. Obtained results are in a good agreement with the results of other works and demonstrate that the method of local Lyapunov exponents has significant stability to noise and has a strong potential to be applied for different nonlinear systems coupled unidirectionally or mutually.

Keywords: Intermittent generalized synchronization of chaos ·
Intermittency characteristics · Local Lyapunov exponents

1 Introduction

Generalized synchronization (GS) is one of the fundamental natural and radiophysics phenomena [1] and it is one of the main types of chaotic synchronization [2]. This phenomenon has a wide application potential, for example, in

Supported by the grant from the President of Russian Federation according to the research project No. MD-21.2020.2.

D. Balandin et al. (Eds.): MMST 2020, CCIS 1413, pp. 161–168, 2021.
https://doi.org/10.1007/978-3-030-78759-2_14

medicine [3] and for secure communication [4]. In the context of the flow dynamical systems

$$\dot{\mathbf{x}} = \mathbf{F}(\mathbf{x})$$
$$\dot{\mathbf{y}} = \mathbf{G}(\mathbf{y}, \mathbf{x}, \mathbf{g}) \tag{1}$$

(where $\mathbf{g}$ is a vector of the coupling parameters, $\mathbf{x}, \mathbf{y}$ are vectors of the system states, $\mathbf{F}, \mathbf{G}$ are vector of the functional relations, $\dot{\mathbf{x}}, \dot{\mathbf{y}}$ are time derivatives) the GS is understood as the existence of a functional relation (the functional in general case) between the states of interacting systems (Eq. (1)).

In this work, as an example, we consider two unidirectionally and mutually coupled chaotic Lorenz oscillators [5] described by the following system of ordinary differential equations:

$$\dot{x}_{1,2} = \sigma(y_{1,2} - z_{1,2}) + \varepsilon_{1,2}(x_{2,1} - x_{1,2}) + N_{1,2}\psi$$
$$\dot{y}_{1,2} = r_{1,2}x_{1,2} - y_{1,2} + x_{1,2}z_{1,2} \tag{2}$$
$$\dot{z}_{1,2} = -b_{1,2}z_{1,2} + x_{1,2}y_{1,2}$$

where $\sigma = 10.0, b_1 = 2, b_2 = 8/3, r_1 = 40, r_1 = 35$. In the case of unidirectional coupling $\varepsilon_1 = 0, \varepsilon_2 = \varepsilon$, in the case of mutual coupling $\varepsilon_1 = \varepsilon_2 = \varepsilon$, $N_{1,2}\psi$ are the noise terms, where ψ is a random number, corresponding to the normal distribution with a mean of 0 and a standard deviation of 1. The noise was added only to the second system, i.e. $N_1 = 0$, $N_2 = N$. The noise amplitude depends on real number N.

The system parameters were set in such a way that for any choice of initial conditions the regime of intermittent GS is observed after the transient is finished [5]. Specifically, in the system (2) a jump intermittency arises due to the complex attractor topologies of the interacting subsystems [5,6]. It should be noted that the equations were solved using general Runge-Kutta integration scheme but the noise was added after each integration step h to the first equation of the second interacting subsystem as $N\psi h$, where $h = 0.003$ [7].

So, for the values of the coupling parameter that are slightly less than a certain critical one instead of a completely asynchronous behavior, as it could be expected, the intermittent GS regime is observed. In such regime certain intervals of time corresponding to the synchronous (in terms of the GS) states of the systems (laminar phases) alternate with the asynchronous oscillations of the systems (turbulent phases). With an increase of the coupling parameter, as the critical value is approached, the mean length of the laminar phases tends to infinity, and the average duration of the turbulent phases tends to zero. Therefore, it is customary to consider the statistical characteristics of the laminar phases due to their considerable amount.

There are several methods to study the intermittent GS, for instance, the auxiliary system approach [8]. Unfortunately, such approach is not applicable in the case of the mutually coupled interacting systems [9]. Therefore, it is necessary to develop the universal methods that could be used to study the intermittent GS regime of coupled nonlinear systems both in the cases of unidirectional and mutual couplings. It should be noted, that there are several attempts allowing to solve such problem. In particular, in Ref. [5,6] the method based on the analysis of the location of representation points has been proposed. The principal disadvantage of such method is the possibility of its application only to the systems with two-sheeted topology of attractors. More universal method proposed by us previously [10,11] is based on the calculation of local Lyapunov exponents. The goal of the present paper is the study of the possibility of the use of local Lyapunov exponents for the analysis of the intermittent GS regime in the presence of noise.

2   Determination of Intermittency Characteristics Using Local Lyapunov Exponents

The main characteristics of intermittency include distributions of the laminar phase lengths for fixed values of the coupling parameter and dependence of the mean length of the laminar phases on the criticality parameter.

To determine these characteristics the separation of laminar phases is required. In this work we propose the method based on calculation of local Lyapunov exponents [13,14]. The Benettin algorithm [12] with Gram-Schmidt orthogonalization is used to acquire the exponents. We use the following formula where the accumulation time is limited:

$$\Lambda \cong \frac{1}{k\tau} \sum_{j=1}^{k} \ln \frac{\|\tilde{x}_j\|}{\epsilon}, \tag{3}$$

where $\tilde{x}_j$ is the perturbation vector at the jth iteration of the algorithm, k is the number of algorithm iterations, τ is a dimensionless time interval between renormalizations, $\|...\|$ is the Euclidean norm, and $\epsilon = 1$ is the initial norm of the perturbation vectors.

Firstly, an orthonormal basis of unit perturbation vectors is specified. Then, in equal intervals, the transformation of their measure (in this case, the length) is calculated, then renormalization and orthogonalization take place and everything mentioned together constitutes an iteration. The process is repeated until the value of $k\tau$ becomes more or equal to a certain value of an accumulation time T. After that, the iterations which did not happen within the dimensionless time interval $[T - t,\ t]$, where t is currently observed dimensionless time value, are disposed. This makes it possible to assess the temporal dynamics of the coupled systems.

Lyapunov exponents characterize the dynamics of the system. Positive values refer to the chaotic part, i.e. the speed of scattering of the phase trajectories initially close along a certain direction; the negatives ones correspond to a periodic motion. Zero Lyapunov exponents are associated with the shift along the time direction and, thus, attractors always have at least one of them. The number of the directions and, accordingly, the exponents coincide with the number of variables in a full system of differential equations describing the states of the systems. However, the Lyapunov exponents do not depend on the directions but rather characterize the systems at all.

The algorithm for identifying the characteristic phases of the behavior from time series is based on the amplitude method. A certain threshold value Δ of the investigated quantity, delimiting the areas of this quantity into the corresponding phases, is set. In this case, at the values less than the threshold a laminar phase is observed, and at values exceeding it a turbulent one is detected. Since during evaluation of local Lyapunov exponents the accumulation time is limited and, therefore, the accuracy of their calculation is quite low, this threshold should be set a little exceeding zero ($\Delta = 0.07$). The threshold value itself also depends on the magnitude of the accumulation interval $T = k\tau$ and decreases as the last one increases. The interval was fixed at 400, although it is possible to obtain characteristics using other values in a wide range of intervals. The choice of the value depends on the systems under study. It should be much greater than an average system period of time in order to minimize local Lyapunov exponents fluctuation. However, too big values lead to a loss of information and, hence, to an increase of the characteristic estimation error. Also, in order to minimize the estimation errors we ignore ultra short laminar phases the length of which is 10 or less, whereas the turbulent phases of lengths not exceeding 25 were considered as parts of the laminar phases.

To understand the further results we should firstly consider the dependencies of the critical parameter value, related to the onset of the GS regime in the cases of unidirectional and mutual coupling, on the noise intensity parameter N (see Fig. 1).

The results indicate that at the beginning with the growth of N up to a certain value ($N \approx 900$ in case of directional coupling and $N \approx 350$ in case of mutual coupling) the critical parameter value also increases[1]. However, after that it tends to decrease until it reaches 0 value indicating that the GS regime arises almost immediately after the transient process dies out. The cause of this behavior is that the noise influence completely hides all information related to the initial GS. In such case the noise-induced synchronization regime is observed [15]. Comparison between the results for both coupling cases allows to conclude that for N in the range of $[0, 200]$ the GS onset in the case of the mutual coupling occurs for a smaller ε value than in the case of the unidirectional coupling that is in a good accordance with the known works. So, this range was chosen for further studies.

---

[1] For relatively small values of the noise intensity comparable with the amplitude of own oscillations of the system the boundary value of the GS regime onset does not change dramatically, and in the case of unidirectional coupling even remains almost constant.

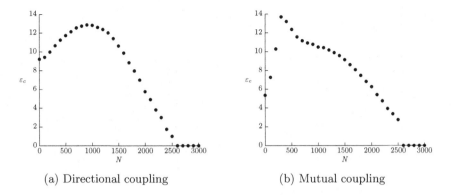

(a) Directional coupling   (b) Mutual coupling

Fig. 1. a, b - Dependencies of the critical coupling parameter value on the noise intensity parameter N in the case of unidirectional and mutual coupling, respectively

3   Estimation of Statistical Characteristics of the Intermittent GS

The first statistical characteristic that has been estimated is the distributions of the laminar phase lengths for fixed values of the coupling parameter (see Fig. 2). In the regime of jump intermittency such distributions should obey the exponential law [5].

The numerically obtained distributions were normalized on the number of laminar phases and their accumulation interval (1000 for a, b, c and 200 for d, do not mix it up with the one related to the local Lyapunov exponents). The average relative deviation of the calculated characteristics from the theoretical exponential law does not exceed 20%. The obtained distributions indicate that with the noise intensity N increasing the deviation of experimental data from the theoretical curves also increases. This effect is greater in the case of mutual coupling since the critical coupling parameter value, as it was shown in Fig. 1, changes faster with N increasing than in the unidirectional case. So, the lower values of the coupling parameter become farther from the critical point, at that the mean length of the laminar phases decreases.

The correctness of the statement mentioned above was confirmed by the calculation of the other intermittency characteristic which is the dependence of the mean length of the laminar phases on the coupling parameter ε (see Fig. 3).

It is clearly seen that for all considered values of the noise intensity the mean length of the laminar phases decreases faster in the case of the mutual coupling in comparison with the unidirectional case. At that, in both cases the dependencies of the studied characteristic on the coupling parameter ε obey close to exponential law

$$L \approx C \exp(k\delta) \times \exp\left(\frac{2}{D}\left(\frac{1}{12} + \frac{\delta}{\sqrt{3}} + \frac{2(-\delta)^{3/2}}{3\sqrt[4]{3}} + \frac{\delta^2}{12}\right)\right). \tag{4}$$

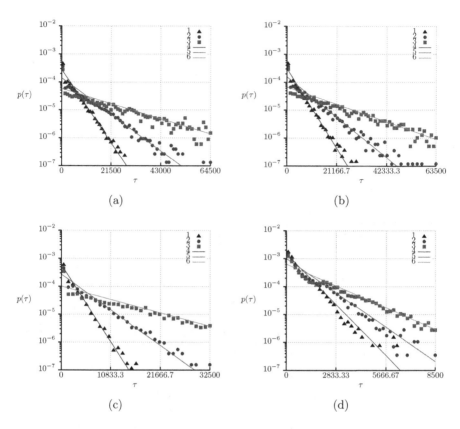

Fig. 2. Distributions of the laminar phase lengths for fixed values of the coupling parameter in the case of unidirectional (a, b) and mutual (c, d) couplings, respectively. The noise intensity parameter $N = 25$ (a, c) and 50 (b, d). The numbers from 1 to 3 represent results obtained for the coupling parameter $\varepsilon = \{9.5, 9.75, 10.0\}$ (a) and $\varepsilon = \{5.7, 5.9, 6.1\}$ (b). The lines 4, 5 and 6 represent approximations in a form $\ln(x) = \ln(1/T) - x/T$, where the mean length of the laminar phases $T \approx \{2973, 5967, 14479\}$ (a), $T \approx \{2687, 5287, 12225\}$ (b), $T \approx \{1433, 3082, 8208\}$ (c), $T \approx \{565, 836, 1316\}$ (d), respectively.

(where $\delta = \varepsilon - \varepsilon_c$, C, k and D are positive constants) in full accordance with the theory developed in [5].

Thus, the method based on calculation of local Lyapunov exponents makes it possible to estimate the characteristics of intermittency with a fairly good accuracy in a sufficiently large range of the coupling parameter values for different values of the noise intensity, both in the cases of unidirectional and mutual couplings. In spite of the noise presence, the obtained characteristics do not change qualitatively, i.e. for all considered values of the noise intensity the jump intermittency is observed.

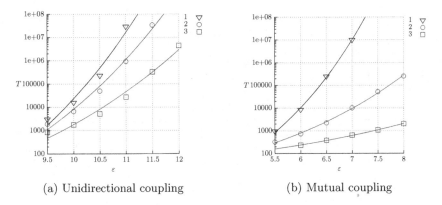

(a) Unidirectional coupling   (b) Mutual coupling

Fig. 3. Dependencies of the mean length of the laminar phases on the coupling parameter ε for different values of the noise intensity parameter N in the cases of unidirectional (a) and mutual (b) couplings. The numbers from 1 to 3 represent results obtained for the values of the noise intensity $N = \{0, 100, 200\}$ in the case of unidirectional coupling (a) and $N = \{20, 60, 120\}$ in the case of mutual coupling (b). Solid lines correspond to the theoretical approximations in the form

$$\ln g(x) = a_2 + \exp b_2 x + 2d \left(\frac{1}{12} + \frac{x}{\sqrt{3}} + \frac{2(-x^{3/2})}{3\sqrt[4]{3}} + \frac{x^2}{12} \right)$$

4   Conclusion

To sum up, using the calculation of local Lyapunov exponents the influence of a stationary noise on general statistical characteristics of intermittent GS of chaos was estimated. It turns out that the characteristics depend greater on the noise intensity in the case of mutual coupling than in the case of the unidirectional one, and it is getting stronger when the coupling parameter tends to the critical value. Nevertheless, there is a wide range of the coupling parameter values where the characteristics do not change rapidly. The numerical results are in a good agreement with the theoretical laws and the other works' findings [5,6]. The work demonstrates that the proposed method has a reliable stability to stationary noise in the case of unidirectional and mutual couplings. Therefore, taking into account the universality of the local Lyapunov exponents we can state that the method has strong potential to be applied for various actions with nonlinear systems under the noise influence. For example, the results can be used when studying interacting systems under noise presence or developing a scheme for secure information transmission.

References

1. Boccaletti, S., Kurths, J., Osipov, G., et al.: The synchronization of chaotic systems. Phys. Rep. **36**(6), 1–101 (2002)

2. Rulkov, N.F., Sushchik, M.M., Tsimring, L.S., Abarbanel, H.D.I.: Generalized synchronization of chaos in directionally coupled chaotic systems. Phys. Rev. E **51**(2), 980–994 (1995)

3. Moskalenko, O.I., Koronovskii, A.A., Hramov, A.E.: Generalized synchronization of chaos for secure communication: remarkable stability to noise. Phys. Lett. A **374**(29), 2925–2931 (2010)

4. Hramov, A.E., Koronovskii, A.A., Ponomarenko, V.I., Prokhorov, M.D.: Detecting synchronization of self-sustained oscillators by external driving with varying frequency. Phys. Rev. E **73**, 026208 (2006)

5. Koronovskii, A.A., Moskalenko, O.I., Pivovarov, A.A., Khanadeev, V.A., Hramov, A.E., Pisarchick, A.N.: Jump intermittency as a second type of transition to and from generalized synchronization. Phys. Rev. E. **102**, 012205 (2020)

6. Moskalenko, O.I., Koronovskii, A.A., Khanadeev, V.A.: Intermittency at the boundary of generalized synchronization in mutually coupled systems with complex attractor topology. Tech. Phys. **64**(3), 302–305 (2019). https://doi.org/10.1134/S1063784219030198

7. Nikitin, N.N., Pervachev, S.V., Razevig, V.D.: About solution of stochastic differential equations of follow-up systems. Autom. Telemech. **4**, 133–137 (1975). (in Russian)

8. Abarbanel, H.D.I., Rulkov, N.F., Sushchik, M.M.: Generalized synchronization of chaos: the auxiliary system approach. Phys. Rev. E. **53**, 4528–4535 (1996)

9. Moskalenko, O.I., Koronovskii, A.A., Hramov, A.E.: Inapplicability of an auxiliary system approach to chaotic oscillators with mutual-type coupling and complex networks. Phys. Rev. E **87**, 064901(2013)

10. Moskalenko, O.I., Evstifeev, E.V., Koronovskii, A.A.: A method of determining the characteristics of intermittent generalized synchronization based on the calculation of local Lyapunov exponents. Tech. Phys. Lett. **46**(8), 792–795 (2020). https://doi.org/10.1134/S1063785020080246

11. Koronovskii, A.A., Moskalenko, O.I., Pivovarov, A.A., Evstifeev, E.V.: Intermittent route to generalized synchronization in bidirectionally coupled chaotic oscillators. Chaos **30**, 083133 (2020)

12. Benettin, G., Galgani, L., Giorgilli, A., Strelcyn, J.M.: Lyapunov characteristic exponents for smooth dynamical systems and for Hamiltonian systems; a method for computing all of them. Part 1: Theory. Meccanica **15**(1), 9–20 (1980)

13. Abarbanel, H.D.I., Brown, R., Kennel, M.B.: Variation of Lyapunov exponents on a strange attractor. J. Nonlinear Sci. **1**, 175–199 (1991)

14. Hramov, A.E., Koronovskii, A.A., Kurovskaya, M.K.: Zero Lyapunov exponent in the vicinity of the saddle-node bifurcation point in the presence of noise. Phys. Rev. E. **78**, 036212 (2008)

15. Toral, R., Mirasso, C.R., Hernández-Garsia, E., Piro, O.: Analytical and numerical studies of noise-induced synchronization of chaotic systems. Chaos **11**(3), 665–673 (2001)

Application of the Kovacic Algorithm to the Problem of Rolling of a Heavy Homogeneous Ball on a Surface of Revolution

Alexander S. Kuleshov$^{(\boxtimes)}$ and Darya V. Solomina

Department of Mechanics and Mathematics, Lomonosov Moscow State University,
Moscow 119934, Russian Federation
kuleshov@mech.math.msu.su

Abstract. The problem of rolling without sliding of a homogeneous ball on a fixed surface under the action of gravity is a classical problem of nonholonomic system dynamics. Usually, when considering this problem, following the E. J. Routh approach [1] it is convenient to define explicitly the equation of the surface, on which the ball's centre is moving. This surface is equidistant to the surface, over which the contact point is moving. From the classical works of E. J. Routh [1] and F. Noether [2] it was known that if the ball rolls on a surface such that its centre moves along a surface of revolution, then the problem is reduced to solving the second order linear differential equation. However it is impossible to find the general solution of this differential equation for an arbitrary surface of revolution. Therefore it is interesting to study for which surface of revolution the corresponding second order linear differential equation admits the explicit solution, for example, Liouvillian solution. To solve this problem it is possible to apply the Kovacic algorithm [3] to the corresponding second order linear differential equation. In this paper we present our own method to derive the corresponding second order linear differential equation. In the case when the centre of the ball moves along the paraboloid of revolution we prove that the corresponding second order linear differential equation admits a Liouvillian solution.

Keywords: Rolling without sliding · Homogeneous ball · Surface of revolution · Kovacic algorithm

1 Introduction

Investigation of various problems of classical mechanics and mathematical physics is reduced to solving the second–order linear differential equation with

This work was supported financially by the Russian Foundation for Basic Research (grant No. 19-01-00140 and No. 20-01-00637).

D. Balandin et al. (Eds.): MMST 2020, CCIS 1413, pp. 169–177, 2021.
https://doi.org/10.1007/978-3-030-78759-2_15

variable coefficients. In 1986, the American mathematician J. Kovacic proposed an algorithm for solving the second–order linear differential equation with rational coefficients in the case where the solution can be expressed in terms of so-called Liouvillian functions [3]. Liouvillian functions are constructed from rational functions by algebraic operations, taking exponentials and integration [3,4]. If a second–order linear differential equation has no Liouvillian solutions, the Kovacic algorithm also allows one to ascertain this fact. Therefore, the Kovacic algorithm is a very effective method for investigation the problems which solution are reduced to the integration the second–order linear differential equation. However, this algorithm is not very known for the specialists in mechanics and mechanical engineering. The goal of this paper is to avoid this problem and to made the Kovacic algorithm more popular for the investigation of various problems, where we need to solve the second–order linear differential equation. In this paper we discuss the application of the Kovacic algorithm to the problem of motion of a heavy homogeneous ball on a fixed perfectly rough surface of revolution.

The paper is organized as follows. In the Sect. 2 we present the detailed formulation of the problem and we give our own way to reduce the problem to the integration the second–order linear differential equation. In the Sect. 3 we prove that the problem of motion of a heavy homogeneous ball on a surface of revolution such that the centre of a ball moves along a paraboloid of revolution in integrated in terms of Liouvillian functions.

2   Problem Formulation. General Equations of Motion

Let us consider the problem of motion of a heavy homogeneous ball on any fixed perfectly rough surface under the action of any forces whose resultant passes through the centre of the ball [1]. Let G be the centre of gravity of the ball and let the moving axes GC, GA, GB be respectively a normal to the surface and some two lines at right angles to be afterwards chosen at our convenience. Let e_1, e_2, e_3 be the unit vectors of these axes GA, GB and GC respectively. Let $\Omega = \theta_1 e_1 + \theta_2 e_2 + \theta_3 e_3$ be the angular velocity of these axes; $v_G = u e_1 + v e_2 + w e_3$ be the velocity of G (then $w = 0$ since the ball is always in contact with the supporting surface); $\omega = \omega_1 e_1 + \omega_2 e_2 + \omega_3 e_3$ be the angular velocity of the body about these axes. Let $R = F e_1 + F' e_2 + R e_3$ be the reaction, acting on the ball from the surface and let $P = X e_1 + Y e_2 + P e_3$ be the impressed force on the centre of gravity of the ball. Let m be the mass of the ball, a – its radius, J – the moment of inertia of the ball about a diameter. We shall suppose that the ball rolls on the convex side of the fixed surface and that the positive direction of the axis GC is drawn outwards from the surface. Then the equations of motion of the ball can be written in vector form:

$$m\dot{v}_G + [\Omega \times v_G] = P + R, \tag{1}$$

$$J\dot{\omega} + [\Omega \times J\omega] = \left[\overrightarrow{GK} \times R\right]. \tag{2}$$

Equations (1) and (2) represents the behavior of momentum and angular momentum of the ball respectively. Here $\overrightarrow{GK} = -ae_3$ is the radius – vector of the ball's point of contact with the surface relative to its centre of gravity. Since the point of contact of the sphere and surface is at rest we have

$$v_G + \left[\omega \times \overrightarrow{GK}\right] = 0. \tag{3}$$

In scalar form Eqs. (3)–(4) can be written as follows:

$$m\dot{u} - m\theta_3 v = X + F, \quad m\dot{v} + m\theta_3 u = Y + F', \quad m\theta_1 v - m\theta_2 u = P + R; \tag{4}$$

$$J\dot{\omega}_1 + J\theta_2\omega_3 - J\theta_3\omega_2 = F'a, \quad J\dot{\omega}_2 + J\theta_3\omega_1 - J\theta_1\omega_3 = -Fa, \quad \dot{\omega}_3 + \theta_1\omega_2 - \theta_2\omega_1 = 0; \tag{5}$$

$$u - a\omega_2 = 0, \quad v + a\omega_1 = 0. \tag{6}$$

Eliminating F, F', ω_1, ω_2 from the Eqs. (4)–(6) we get:

$$\dot{u} - \theta_3 v = \frac{a^2 X}{J + ma^2} + \frac{Ja\theta_1\omega_3}{J + ma^2}, \quad \dot{v} + \theta_3 u = \frac{a^2 Y}{J + ma^2} + \frac{Ja\theta_2\omega_3}{J + ma^2}. \tag{7}$$

The meaning of the Eq. (7) may be found as follows. They are the two equations of motion of the centre of the ball, which we should have obtained if the given surface had been smooth and the centre G had been acted on by acceleration forces

$$\frac{Ja\theta_1\omega_3}{J + ma^2} \quad \text{and} \quad \frac{Ja\theta_2\omega_3}{J + ma^2}$$

along the axes GA, GB and by the same impressed forces as before reduced in the ratio

$$\frac{a^2}{J + ma^2}.$$

The centre G of the ball moves along a surface formed by producing all the normals to the given surface a constant length equal to the radius of the ball. Let us take the axes GA, GB to be tangents to the lines of curvature of this surface. Let us find expression for the angular velocity $\boldsymbol{\Omega}$ of the chosen moving coordinate system GA, GB, GC in terms of the components u and v of the velocity v_G of the ball. We will assume that the surface along which the centre of the ball moves is given with respect to some fixed coordinate system by the equation

$$r = r(q_1, q_2), \tag{8}$$

where q_1 and q_2 are gaussian coordinates on this surface. We shall assume that a coordinate grid on the surface (8) consists of curvature lines whose directions at every point are given by unit vectors:

$$e_1 = \frac{1}{h_1}\frac{\partial r}{\partial q_1}, \quad e_2 = \frac{1}{h_2}\frac{\partial r}{\partial q_2}, \quad (e_i \cdot e_j) = \delta_{ij}. \tag{9}$$

Here we have denoted by h_1, h_2 the Lame's parameters

$$h_i(q_1, q_2) = \left| \frac{\partial \boldsymbol{r}}{\partial q_i} \right|, \quad i = 1, 2.$$

The vector $\boldsymbol{e}_3 = [\boldsymbol{e}_1 \times \boldsymbol{e}_2]$ is a normal vector to the surface (8) at point (q_1, q_2). The velocity of centre of the ball $\boldsymbol{v}_G$ may be written as follows:

$$\boldsymbol{v}_G = \frac{d\boldsymbol{r}}{dt} = \frac{\partial \boldsymbol{r}}{\partial q_1} \dot{q}_1 + \frac{\partial \boldsymbol{r}}{\partial q_2} \dot{q}_2 = u\boldsymbol{e}_1 + v\boldsymbol{e}_2.$$

Therefore we have the following equation connecting the velocities u and v with the coordinates q_1, q_2 and their derivatives

$$u = h_1 \dot{q}_1, \quad v = h_2 \dot{q}_2. \tag{10}$$

Let $k_i(q_1, q_2)$, $i = 1, 2$ be principal curvatures of the surface (8). Then we have the following equations:

$$\frac{\partial \boldsymbol{e}_3}{\partial q_1} = -h_1 k_1 \boldsymbol{e}_1, \quad \frac{\partial \boldsymbol{e}_3}{\partial q_2} = -h_2 k_2 \boldsymbol{e}_2. \tag{11}$$

Equations (11) follow from the Rodrigues's theorem well known in differential geometry (see e.g. [5]), where we have to additionally account for the fact that our coordinate grid on the surface (8) is orthogonal and consists of curvature lines. Taking into account (9) and (11) it is easy to derive the following equations:

$$\frac{\partial \boldsymbol{e}_1}{\partial q_1} = -\frac{1}{h_2} \frac{\partial h_1}{\partial q_2} \boldsymbol{e}_2 + h_1 k_1 \boldsymbol{e}_3, \quad \frac{\partial \boldsymbol{e}_1}{\partial q_2} = \frac{1}{h_1} \frac{\partial h_2}{\partial q_1} \boldsymbol{e}_2,$$

$$\frac{\partial \boldsymbol{e}_2}{\partial q_1} = \frac{1}{h_2} \frac{\partial h_1}{\partial q_2} \boldsymbol{e}_1, \quad \frac{\partial \boldsymbol{e}_2}{\partial q_2} = -\frac{1}{h_1} \frac{\partial h_2}{\partial q_1} \boldsymbol{e}_1 + h_2 k_2 \boldsymbol{e}_3. \tag{12}$$

The angular velocity of the coordinate system GA, GB, GC can be found by the well known formula:

$$\boldsymbol{\Omega} = (\dot{\boldsymbol{e}}_2 \cdot \boldsymbol{e}_3) \boldsymbol{e}_1 + (\dot{\boldsymbol{e}}_3 \cdot \boldsymbol{e}_1) \boldsymbol{e}_2 + (\dot{\boldsymbol{e}}_1 \cdot \boldsymbol{e}_2) \boldsymbol{e}_3,$$

where

$$\dot{\boldsymbol{e}}_i = \frac{d\boldsymbol{e}_i}{dt} = \frac{\partial \boldsymbol{e}_i}{\partial q_1} \dot{q}_1 + \frac{\partial \boldsymbol{e}_i}{\partial q_2} \dot{q}_2, \quad i = 1, 2, 3.$$

Taking into account (11)–(12) we obtain for the angular velocity $\boldsymbol{\Omega}$ the following expression

$$\boldsymbol{\Omega} = h_2 k_2 \dot{q}_2 \boldsymbol{e}_1 - h_1 k_1 \dot{q}_1 \boldsymbol{e}_2 + \left(\frac{\dot{q}_2}{h_1} \frac{\partial h_2}{\partial q_1} - \frac{\dot{q}_1}{h_2} \frac{\partial h_1}{\partial q_2} \right) \boldsymbol{e}_3.$$

This expression can be rewritten in the form

$$\boldsymbol{\Omega} = k_2 v \boldsymbol{e}_1 - k_1 u \boldsymbol{e}_2 + \frac{1}{h_1 h_2} \left(\frac{\partial h_2}{\partial q_1} v - \frac{\partial h_1}{\partial q_2} u \right) \boldsymbol{e}_3.$$

if we take into account Eq. (10). Therefore we have the following expressions for the components θ_1, θ_2, θ_3 of the angular velocity $\boldsymbol{\Omega}$:

$$\theta_1 = k_2 v, \quad \theta_2 = -k_1 u, \quad \theta_3 = \frac{1}{h_1 h_2} \left(\frac{\partial h_2}{\partial q_1} v - \frac{\partial h_1}{\partial q_2} u \right). \tag{13}$$

We suppose now that the surface along which the centre of the ball moves is a surface of revolution, given with respect to some fixed coordinate system by the equation

$$\boldsymbol{r} = \begin{pmatrix} \rho(q_1) \cos q_2 \\ \rho(q_1) \sin q_2 \\ \zeta(q_1) \end{pmatrix}. \tag{14}$$

In this case the Lame's parameters h_1 and h_2 take the form:

$$h_1 = h_1(q_1) = \sqrt{\left(\frac{d\rho}{dq_1} \right)^2 + \left(\frac{d\zeta}{dq_1} \right)^2}, \quad h_2 = h_2(q_1) = \rho(q_1), \tag{15}$$

and the principal curvatures k_1 and k_2 may be written as follows:

$$k_1 = k_1(q_1) = \frac{\left(\frac{d^2\zeta}{dq_1^2} \frac{d\rho}{dq_1} - \frac{d\zeta}{dq_1} \frac{d^2\rho}{dq_1^2} \right)}{\left(\left(\frac{d\rho}{dq_1} \right)^2 + \left(\frac{d\zeta}{dq_1} \right)^2 \right)^{\frac{3}{2}}}, \quad k_2 = k_2(q_1) = \frac{\frac{d\zeta}{dq_1}}{\rho \sqrt{\left(\frac{d\rho}{dq_1} \right)^2 + \left(\frac{d\zeta}{dq_1} \right)^2}}. \tag{16}$$

In this case the meridians and parallels are the lines of curvature. Let the axis Z of upward vertical be symmetry axis of the considered surface of revolution. Except the coordinates q_1 and q_2 we introduce the Euler angles θ, ψ and φ, where the angle the axis GC makes with the axis of Z equals θ and ψ is the angle the plane containing Z and GC makes with any fixed vertical plane. We suppose, that the components θ_1, θ_2, θ_3 of the angular velocity $\boldsymbol{\Omega}$ are defined by the Euler kinematic equations:

$$\theta_1 = \dot{\psi} \sin\theta \sin\varphi + \dot{\theta} \cos\varphi, \quad \theta_2 = \dot{\psi} \sin\theta \cos\varphi - \dot{\theta} \sin\varphi, \quad \theta_3 = \dot{\psi} \cos\theta + \dot{\varphi},$$

in which we put $\varphi = -\pi/2$. Therefore we have

$$\theta_1 = -\dot{\psi} \sin\theta, \quad \theta_2 = \dot{\theta}, \quad \theta_3 = \dot{\psi} \cos\theta. \tag{17}$$

Comparing these equations with Eq. (13) we find:

$$-\dot{\psi} \sin\theta = k_2 h_2 \dot{q}_2, \quad \dot{\theta} = -k_1 h_1 \dot{q}_1, \quad \dot{\psi} \cos\theta = \frac{1}{h_1} \frac{dh_2}{dq_1} \dot{q}_2. \tag{18}$$

From the second equation of the system (18) the connection between θ and q_1 is determined. Therefore we can assume that the surface (14) is defined by the variables θ and q_2, i.e.

$$\rho|_{q_1 = q_1(\theta)} = \sigma(\theta), \quad \zeta|_{q_1 = q_1(\theta)} = \tau(\theta). \tag{19}$$

The Lame parameters, calculated according to (15), are determined as follows:

$$h_1 = h_1(\theta) = \sqrt{\left(\frac{d\sigma}{d\theta}\right)^2 + \left(\frac{d\tau}{d\theta}\right)^2}, \quad h_2 = h_2(\theta) = \sigma(\theta), \tag{20}$$

and the principal curvatures $k_1 = k_1(\theta)$ and $k_2 = k_2(\theta)$ are calculated by the formulas:

$$k_1 = k_1(\theta) = \frac{\left(\frac{d^2\tau}{d\theta^2}\frac{d\sigma}{d\theta} - \frac{d\tau}{d\theta}\frac{d^2\sigma}{d\theta^2}\right)}{\left(\left(\frac{d\sigma}{d\theta}\right)^2 + \left(\frac{d\tau}{d\theta}\right)^2\right)^{\frac{3}{2}}}, \quad k_2 = k_2(\theta) = \frac{\frac{d\tau}{d\theta}}{\sigma\sqrt{\left(\frac{d\sigma}{d\theta}\right)^2 + \left(\frac{d\tau}{d\theta}\right)^2}}. \tag{21}$$

From the second equation of the system (18) we obtain, taking into account (10), that

$$u = -\frac{\dot\theta}{k_1}. \tag{22}$$

From Eq. (6) we get

$$\omega_1 = -\frac{v}{a}, \quad \omega_2 = \frac{u}{a} = -\frac{\dot\theta}{ak_1}.$$

Therefore, from the third equation of the system (5) we obtain:

$$\dot\omega_3 = \theta_2\omega_1 - \theta_1\omega_2 = \frac{v\dot\theta}{ak_1}(k_2 - k_1). \tag{23}$$

Equation (23) can be rewritten as follows:

$$\frac{d\omega_3}{d\theta} = \frac{v}{ak_1}(k_2 - k_1). \tag{24}$$

Now we suppose the ball rolls on a surface under the action of gravity. Then we have

$$Y = 0, \quad \theta_3 = \frac{1}{h_1h_2}\frac{dh_2}{d\theta}v$$

and the second equation of the system (7) takes the form:

$$\frac{dv}{d\theta} - \frac{v}{h_1h_2k_1}\frac{dh_2}{d\theta} = \frac{Ja}{J+ma^2}\omega_3. \tag{25}$$

Differentiating repeatedly (25) and taking into account (24) we have

$$\frac{d}{d\theta}\left(\frac{dv}{d\theta} - \frac{v}{h_1h_2k_1}\frac{dh_2}{d\theta}\right) = \frac{J}{J+ma^2}\frac{v}{k_1}(k_2 - k_1). \tag{26}$$

Thus, the problem of rolling of a ball on a fixed perfectly rough surface under the action of gravity, under the assumption that the ball's centre moves along a given surface of revolution, is reduced to integration the second order linear differential equation (26) with respect to the ball's velocity component v. Therefore, it is interesting to study for which surfaces of revolution equation (26) is integrable in Liouvillian functions. For study the problem of existence of Liouvillian solution of a given second order linear differential equation the Kovacic algorithm is usually used [3]. Below we prove that the problem of rolling of a heavy homogeneous ball on a fixed perfectly rough surface is integrable in Liouvillian functions when the centre of the ball moves along a paraboloid of revolution.

3   Rolling of a Ball on a Paraboloid of Revolution

Let the perfectly rough surface on which the ball rolls be such that the centre of the ball moves along a paraboloid of revolution. We write equation of the paraboloid in the form (14):

$$r = \begin{pmatrix} Rq_1 \cos q_2 \\ Rq_1 \sin q_2 \\ -\frac{Rq_1^2}{2} \end{pmatrix}.$$

In the considered case

$$\rho(q_1) = Rq_1, \quad \zeta(q_1) = -\frac{Rq_1^2}{2}.$$

Here R is a parameter having the dimension of length. The Lame's parameters h_1 and h_2 calculated by (15) have a form:

$$h_1 = R\sqrt{1 + q_1^2}, \quad h_2 = Rq_1,$$

and the principal curvatures k_1 and k_2, calculated by (16), have a form:

$$k_1 = -\frac{1}{R(1 + q_1^2)^{\frac{3}{2}}}, \quad k_2 = -\frac{1}{R\sqrt{1 + q_1^2}}.$$

From the second equation of the system (18) we find the connection between variables q_1 and θ:

$$\dot{\theta} = \frac{\dot{q}_1}{1 + q_1^2}.$$

Therefore we have

$$q_1 = \tan\theta. \tag{27}$$

Taking into account (27), we can suppose, that the expressions for $\rho(q_1)$ and $\zeta(q_1)$ are rewritten as follows:

$$\sigma(\theta) = R\tan\theta, \quad \tau(\theta) = -\frac{R}{2}\tan^2\theta.$$

As a result we obtain the following expressions for the Lame's parameters h_1 and h_2 and the principal curvatures k_1 and k_2:

$$h_1 = \frac{R}{\cos^3 \theta}, \quad h_2 = R \tan \theta, \quad k_1 = -\frac{\cos^3 \theta}{R}, \quad k_2 = -\frac{\cos \theta}{R}.$$

The second order linear differential equation (26) can be represented in the form:

$$\frac{d}{d\theta} \left(\frac{dv}{d\theta} + \frac{v}{\sin \theta \cos \theta} \right) = \frac{J}{J + ma^2} \frac{\sin^2 \theta}{\cos^2 \theta} v. \tag{28}$$

Thus, the problem of motion of a heavy homogeneous ball on a perfectly rough surface such that the centre of the ball moves along a paraboloid of revolution is reduced to integration the second order linear differential equation (28). Let us change the independent variable in Eq. (28) by formula $x = \cos^2 \theta$ and denote:

$$\frac{J}{J + ma^2} = n^2 < 1.$$

Then Eq. (28) is reduced to the equation with rational coefficients:

$$\frac{d^2 v}{dx^2} + \frac{1}{x - 1} \frac{dv}{dx} - \frac{\left(n^2 x^2 + 2\left(1 - n^2\right) x + n^2 - 1\right)}{4x^2 \left(x - 1\right)^2} v = 0. \tag{29}$$

Since Eq. (29) is the second–order linear differential equation with rational coefficients, we can apply the Kovacic algorithm to this differential equation. The goal of this algorithm is to find a solution of the differential equation

$$\frac{d^2 v}{dx^2} + a\left(x\right) \frac{dv}{dx} + b\left(x\right) v = 0, \tag{30}$$

where $a\left(x\right)$ and $b\left(x\right)$ are rational functions of one (in general case complex) variable x. The first step of the algorithm is to reduce the differential equation (30) to a simpler form, using the following formula

$$y\left(x\right) = v\left(x\right) \exp\left(\frac{1}{2} \int a\left(x\right) dx\right). \tag{31}$$

Then Eq. (30) takes the form

$$\frac{d^2 y}{dx^2} = R\left(x\right) y, \quad R\left(x\right) = \frac{1}{2} \frac{da}{dx} + \frac{a^2}{4} - b, \tag{32}$$

where $R\left(x\right)$ is also rational function of one variable x. The Kovacic algorithm allows one to find explicitly the solution of Eq. (32), expressed in terms of Liouvillian functions.

Applying to the differential equation (29) the transformation of the form (31), we reduce it to the differential equation

$$\frac{d^2 y}{dx^2} = \frac{n^2 - 1}{4x^2} y. \tag{33}$$

Application of the Kovacic algorithm to the second order linear differential equation (33) shows that the general solution of this equation can be represented in the form:

$$y = C_1 x^{\frac{1+n}{2}} + C_2 x^{\frac{1-n}{2}},$$

where C_1 and C_2 are arbitrary constants. Therefore, the general solution of Eq. (33) is expressed in terms of Liouvillian functions. For the differential equation (28), we corresponding general solution has the form:

$$v(\theta) = \frac{\cos\theta}{\sin\theta} \left(K_1 (\cos\theta)^n + K_2 (\cos\theta)^{-n} \right),$$

where K_1 and K_2 are arbitrary constants. Thus the problem of motion of a heavy homogeneous ball of a surface of revolution such that the centre of the ball moves along the paraboloid of revolution is integrable in Liouvillian functions.

4 Conclusions

In this paper we apply the Kovacic algorithm to the problem of rolling of a heavy homogeneous ball on a fixed surface such that the centre of the ball moves along a given surface of revolution. This problem is reduced to solving the second–order linear differential equation with respect to the projection of velocity of the ball's centre onto the tangent to the parallel of the corresponding of revolution and we present here our own method to derive the corresponding equation. In the case when the centre of the ball moves along the paraboloid of revolution we are presenting the corresponding linear differential equation in explicit form and reduce its coefficients to a form of rational functions. Using the Kovacic algorithm we prove that the general solution of the corresponding second–order linear differential equation is expressed in terms of Liouvillian functions for all values of parameters of the problem.

References

1. Routh, E.J.: The Advanced Part of a Treatise on the Dynamics of a System of Rigid Bodies: Being Part II of a Treatise on the Whole Subject. Cambridge University Press, Cambridge (2013). https://doi.org/10.1017/CBO9781139237284
2. Noether, F.: Über rollende Bewegung einer Kugel auf Rotationsflächen. Teubner, Leipzig (1909)
3. Kovacic, J.: An algorithm for solving second order linear homogeneous differential equations. J. Symb. Comput. **2**(1), 3–43 (1986). https://doi.org/10.1016/S0747-7171(86)80010-4
4. Kaplansky, I.: An Introduction to Differential Algebra. Hermann, Paris (1957)
5. Do Carmo, M.P.: Differential Geometry of Curves and Surfaces. Prentice-Hall, Englewood Cliffs (1976)

Computation in Optimization and Optimal Control

Optimal Rotor Stabilization
in an Electromagnetic Suspension System Using
Takagi-Sugeno Fuzzy Models

Aleksey V. Mukhin$^{(\boxtimes)}$ ⓘ

Lobachevsky State University, Nizhny Novgorod, Russia

Abstract. The paper presents the results of solving the problem of designing stabilizing output controllers for an electromagnetic suspension system based on the use of Takagi-Sugeno fuzzy models. Two major problems were considered: the construction of stabilizing controllers and the construction of optimal controllers based on the quadratic performance criterion. To calculate the controllers, an original nonlinear mathematical model of the plant was replaced by an equivalent fuzzy model, which is a set of linear subsystems. For the synthesis of control laws, the apparatus of linear matrix inequalities was used. The calculated controllers for the fuzzy system were substituted into the original nonlinear object. The results of the mathematical modeling showed that using Takagi-Sugeno fuzzy models, it is possible to construct both a stabilizing controller and an optimal controller based on a quadratic performance criterion. The calculated controllers ensured rotor stabilization in a fairly wide range of initial deviations, up to the maximum possible values. Based on obtained results, it can be concluded that the presented approach, realized on the use of Takagi-Sugeno fuzzy models, allows one to describe the rotor dynamics in an electromagnetic suspension system in a wide range of initial disturbances.

Keywords: Electromagnetic suspension · Takagi-Sugeno fuzzy models · Linear matrix inequalities

1 Introduction

The operation principle of an electromagnetic suspension is based on the phenomenon of magnetic levitation. Thanks to this, it becomes possible to overcome gravity without contact and to provide the rotor hanging in active magnetic bearings. The obvious advantage of such systems is, first of all, the absence of mechanical contact and, as a consequence, the absence of friction. Due to this advantage it is possible to increase significantly the service life and efficiency compared to traditional mechanical counterparts.

Control of a rotor in an electromagnetic suspension is an important and pressing problem associated with the wide practical application of electromagnetic bearings. Electromagnetic bearings are of great interest for a whole range of various fields of industry and technology, as well as some areas of medicine [1–4].

© Springer Nature Switzerland AG 2021
D. Balandin et al. (Eds.): MMST 2020, CCIS 1413, pp. 181–192, 2021.
https://doi.org/10.1007/978-3-030-78759-2_16

Control is realized, as a rule, by changing the magnitude of the magnetic field created by the electromagnet. For designing the controllers, the most common approach is based on the use of linearized models [5–9]. Despite the easiness and convenience, an obvious drawback of linearized models is their limited applicability. Linearized models work only in the neighborhood of the equilibrium position, with small initial deviations. In fact, the initial disturbances of an object can go far beyond the limits of applicability of linearized models. As a result, such models cannot fully describe the dynamics of the investigated plant, and the calculated controllers are efficient only under small initial disturbances. One of the ways to take into account nonlinearities and build nonlinear controllers can be the use of fuzzy logic and fuzzy systems based on it. Takagi-Sugeno fuzzy controller design using the negative absolute eigenvalue approach for a class nonlinear and unstable system including electromagnetic suspension was described in [10].

The article presents the results of solving the rotor control problem in an electromagnetic suspension based on the use of continuous fuzzy Takagi-Sugeno models [11]. The control problem was considered under the assumption that the measured variable is the vertical displacement of the rotor. In the context of this problem, two approaches were used: the construction of a stabilizing controller and the construction of an optimal controller with a given quadratic performance criterion. For the synthesis of control laws, the apparatus of linear matrix inequalities was used [12, 13].

The article is organized as follows: the first section presents the derivation of a fuzzy mathematical model. The second section is the formulation of control problems. In the third section, systems of linear matrix inequalities are presented, as well as a procedure for their numerical implementation. The fourth section contains the results of mathematical modeling.

2 Fuzzy Object Model

The rotor in an electromagnetic suspension is in the field of action of two forces: gravity and magnetic attraction. According to Newton's second law, when these forces are equal, the body will be stationary. The rotor dynamics in the suspension is described by the following system of equations [5]

$$\dot{x}_1 = x_2$$
$$\dot{x}_2 = \frac{1}{2}\left[\frac{(1+x_3)^2}{(1-x_1)^2} - 1\right]$$
$$\dot{x}_3 = -\frac{(1+x_3)}{(1-x_1)}x_2 - a(1-x_1)x_3 + (1-x_1)u \qquad (1)$$

where $x = (x_1, x_2, x_3)^T \in R^{n_x}$ is a state system;
$u \in R^{n_u}$ is the control;
a is a constant parameter.

The dimensionless variable x_1 corresponds to the vertical movement of the rotor, x_2 corresponds to the speed of movement, and x_3 describes the current in the electromagnet circuit.

For the derivation the Takagi-Sugeno fuzzy model it is necessary to reduce the object (1) to the following form

$$\dot{x} = F(\sigma)x + B(\sigma)u, \, x(0) = x_0 \tag{2}$$

where $F(\sigma) \in R^{n_x \times n_x}, \sigma = \sigma(x) \in R^{n_\sigma}$;
$B(\sigma) \in R^{n_x \times n_u}$.

The elements of the matrices $F(\sigma)$ and $B(\sigma)$ must be continuous nonlinear functions $\sigma(x)$. To transform object (1) to form (2), a new phase variable x_4 was introduced, which is equals to

$$x_4 = \frac{1 + x_3}{1 - x_1} \tag{3}$$

After differentiating expression (3), it is obtained the following linear equation

$$\dot{x}_4 = \frac{\dot{x}_3 + x_4 x_2}{1 - x_1} = -ax_3 + u, \tag{4}$$

Then, the system (1) takes the form

$$\dot{x}_1 = x_2,$$
$$\dot{x}_2 = \frac{1}{2}\left[x_4^2 - 1\right],$$
$$\dot{x}_3 = -x_4 x_2 - a(1 - x_1)x_3 + (1 - x_1)u$$
$$\dot{x}_4 = -ax_3 + u \tag{5}$$

where $x^* = (x_1, x_2, x_3, x_4)^T \in R^{n_x^*}$.

The conversion from (1) to (5) means a mapping of the form R^{n_x} to $R^{n_x^*}$. The matrices $F(\sigma)$ and $B(\sigma)$ are written as.

$$F(\sigma) = \begin{pmatrix} 0 & 1 & 0 & 0 \\ 0 & 0 & 0 & \sigma_1 \\ \sigma_2 & \sigma_3 & \sigma_4 & \sigma_5 \\ 0 & 0 & -a & 0 \end{pmatrix}, \, B(\sigma) = \begin{pmatrix} 0 \\ 0 \\ \sigma_6 \\ 1 \end{pmatrix} \tag{6}$$

The corresponding nonlinear functions are defined as $\sigma_1 = \frac{1}{2}\left(x_4 - \frac{1}{x_4}\right)$, $\sigma_2 = ax_3$, $\sigma_3 = -x_4$, $\sigma_4 = a(x_1 - 1)$, $\sigma_5 = -x_2$, $\sigma_6 = -\frac{\sigma_4}{a}$.

Let us define a subset $\Omega = \{a_{i1} \le x_i < a_{i2}, i = \overline{1, n_x^*}\}$, in which a nonlinear object (5) will be considered. In order to ensure the continuity of the function σ_1, it is necessary to make the following change in the second equation of system (5)

$$\dot{x}_2 = \frac{1}{2}\left[x_4^2 - 1\right] = \frac{1}{2}(x_4 - 1)(x_4 + 1) = \frac{1}{2}x_4^*\left(x_4^* + 2\right) \tag{7}$$

Omitting the asterisk in the new variable, the system takes the following form

$$\dot{x}_1 = x_2,$$

$$\dot{x}_2 = \frac{1}{2}x_4(x_4 + 2),$$

$$\dot{x}_3 = -(x_4 + 1)x_2 - a(1 - x_1)x_3 + (1 - x_1)u,$$

$$\dot{x}_4 = -ax_3 + u, \tag{8}$$

the corresponding membership functions of the form

$$M_{i1,2} = \frac{\pm \sigma_i \mp \sigma_i^{min,max}}{\sigma_i^{max} - \sigma_i^{min}}, i = \overline{1, n_\sigma} \tag{9}$$

The range of values of each membership function forms its own normalized fuzzy set. Let's form a base of fuzzy rules for the plant (8)

$$R^i : IF\ \sigma_1\ is\ M_{11}\ and\ \dots\ and\ \sigma_{n_\sigma}\ is\ M_{n_\sigma 1} \tag{10}$$
$$THEN\dot{x}(t) = A_i x(t) + B_i u(t)$$

where R^i is the fuzzy rule $(i = \overline{1, r})$;
$r = 2^{n_\sigma}$ is the rule number.

For each rule there is corresponding linear system with matrices A_i and B_i, which are defined from $F(\sigma)$ depending on values σ_i. Then, the fuzzy model of the nonlinear system (8) is represented as a weighted sum of all linear subsystems

$$\dot{x} = \sum_{i=1}^{r} h_i(\sigma)[A_i x + B_i u] \tag{11}$$

where $h_i(\sigma) = \prod_{j=1}^{n_\sigma} M_{i1,2}^j(\sigma_j)$.

The obtained continuous fuzzy model (11) represents a nonlinear plant (8) on the considered subset Ω. Before proceeding to the problem formulation, let us make one simplification. Since the largest number of nonlinear functions is concentrated in the third equation of system (8), the linearization of this equation will significantly reduce the number of rules and thereby simplify the entire further theoretical analysis. After linearization of this equation in the neighborhood of the equilibrium position, system (8) takes the form

$$\dot{x}_1 = x_2,$$

$$\dot{x}_2 = \frac{1}{2}x_4(x_4 + 2),$$

$$\dot{x}_3 = -x_2 - ax_3 + u,$$

$$\dot{x}_4 = -ax_3 + u, \tag{12}$$

The only nonlinear function σ_1 is defined as

$$\sigma_1 = \frac{1}{2}(x_4 + 2) \tag{13}$$

The final fuzzy model of the simplified nonlinear system (12) is written in the following form

$$\dot{x} = \sum_{i=1}^{r} M_i A_i x + Bu \tag{14}$$

3   Formulation of the Problems

Let us consider two control problems: the first is the construction of a stabilizing fuzzy controller, the second is a fuzzy controller with a quadratic performance criterion. The vertical displacement of the rotor x_1 is considered as the measured variable. Corresponding matrix equation of the measured output

$$y = C_2 x \tag{15}$$

where $y \in R^{n_y}$ is the measured output;

$$C_2 = \begin{pmatrix} 1 & 0 & 0 & 0 \end{pmatrix}.$$

The control law of a linear system by the measured output in the form of a linear dynamic controller has the following form [12]

$$\dot{x}_r(t) = A_{ri} x_r(t) + B_{ri} y$$
$$(t) = C_{ri} x_r(t) + D_{ri} y \tag{16}$$

where $x_r \in R^{n_x^*}$ is a controller state.

In the case of a fuzzy system, the original plant is represented as a set of linear subsystems, each of which is characterized by its own matrices. Let us write a fuzzy control model in the form of a fuzzy dynamic controller for system (14)

$$\dot{x}_r = \sum_{i=1}^{r} M_i A_{ri} x_r + \sum_{i=1}^{r} M_i B_{ri} y$$
$$u = \sum_{i=1}^{r} M_i C_{ri} x_r + \sum_{i=1}^{r} M_i D_{ri} y \tag{17}$$

In addition, for the second problem, it is necessary to introduce the target output equation $z(t) \in R^{n_z}$

$$z(t) = Cx(t) + Du(t) \tag{18}$$

As a quadratic criterion, it is considered the following functional

$$\|z(t)\|^2 = \int_0^{\infty} \left(\sum_{i=2}^{n_x^*} x_i^2(t) + u^2(t) \right) dt \tag{19}$$

The problem is to calculate a γ-optimal fuzzy controller of the form (17), which will satisfy the following inequalities [12]

$$\inf_u \|z(t)\|^2 < \gamma^2 |x_0|^2, \forall x_0 \neq 0 \tag{20}$$

where γ is the minimum possible positive parameter.

Let us reduce Eq. (18) to the following form

$$z(t) = (C + DD_{ri} C_2)x + DC_{ri} x_r = C_{ci} x_c \tag{21}$$

Then, the closed-loop fuzzy system takes the following form

$$\dot{x}_c = \sum_{i=1}^{r} M_i A_{ci} x_c$$

$$z = \sum_{i=1}^{r} M_i C_{ci} x_c \tag{22}$$

where $x_c = (x x_r)^T$.

Closed-loop subsystem matrices are defined as follows

$$A_{ci} = \begin{pmatrix} A_i + B D_{ri} C_2 & B C_{ri} \\ B_{ri} C_2 & A_{ri} \end{pmatrix} \tag{23}$$

$$C_{ci} = [(C + D D_{ri} C_2) D C_{ri}] \tag{24}$$

Thus, it is required to calculate a stabilizing type controller and a γ-optimal controller according to criterion (20).

4 The Solving Methodology

To synthesize control laws, the apparatus of linear matrix inequalities [12], adapted for the case of fuzzy systems was used.

4.1 Calculation of a Stabilizing Fuzzy Controller

In order to ensure the stability of the closed-loop system (22) with matrices (23), it is necessary and sufficient to find such a quadratic function matrix $X = X^T > 0$, which satisfies the following inequalities

$$A_{ci}^T X + X A_{ci} < 0, i = \overline{1, r} \tag{25}$$

We introduce the matrix of parameters of the controllers Θ_i and represent the matrices of closed subsystems in the following form

$$A_{ci} = A_{0i} + B_0 \Theta_i C_0 \tag{26}$$

where $A_{0i} = \begin{pmatrix} A_i & 0_{n_x^* \times n_x^*} \\ 0_{n_x^* \times n_x^*} & 0_{n_x^* \times n_x^*} \end{pmatrix}$, $B_0 = \begin{pmatrix} 0_{n_x^* \times n_x^*} & B \\ I & 0_{n_x^* \times n_u} \end{pmatrix}$, $\Theta_i = \begin{pmatrix} A_{ri} & B_{ri} \\ C_{ri} & D_{ri} \end{pmatrix}$, $C_0 = \begin{pmatrix} 0_{n_x^* \times n_x^*} & I \\ C_2 & 0_{n_y \times n_x^*} \end{pmatrix}$.

Substitute (26) into (25) and write the resulting inequalities in the form

$$A_{0i}^T X + X A_{0i} + C_0^T \Theta_i^T B_0^T X + X B_0 \Theta_i C_0 < 0 \tag{27}$$

Next, it is necessary to reduce inequalities (27) to a system of linear matrix inequalities of the form

$$\Psi_i + P^T \Theta_i^T Q + Q^T \Theta_i P < 0, i = \overline{1, r} \tag{28}$$

where $\Psi_i = A_{0i}^T X + X A_{0i}$, $P = C_0$, $Q = B_0^T X$.

According to the elimination lemma, inequalities (28) are solvable with respect to the Θ_i if and only if [12]

$$W_C^T (A_i^T X_1 1 + X_1 1 A_i) W_C < 0, X_1 1 = X_1 1^T > 0$$
$$W_{B^T}^T \left(Y_{11} A_i^T + A_i Y_{11} \right) W_{B^T} \left(0, Y_{11} = Y_{11}^T \right) 0 \tag{29}$$

where W_C и W_{B^T} are matrices whose columns are based for the null spaces of C and B^T, respectively;

X_{11} и Y_{11} are upper-left hand submatrices of inverse matrices X and Y, respectively.

In order to implement the condition $XY = I$, it is necessary and sufficient that X_{11} and Y_{11} satisfy the inequality

$$\begin{pmatrix} X_{11} & I \\ I & Y_{11} \end{pmatrix} > 0 \tag{30}$$

Thus, the object is stabilized if and only if there are two positive definite symmetric submatrices X_{11} and Y_{11} satisfying the linear matrix inequalities (29) and (30). If the submatrices X_{11} and Y_{11} have been found, then the general matrix X can be restored by the formula [12].

$$X = \begin{pmatrix} X_{11} & X_{11} - Y_{11}^{-1} \\ X_{11} - Y_{11}^{-1} & X_{11} - Y_{11}^{-1} \end{pmatrix} \tag{31}$$

After the matrix X is has been found, the matrices of the parameters of the controllers Θ_i are calculated from inequalities (28) and then substituted into the fuzzy dynamic controller (17).

4.2 Calculation of a Fuzzy γ-Optimal Controller

The problem of γ-optimal fuzzy control consists in calculating such a fuzzy controller that ensures that condition (20) is satisfied for all linear subsystems. In order to implement this condition, the norms of the transfer matrices of a closed-loop plant must satisfy the conditions

$$\|H_{ci}\|_2 < \gamma |x_0| \quad \forall x_0 \neq 0 \tag{32}$$

where $H_{ci} = [C + DD_{ri} C_2 DC_{ri}] (sI - A_{ci})^{-1} x_c^0$.

The main idea of calculating controllers of this type is to reduce the matrices of closed-loop subsystems A_{ci} to linear matrix inequalities of the form (28). The subsequent steps are similar to the previous case. First, the matrices Ψ_i are calculated, and then the sought matrices of controllers are found from (28). Since the matrices Ψ_i contain a common unknown positive definite symmetric block matrix, it is necessary to first calculate this matrix. Since the ranks of the matrices P and Q are less than the rank of the matrices Ψ_i, then for the system of inequalities (28) is solvable when [12]

$$W_P^T \Psi_i W_P < 0, \; W_Q^T \Psi_i W_Q < 0 \tag{33}$$

where W_P и W_Q are matrices whose columns are based for the null spaces of P and Q, respectively.

By performing a series of matrix transformations, taking into account the block structure of matrices, inequalities (33) can be reduced to the following inequalities

$$\begin{pmatrix} W_{C_2} & 0 \\ 0 & I \end{pmatrix}^T \begin{pmatrix} A_i^T X_{11} + X_{11}A_i & C^T \\ C & -\gamma I \end{pmatrix} \begin{pmatrix} W_{C_2} & 0 \\ 0 & I \end{pmatrix} < 0 \tag{34}$$

$$N^T \begin{pmatrix} Y_{11}A_i^T + A_i^T Y_{11} & Y_{11}C^T \\ CY_{11} & -\gamma I \end{pmatrix} N < 0$$

where $i = \overline{1, r}$;

W_{C_2} и N are matrices whose columns are based for the null spaces of C_2 and $(B^T D^T)$, respectively;

For the existence of γ-optimal full-order controllers, it is necessary and sufficient that there exist two positive definite symmetric matrices X_{11} and Y_{11} satisfying inequalities (34), as well as inequalities

$$\begin{pmatrix} X_{11} & I \\ I & Y_{11} \end{pmatrix} \geq 0 \tag{35}$$

$$\begin{pmatrix} Y_{11} & I \\ I & \gamma I \end{pmatrix} > 0 \tag{36}$$

If the matrices X_{11} and Y_{11} have been found, then the matrix Y can be calculated from the formula (31). If the matrix X is has been found, the matrices of the parameters of the controllers Θ_i are calculated from inequalities (28) and then substituted into the fuzzy dynamic controller (17).

Linear matrix inequalities and systems of inequalities define nonlinear but convex constraints [12]. Therefore, convex optimization methods can be used for numerical implementation. The solvability of the system of inequalities (29) is reduced by minimizing the parameter t for which a linear matrix inequality of the form $F(x) - tI \leq 0$ is valid. The system of inequalities (34–36) is the problem of the solvability of a linear function with constraints given by a system of linear matrix inequalities. For the numerical implementation of linear matrix inequality systems, the interior-point method was used, implemented in the form of standard commands of the MATLAB software package [14].

5 Results

The calculated controllers were substituted in turn into the original nonlinear plant (1) with $a = 7.5$, closed by a fuzzy controller (17). Graphs of transient processes in the closed-loop system with a stabilizing type fuzzy controller for different initial values are shown in Figs. 1 and 2. Graphs of transient processes in the closed-loop system with a fuzzy γ-optimal controller for the same initial values are shown in Figs. 3 and 4. The initial values of the controllers were zero.

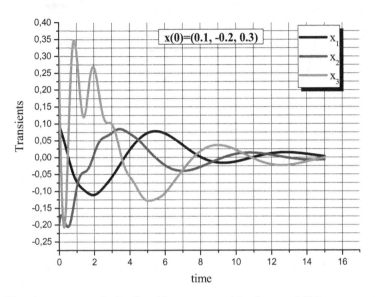

Fig. 1. Transient processes in the closed-loop system with a fuzzy stabilizing type controller

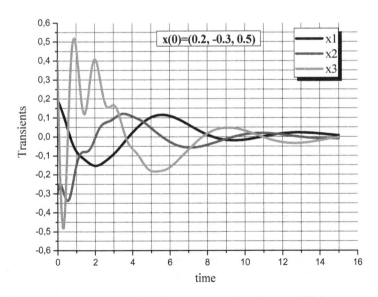

Fig. 2. Transient processes in the closed-loop system with a fuzzy stabilizing type controller

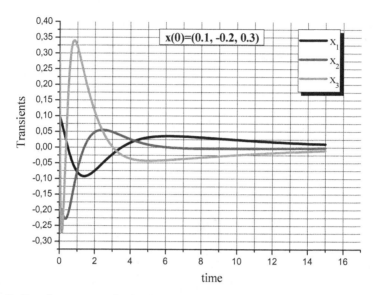

Fig. 3. Transient processes in the closed-loop system with a fuzzy γ-optimal controller

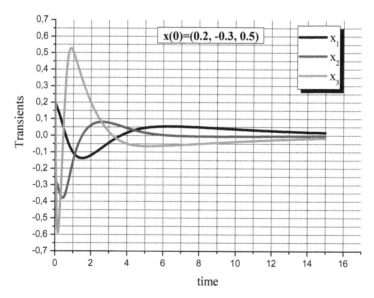

Fig. 4. Transient processes in the closed-loop system with a fuzzy γ-optimal controller

The results of mathematical modeling showed that the calculated fuzzy controllers of both types make it possible to stabilize the rotor in a wide range of initial disturbances, up to the maximum possible values. Based on the results obtained, it can be concluded that the calculated fuzzy controllers allow control the object in wide range of initial disturbances.

6 Conclusion

The article presents the results of solving the control rotor problem in an electromagnetic suspension according to the measured output of full order. A fuzzy model of an object is derived, which is equivalent to the original nonlinear model, as well as a fuzzy control model. Numerical calculations of fuzzy controllers and mathematical modeling of transient processes are performed. Based on the obtained results, it can be concluded that by measuring only the movement of the rotor, it is possible to calculate both a stabilizing controller and a γ-optimal controller with a quadratic quality criterion. The calculated fuzzy controllers ensured rotor stabilization under any initial disturbances, up to the maximum possible values. Thus, the presented fuzzy models, in contrast to the linearized models, make it possible to control the dynamics and stabilize the rotor in an electromagnetic suspension in a wide range of initial disturbances.

Acknowledgments. The author thanks his scientific adviser, Professor of Differential Equations Department of Lobachevsky State University D.V. Balandin for advice, as well as valuable and useful comments.

References

1. Zhuravlev, Yu.N.: Active Magnetic Bearings. Theory, Calculation, Application. Politechnica, St. Peterburg (2003)
2. Schweitzer, G., Maslen, E.: Magnetic Bearings Theory, Design, and Application to Rotating Rachinery. Springer, Berlin (2009). https://doi.org/10.1007/978-3-642-00497-1
3. Grinvald, V.M., Kusmin, G.S., Masloboev, Y., Selishchev, S.V., Telyshev, D.V.: First domestic ventricular assistant device AVK-N "Sputnik" on basis of implantable blood pump. Izvestiya Vysshikh Uchebnykh Zavedenii. Elektronika 20(5), 516–521 (2015)
4. Masuzawa, T., Osa, M., Mapley, M.: Motor design and impeller suspension. Mech. Circulatory Respiratory Support 11, 335–377 (2017)
5. Balandin, D.V., Biryukov, R.S., Kogan, M.M., Fedyukov, A.A.: Optimal stabilization of bodies in electromagnetic suspensions without measurements of their locations. J. Comput. Syst. Sci. Int. 56, 351–363 (2017). https://doi.org/10.1134/S1064230717020046
6. Gruber, W., Pichler, M., Rothbock, M., Amrhein, W.: Self-sensing active magnetic bearing using 2-level PWM current ripple demodulation. In: Proceedings 7th International Conference on Sensing Technology. Wellington, pp. 591–595 (2013)
7. Gluck, T., Kemmetmuller, W., Tump, C., Kugi, A.: Resistance estimation algorithm for self-sensing magnetic levitation systems. In: Proceedings 5th IFAC Symposium on Mechatronic Systems, Boston, pp. 32–37 (2010)
8. Kumar, V., Jerome, J.: LQR based optimal tuning of PID controller for trajectory tracking of magnetic levitation system. Procedia Eng. 64, 254–264 (2013). https://doi.org/10.1016/j.proeng.2013.09.097
9. Yifei, Y., Huangqiu, Z.: Optimal control and H_∞ output feedback design options for active magnetic bearing spindle position regulation. J. Networks 8, 1624–1631 (2013). https://doi.org/10.4304/jnw.8.7.1624-1631
10. Gandhi, R.V., Adhyaru, D.M.: Takagi-Sugeno fuzzy regulator design for nonlinear and unstable systems using negative absolute eigenvalue approach. IEEE/CAA J. Autom. Sinica. 7, 1–12 (2019). https://doi.org/10.1109/JAS.2019.1911444

11. Takagi, T., Sugeno, M.: Fuzzy identification of systems and its applications to modeling and control. IEEE Trans. Syst. Man Cybernet. **15**(116), 116–132 (1985)
12. Balandin, D.V., Kogan, M.M.: Synthesis of control laws based on linear matrix inequalities. Autom. Remote Control **68**(3), 371–385 (2007)
13. Tanaka, K., Wang, H.O.: Fuzzy Control Systems Design and Analysis: A Linear Matrix Inequality Approach, pp. 2–3. John Wiley & Sons, New York (2001)
14. Gahinet, P., Nemirovski, A., Laub, A.J., et al.: The LMI Control Toolbox for Use with Matlab. User's Guide. The MathWorks, Inc., Natick (1995)

Analysis of Indicators of Optimal Functioning of a Network of Hospitals, Taking into Account Errors of Observation Results

Mikhail Andreevich Fedotkin⬤ and Ekaterina Vadimovna Proydakova⁽⊠⁾⬤

Lobachevsky State University of Nizhni Novgorod (UNN), Nizhny Novgorod, Russia
unn@unn.ru
http://www.unn.ru

Abstract. In Russia, over the past few years, there has been a tendency for an increase in the number of elderly citizens and the incidence of this category of persons. The situation that has arisen requires an increase in the efficiency of the existing system of medical institutions in the context of limited financial and other resources in health care. This problem cannot be solved without an analytical study aimed at economic optimisation of costs in each medical institution and in the network of medical institutions of the subject as a whole. In this paper, we use a representation of the process of functioning of a network of medical institutions in the form of a control cybernetic system. The authors, using the cybernetic approach, have synthesized a mathematical model of the functioning of a typical medical institution of the network in the form of a control system. The created mathematical model allows for a single measurement with a given accuracy to generate the values of the main indicators of the work of a medical institution during each reporting period and, thus, to obtain additional statistics of any volume for these indicators. In the work, on the basis of the additional statistics obtained, an analysis of the optimal distribution of resources between medical institutions of a network of a particular subject is carried out, using the example of Nizhny Novgorod.

Keywords: Control cybernetic system · Mathematical model · Functioning of a medical institution · Basic indicators · Measurement accuracy · Realization of a random variable · Sample values · Optimization.

1 Introduction

In works [1–4], a method is proposed for constructing and studying mathematical models of various kinds of systems and control processes based on the cybernetic

This work was carried out with the financial support of the Russian Foundation for Basic Research, project No. 18-413-520005.

D. Balandin et al. (Eds.): MMST 2020, CCIS 1413, pp. 193–204, 2021.
https://doi.org/10.1007/978-3-030-78759-2_17

approach of Lyapunov – Yablonsky. In this article, which is a direct continuation of the listed works, the synthesis of a mathematical model of the functioning of a medical institution in the form of a control cybernetic system is carried out, as well as an analytical and numerical analysis of its activities, taking into account the errors of observation results. The cybernetic approach allows for a single measurement of the main indicators of the performance of a particular medical institution during each reporting year to obtain additional statistics on the main indicators of any final volume. Using additional statistics, the quality and dynamics of the functioning of a medical institution were studied. This paper also proposes a rationale for the selection of criteria for the effectiveness of the network from the mentioned medical institutions. This allows solving the problem of optimizing the distribution of resources within a specific network of medical institutions.

2   Mathematical Model

Let us denote by the symbol t the number of the reporting year from the set $T \in \{1, 2, \ldots, k\}$ of the discrete functioning of a typical hospital M_j with number j from the set $J \in \{1, 2, \ldots, n\}$. In this notation, the duration of the entire observation period for the network of n hospitals is k. The administration of the medical regional unit and the administration of each medical institution M_j during each reporting year t selects the so-called regulatory management $u(t)$ from a set R. Normative management $u = u(t)$ unambiguously determines for a typical medical institution M_j during the reporting year t the number of $x_j^{(u)}(t)$ beds used and the accuracy $\varepsilon = \varepsilon(t)$ measurements of all parameters and results of network operation. For measurement accuracy, the natural constraint $0 < \varepsilon(t) < 1$ is valid. Based on the results of the hospital operation M_j in the reporting year t and when using $x_j^{(u)}(t)$ beds, only one observation is recorded as the actual values of various m items of financial costs $A_{1,j}^{(u)}(t), A_{2,j}^{(u)}(t), \ldots, A_{m,j}^{(u)}(t)$, so and actual values of the number $Q_j^{(u)}(t)$ of the treated patients, the number of $G_j^{(u)}(t)$ deceased, the number of $L_j^{(u)}(t)$ bed days. The name of all cost items is given in [2]. In particular, the item $A_{m,j}^{(u)}(t)$ plays an important role, as it determines the total amount of financing of expenses for all cost items of the hospital numbered j in the reporting year t. On the basis of only one measurement of these indicators, it is difficult to evaluate the operation of a network of n nursing departments and thereby solve the problem of optimizing the process of medical care. Therefore, it is necessary to develop a methodology for obtaining sample statistics of a given volume for the value of the cost of cost items, for the number of patients treated, for the number of patients who died and for the number of bed-days. To solve this problem in [2, 3] with normative control $u = u(t) \in R$, for a given accuracy $\varepsilon = \varepsilon(t)$ for a medical institution M_j for each reporting year t proposed an algorithm for constructing and a method for studying the following normally distributed random variables: a) the cost $A_{i,j}^{(u;\varepsilon)}(\omega; t)$ the cost of the article with the number $i \in I = \{1, 2, \ldots, m\}$; b) the number of $Q_j^{(u;\varepsilon)}(\omega; t)$ treated patients; c) the number of $G_j^{(u;\varepsilon)}(\omega; t)$ deceased patients; d) the numbers $L_j^{(u;\varepsilon)}(\omega; t)$ bed-days. Here the real number ω is the value or realization of the random variable ξ with a uniform distribution law on the segment $[0, 1]$. By generating a set of values of the random variable ξ, it is possible to obtain a sample of a given volume for the cost of cost items, for the number of patients treated, for the number of patients who died and for the number of bed days.

For the optimal distribution of material and financial resources between the medical institutions of the network during each reporting year $t \in T = \{1, 2, \ldots, k\}$, it is necessary to choose the control $r(t) \in R$. Management $r(t)$ significantly affects the main parameters and performance indicators of the health care network, for example, it uniquely determines the number of $x_j^{(r)}(t)$ beds used by the medical institution M_j in the reporting year t ... Let us denote by the symbol $[x]$ the closest integer to the real number x. Then, under control $r = r(t)$ and for a given value of $\omega \in [0, 1]$, relation (1) allows calculating with a given accuracy ε the values $A_{i,j}^{(r;\varepsilon)}(\omega; t)$ of each cost item, values $Q_j^{(r;\varepsilon)}(\omega; t)$ number of patients treated, values $G_j^{(r;\varepsilon)}(\omega; t)$ the number of deaths and the values $L_j^{(r;\varepsilon)}(\omega; t)$ the number of bed days for the medical institution M_j for the reporting year t.

$$A_{i,j}^{(r;\varepsilon)}(\omega; t) = A_{i,j}^{(u;\varepsilon)}(\omega; t) x_j^{(r)}(t) / x_j^{(u)}(t),$$

$$Q_j^{(r;\varepsilon)}(\omega; t) = [Q_j^{(u;\varepsilon)}(\omega; t) x_j^{(r)}(t) / x_j^{(u)}(t)], \tag{1}$$

$$G_j^{(r;\varepsilon)}(\omega; t) = [G_j^{(u;\varepsilon)}(\omega; t) x_j^{(r)}(t) / x_j^{(u)}(t)],$$

$$L_j^{(r;\varepsilon)}(\omega; t) = [L_j^{(u;\varepsilon)}(\omega; t) x_j^{(r)}(t) / x_j^{(u)}(t)].$$

In particular, in papers [2,3], using formulas (1) for $r(t) = u(t)$ and for different values of the parameters $\varepsilon = \varepsilon(t)$ and ω, the characteristics $A_{i,j}^{(u;\varepsilon)}(\omega; t)$, $Q_j^{(u;\varepsilon)}(\omega; t)$, $G_j^{(u;\varepsilon)}(\omega; t)$, $L_j^{(u;\varepsilon)}(\omega; t)$ and studied their properties. For fixed values of the parameters, we denote ε and ω through the symbols $a_{i,j}^{(u;\varepsilon)}(\omega; t)$, $q_j^{(u;\varepsilon)}(\omega; t)$, $g_j^{(u;\varepsilon)}(\omega; t)$, $l_j^{(u;\varepsilon)}(\omega; t)$ are constants that are determined from relations (2).

$$a_{i,j}^{(u;\varepsilon)}(\omega; t) = A_{i,j}^{(u;\varepsilon)}(\omega; t) / x_j^{(u)}(t),$$

$$q_j^{(u;\varepsilon)}(\omega; t) = Q_j^{(u;\varepsilon)}(\omega; t) / x_j^{(u)}(t), \tag{2}$$

$$g_j^{(u;\varepsilon)}(\omega; t) = G_j^{(u;\varepsilon)}(\omega; t) / x_j^{(u)}(t),$$

$$l_j^{(u;\varepsilon)}(\omega; t) = L_j^{(u;\varepsilon)}(\omega; t) / x_j^{(u)}(t).$$

For the hospital M_j, constant values or standard coefficients of the form $a_{i,j}^{(u;\varepsilon)}(\omega; t)$, $q_j^{(u;\varepsilon)}(\omega; t)$, $g_j^{(u;\varepsilon)}(\omega; t)$ and $l_j^{(u;\varepsilon)}(\omega; t)$ per bed are the cost density articles with the number i, the density of the number of patients treated, the density of the number of patients who died and, accordingly, the density of bed-days. Management $r(t)$ uniquely determines the number of beds $x_j^{(r)}(t)$ used by the medical institution M_j in each reporting year t. Therefore, in what follows, when posing and solving the problem of optimizing the operation of a medical network, by the mathematical description of the control $r(t)$ we mean an integer n - dimensional vector $x^{(r)}(t) = (x_1^{(r)}(t), x_2^{(r)}(t), \ldots, x_n^{(r)}(t)) \in X^{(n)}(t)$.

For the sake of simplicity, except for the case $r(t) = u(t)$, we will omit the symbol r for superscripts and write the control $x(t) = (x_1(t), x_2(t), \ldots, x_n(t))$ instead of control $x^{(r)}(t) = (x_1^{(r)}(t), x_2^{(r)}(t), \ldots, x_n^{(r)}(t))$. Due to this, the value $A_{i,j}^{(r;\varepsilon)}(\omega; t), Q_j^{(r;\varepsilon)}(\omega; t), G_j^{(r;\varepsilon)}(\omega; t), L_j^{(r;\varepsilon)}(\omega; t)$ will now be denoted by $A_{i,j}^{(\varepsilon)}(\omega; t)$, $Q_j^{(\varepsilon)}(\omega; t), G_j^{(\varepsilon)}(\omega; t)$ and $L_j^{(\varepsilon)}(\omega; t)$. Using relations (1) and (2), the indicated values will now be calculated using relation (3).

$$A_{i,j}^{(\varepsilon)}(\omega; t) = a_{i,j}^{(u;\varepsilon)}(\omega; t)x_j(t),$$

$$Q_j^{(\varepsilon)}(\omega; t) = [q_j^{(u;\varepsilon)}(\omega; t)x_j(t)], \tag{3}$$

$$G_j^{(\varepsilon)}(\omega; t) = [g_j^{(u;\varepsilon)}(\omega; t)x_j(t)],$$

$$L_j^{(\varepsilon)}(\omega; t) = [l_j^{(u;\varepsilon)}(\omega; t)x_j(t)].$$

Let us proceed directly to the definition of the domain $X^{(n)}(t)$ or the system of restrictions on the control variables $x_1(t), x_2(t), \ldots, x_n(t)$. This system includes the following types of restrictions.

1. Limitations on the number of $x_j(t)$ beds in the M_j hospital:

$$x_{j,min}(t) \leq x_j(t) \leq x_{j,max}(t), j = 1, 2, \ldots, n. \tag{4}$$

Here the values $x_{j,min}(t)$ and $x_{j,max}(t)$ determine the minimum possible number of beds and the maximum possible number of beds for the medical institution M_j with the number j. These values for the reporting year t are set by the operating conditions for a network of n medical institutions. If $x_{j,p}(t)$ reserve of beds of a medical institution with the number j, which is a known value in the reporting year t, then $x_{j,max}(t) \leq x_{j,p}(t), x_{j,min}(t) \geq 0$.

2. Limit on the total number of beds on the network:

$$\sum_{j=1}^{n} x_j(t) \leq (\sum_{j=1}^{n} x_j(t))_{max}, \tag{5}$$

where the value $(\sum_{j=1}^{n} x_j(t))_{max}$ characterizes the maximum possible number of beds in a network of medical institutions and is set by the operating conditions of all medical institutions in the reporting year t. Directly from (4) follows the relation: $\sum_{j=1}^{n} x_j(t) \leq \sum_{j=1}^{n} x_{j,max}(t)$. If $(\sum_{j=1}^{n} x_j(t))_{max} = \sum_{j=1}^{n} x_{j,max}(t)$, then the condition (5) follows directly from (4) and need not be required.

3. Limitations on the total number of patients treated in the network:

$$\sum_{j=1}^{n} Q^{(u;\varepsilon)}(\omega; t) \leq \sum_{j=1}^{n} q_j^{(u;\varepsilon)}(\omega; t)x_j(t) \leq (\sum_{j=1}^{n} q_j^{(u;\varepsilon)}(\omega; t)x_j(t))_{max}. \tag{6}$$

Here the values $\sum_{j=1}^{n} Q^{(u;\varepsilon)}(\omega;t)$ and $(\sum_{j=1}^{n} q_j^{(u;\varepsilon)}(\omega;t)x_j(t))_{max}$ determine for all medical institutions in the reporting year t the actual number of patients treated and, accordingly, the maximum possible number of patients treated. The value $(\sum_{j=1}^{n} q_j^{(u;\varepsilon)}(\omega;t)x_j(t))_{max} \leq (\sum_{j=1}^{n} x_j(t))_{max} \times 365$ is determined by the conditions of operation of all medical institutions in the reporting year t.

For example, if $(\sum_{j=1}^{n} x_j(t))_{max} = 225$, then $(\sum_{j=1}^{n} q_j^{(u;\varepsilon)}(\omega;t)x_j(t))_{max} \leq$ $\leq 225 \times 365 = 82125$. The quantities $q_1^{(u;\varepsilon)}(\omega;t)$, $q_2^{(u;\varepsilon)}(\omega;t)$,..., $q_n^{(u;\varepsilon)}(\omega;t)$ are the initial information or normative coefficients of the first type for the static optimization model. The value $q_j^{(u;\varepsilon)}(\omega;t)$ characterizes the professional and organizational characteristics of the medical institution with the number j in the reporting year t in restoring the health of patients.

4. Limitation on the total number of deaths in the network:

$$\sum_{j=1}^{n} g_j^{(u;\varepsilon)}(\omega;t)x_j(t) \leq G_j^{(u;\varepsilon)}(\omega;t), \tag{7}$$

where the values $g_1^{(u;\varepsilon)}(\omega;t)$, $g_2^{(u;\varepsilon)}(\omega;t)$, $\ldots, g_n^{(u;\varepsilon)}(\omega;t)$ are also input information or normative coefficients of the second type for a static optimization model. The value $g_j^{(u;\varepsilon)}(\omega;t)$ characterizes the professional and organizational characteristics of the medical institution with the number j in the reporting year t in critical cases of the patient's condition.

5. Limit on the total number of bed-days on the network:

$$\sum_{j=1}^{n} l_j^{(u;\varepsilon)}(\omega;t)x_j(t) \leq (\sum_{j=1}^{n} x_j(t))_{max} \times 365, \tag{8}$$

where the values $l_1^{(u;\varepsilon)}(\omega;t)$, $l_2^{(u;\varepsilon)}(\omega;t)$, $\ldots, l_n^{(u;\varepsilon)}(\omega;t)$ are initial information or normative coefficients of the third type for a static optimization model. The value $l_j^{(u;\varepsilon)}(\omega;t)$ also characterizes the professional and organizational properties of the workload of the medical institution with the number j in the reporting year t.

6. If the number m determines the total number of cost items of each medical institution, then the restrictions on the cost of each cost item numbered i in each reporting year by all medical institutions can be written as:

$$\sum_{j=1}^{n} a_j^{(u;\varepsilon)}(\omega;t)x_j(t) \leq A_{i,plan}(t), i = 1, \ldots, m. \tag{9}$$

In relation (9), the value $A_{i,plan}(t)$ determines in the reporting year the planned costs of all medical institutions for the item with the number i. For fixed $i = 1, 2, \ldots, m$ and $j = 1, 2, \ldots, n$, the value $a_{i,j}^{(u;\varepsilon)}(\omega;t)$ is the initial information or normative factor of the fourth type for the static optimization model. The value $a_{i,j}^{(u;\varepsilon)}(\omega;t)$ characterizes the professional and organizational features of financial costs for item i for hospital numbered j in the reporting year t. At the same time, the total volume $A_m^{(\varepsilon)}(\omega; t)$ financial expenses for all cost items n of the network of medical institutions in the reporting year t is calculated using a formula of the form $A_m^{(\varepsilon)}(\omega; t) = \sum_{j=1}^n a_{m,j}^{(u;\varepsilon)}(\omega; t) x_j(t)$.

In the above constraints using the control variables $x_1(t), x_2(t), \ldots, x_n(t)$ and the normative coefficients $q_j^{(u;\varepsilon)}(\omega; t), g_j^{(u;\varepsilon)}(\omega; t), l_j^{(u;\varepsilon)}(\omega; t), a_{m,j}^{(u;\varepsilon)}(\omega; t)$ formulas for calculating the main indicators $Q^{(\varepsilon)}(\omega; t), G^{(\varepsilon)}(\omega; t)), L^{(\varepsilon)}(\omega; t), A_m^{(\varepsilon)}(\omega; t)$ efficiency of the network of medical institutions. Therefore, these indicators can be approximately found using the following linear functions:

$$Q^{(\varepsilon)}(\omega; t) \approx q^{(\varepsilon)}(\omega; x_1(t), \ldots, x_n(t)) = \sum_{j=1}^n q_j^{(u;\varepsilon)}(\omega; t) x_j(t),$$

$$G^{(\varepsilon)}(\omega; t) \approx g^{(\varepsilon)}(\omega; x_1(t), \ldots, x_n(t)) = \sum_{j=1}^n g_j^{(u;\varepsilon)}(\omega; t) x_j(t), \qquad (10)$$

$$L^{(\varepsilon)}(\omega; t) \approx l^{(\varepsilon)}(\omega; x_1(t), \ldots, x_n(t)) = \sum_{j=1}^n l_j^{(u;\varepsilon)}(\omega; t) x_j(t),$$

$$A_m^{(\varepsilon)}(\omega; t) \approx a^{(\varepsilon)}(\omega; x_1(t), \ldots, x_n(t)) = \sum_{j=1}^n a_{m,j}^{(u;\varepsilon)}(\omega; t) x_j(t).$$

Thus, we can assume that for a network of n hospitals the four linear functions

$$q^{(\varepsilon)}(\omega; x_1(t), x_2(t), \ldots, x_n(t)), g^{(\varepsilon)}(\omega; x_1(t), x_2(t), \ldots, x_n(t)),$$

$$l^{(\varepsilon)}(\omega; x_1(t), x_2(t), \ldots, x_n(t)), a^{(\varepsilon)}(\omega; x_1(t), x_2(t), \ldots, x_n(t))$$

are responsible for the number of patients treated, for the number of patients who died, for the number of bed-days and for the total the amount of financial costs, respectively. For the family of linear functions

$$q^{(\varepsilon)}(\omega; x_1(t), x_2(t), \ldots, x_n(t)), g^{(\varepsilon)}(\omega; x_1(t), x_2(t), \ldots, x_n(t)),$$

$$l^{(\varepsilon)}(\omega; x_1(t), x_2(t), \ldots, x_n(t)), a^{(\varepsilon)}(\omega; x_1(t), x_2(t), \ldots, x_n(t))$$

it is possible to pose and solve various optimization multicriteria problems, which depend on the physical meaning of the target indicators. For example, we will find such a distribution of $x_1^*(t), x_2^*(t), \ldots, x_n^*(t)$ beds in medical institutions for which at least one from the following optimality conditions:

$$q^{(\varepsilon)}(\omega; x_1^*(t), x_2^*(t), \ldots, x_n^*(t))$$
$$= max\left\{q^{(\varepsilon)}(\omega; x_1(t), x_2(t), \ldots, x_n(t)) : (x_1(t), x_2(t), \ldots, x_n(t)) \in X^{(n)}(t)\right\},$$

$$g^{(\varepsilon)}(\omega; x_1^*(t), x_2^*(t), \ldots, x_n^*(t))$$
$$= min\left\{g^{(\varepsilon)}(\omega; x_1(t), x_2(t), \ldots, x_n(t)) : (x_1(t), x_2(t), \ldots, x_n(t)) \in X^{(n)}(t)\right\}.$$

It is desirable to find such a distribution of $x_1^*(t), x_2^*(t), \ldots, x_n^*(t)$ beds, which provides for the reporting year for the entire network of branches nursing care the maximum number of patients treated and the minimum number of deaths. We will call this distribution of beds optimal. The constructed static optimization multi-criteria model of the distribution of the number of beds over medical institutions in each year t is determined by relations (4)–(10) and the optimality conditions. These relations and conditions contain n control variables $x_1(t), x_2(t), \ldots, x_n(t)$ and in the general case at most $(2n + m + 5)$ linear independent restrictions.

3   Optimization of Bed Allocation

It is known that one of the most important criteria for the functioning of a network of medical institutions is the number of patients treated. Consider a one-criterion optimization model for the distribution of the number of beds by medical institutions in each reporting year $t = 1, 2, \ldots, k$, taking into account the measurement errors of actual indicators. To determine and analyze the optimal values of $x_1^*(t), x_2^*(t), \ldots, x_n^*(t)$ control variables in the reporting year only under the condition of maximum of treated patients, it is necessary for fixed values of the parameters ε and ω to know the initial information in the form of values of $A_{1,j}^{(u)}(t), A_{2,j}^{(u)}(t), \ldots, A_{m,j}^{(u)}(t), Q_j^{(u)}(t), G_j^{(u)}(t), L_j^{(u)}(t), A_{1,j}^{(u;\varepsilon)}(\omega; t),$ $A_{2,j}^{(u;\varepsilon)}(\omega; t), \ldots, A_{m,j}^{(u;\varepsilon)}(\omega; t), Q_j^{(u;\varepsilon)}(\omega; t), G_j^{(u;\varepsilon)}(\omega; t), L_j^{(u;\varepsilon)}(\omega; t)$. In this case, the one-criterion optimization model of the distribution of the number of beds by medical institutions in the year t is determined by (4)–(9) and the equality:

$$q^{(\varepsilon)}(\omega; x_1^*(t), x_2^*(t), \ldots, x_n^*(t))$$
$$= max\left\{ q^{(\varepsilon)}(\omega; x_1(t), x_2(t), \ldots, x_n(t)) : (x_1(t), x_2(t), \ldots, x_n(t)) \in X^{(n)}(t) \right\}.$$

The calculation of the optimal values $x_1^*(t), x_2^*(t), \ldots, x_n^*(t)$ of control variables was carried out using the Microsoft Excel based on the data [1] on the functioning for the period 2007–2015 ($t = 1, \ldots, 9$) networks of 5 medical institutions ($j = 1, \ldots, 5$) Nizhny Novgorod. Moreover, $m = 25$, $n = 5$ and $k = 9$. In the calculations, the actual distribution of hospital beds for t was determined by the set $x_1^{(u)}(t) = x_2^{(u)}(t) = x_3^{(u)}(t) = x_4^{(u)}(t) = 50$ and $x_5^{(u)}(t) = 25$. As an example, below is the solution to the optimization problem based on the criterion of the maximum number of treated patients according to the data for 2015, 2014.

Table 1. Optimization problem solved on actual data in 2015.

Hospital	№ 34	№ 24	№ 14	№ 11	№ 37
$x_j^*(t)$	93	10	53	38	42
$a_j(t)$	223599,43	296975,85	376255,98	297359,12	394013,12
$q_j(t)$	10,86	8,58	14,58	8,70	12,24
$l_j(t)$	302,92	309,06	318,24	333,12	300,12
$g_j(t)$	0,16	0,06	0,14	1,00	0,40

From Table 1 it follows that with actual data, the optimal values are: $x_1^*(t) =$
$= 93, x_2^*(t) = 10, x_3^*(t) = 53, x_4^*(t) = 38, x_5^*(t) = 42$. The values obtained as
a result of applying the method of generating new statistics according to the
normal law in 2015 with $\varepsilon = 0,01$ and three different implementations of ω_1, ω_2,
ω_3 are presented in Tables 2, 3 and 4.

Table 2. Optimization problem on the generated data in 2015 with $\varepsilon = 0,01$ and ω_1.

Hospital	№ 34	№ 24	№ 14	№ 11	№ 37
$x_j^*(t)$	91	11	52	38	42
$a_j(t)$	225047,67	298899,36	378692,98	299285,11	396565,13
$q_j(t)$	10,93	8,64	14,67	8,76	12,32
$l_j(t)$	304,88	311,06	320,30	335,28	302,06
$g_j(t)$	0,16	0,06	0,14	1,01	0,40

From Table 2 it follows that calculated according to the statistics generated
at $\varepsilon = 0,01$ and realization ω_1, the optimal distribution of beds has the form:
$x_1^*(t) = 91, x_2^*(t) = 11, x_3^*(t) = 52, x_4^*(t) = 38, x_5^*(t) = 42$.

Table 3. Optimization problem on the generated data in 2015 with $\varepsilon = 0,01$ and ω_2.

Hospital	№ 34	№ 24	№ 14	№ 11	№ 37
$x_j^*(t)$	94	5	53	37	44
$a_j(t)$	224884,93	298683,21	378419,13	299068,68	396278,36
$q_j(t)$	10,92	8,63	14,66	8,75	12,31
$l_j(t)$	304,66	310,84	320,07	335,04	301,85
$g_j(t)$	0,16	0,06	0,14	1,01	0,40

It can be seen from Table 3 that, calculated from the statistics generated at
$\varepsilon = 0,01$ and realization ω_2, the optimal distribution of beds has the following
form: $x_1^*(t) = 94, x_2^*(t) = 5, x_3^*(t) = 53, x_4^*(t) = 37, x_5^*(t) = 44$.

Table 4. Optimization problem on the generated data in 2015 with $\varepsilon = 0,01$ and ω_3.

Hospital	№ 34	№ 24	№ 14	№ 11	№ 37
$x_j^*(t)$	95	5	53	37	44
$a_j(t)$	223490,01	296830,54	376071,87	297213,61	393820,32
$q_j(t)$	10,85	8,58	14,57	8,70	12,23
$l_j(t)$	302,77	308,91	318,08	332,96	299,97
$g_j(t)$	0,16	0,06	0,14	1,00	0,40

From Table 4 it follows that calculated according to statistics generated at $\varepsilon = 0,01$ and realization w_3, the optimal distribution of beds has the form: $x_1^*(t) = 95, x_2^*(t) = 5, x_3^*(t) = 53, x_4^*(t) = 37, x_5^*(t) = 44$.

The values obtained by applying the method of generating new statistics according to the normal law in 2015 with $\varepsilon = 0,1$ and three different implementations of w_1, w_2, w_3 are presented in Tables 5, 6 and 7.

Table 5. Optimization problem on the generated data in 2015 with $\varepsilon = 0,1$ and w_1.

Hospital	№ 34	№ 24	№ 14	№ 11	№ 37
$x_j^*(t)$	102	5	57	40	47
$a_j(t)$	208696,14	277181,90	351177,87	277539,62	367751,47
$q_j(t)$	10,14	8,01	13,61	8,12	11,42
$l_j(t)$	282,73	288,46	297,03	310,92	280,12
$g_j(t)$	0,15	0,06	0,13	0,93	0,37

From Table 5 it can be seen that calculated according to the statistics generated at $\varepsilon = 0,1$ and realization w_1, the optimal distribution of beds is as follows: $x_1^*(t) = 102, x_2^*(t) = 5, x_3^*(t) = 57, x_4^*(t) = 40, x_5^*(t) = 47$.

Table 6. Optimization problem on the generated data in 2015 with $\varepsilon = 0,1$ and w_2.

Hospital	№ 34	№ 24	№ 14	№ 11	№ 37
$x_j^*(t)$	94	6	53	37	43
$a_j(t)$	212381,25	282076,32	357378,90	282440,36	374245,15
$q_j(t)$	10,32	8,15	13,85	8,26	11,63
$l_j(t)$	287,72	293,55	302,27	316,41	285,06
$g_j(t)$	0,15	0,06	0,13	0,95	0,38

From Table 6 it follows that calculated according to statistics generated at $\varepsilon = 0,1$ and realization w_2, the optimal distribution of beds has the form: $x_1^*(t) = 94, x_2^*(t) = 6, x_3^*(t) = 53, x_4^*(t) = 37, x_5^*(t) = 43$.

Table 7. Optimization problem on the generated data in 2015 with $\varepsilon = 0,1$ and w_3.

Hospital	№ 34	№ 24	№ 14	№ 11	№ 37
$x_j^*(t)$	98	10	56	40	44
$a_j(t)$	212381,25	282076,32	357378,90	282440,36	374245,15
$q_j(t)$	10,32	8,15	13,85	8,26	11,63
$l_j(t)$	287,72	293,55	302,27	316,41	285,06
$g_j(t)$	0,15	0,06	0,13	0,95	0,38

From Table 7 it can be seen that calculated according to the statistics generated at $\varepsilon = 0,1$ and implementation ω_3, the optimal distribution of beds has the form: $x_1^*(t) = 98, x_2^*(t) = 10, x_3^*(t) = 56, x_4^*(t) = 40, x_5^*(t) = 44$.

Table 8. Optimization problem solved on actual data in 2014.

Hospital	№ 34	№ 24	№ 14	№ 11	№ 37
$x_j^*(t)$	103	5	53	5	71
$a_j(t)$	173927,94	326254,84	409054,12	355704,48	374803,08
$q_j(t)$	12,08	8,90	14,44	8,28	13,16
$l_j(t)$	341,80	348,94	357,08	347,06	339,68
$g_j(t)$	0,20	0,20	0,22	1,74	1,64

From Table 8 it follows that with actual data, the optimal values are: $x_1^*(t) =$ $= 103, x_2^*(t) = 5, x_3^*(t) = 53, x_4^*(t) = 5, x_5^*(t) = 71$.

The values obtained as a result of applying the method of generating new statistics according to the normal law with $\varepsilon = 0,01$ and three different implementations of ω_1, ω_2, ω_3 are presented in Tables 9, 10 and 11.

Table 9. Optimization problem on the generated data in 2014 with $\varepsilon = 0,01$ and ω_1.

Hospital	№ 34	№ 24	№ 14	№ 11	№ 37
$x_j^*(t)$	110	7	58	5	72
$a_j(t)$	161230,94	302437,75	379192,57	329737,53	372066,96
$q_j(t)$	11,20	8,25	13,39	7,68	13,06
$l_j(t)$	316,85	323,47	331,01	321,72	337,20
$g_j(t)$	0,19	0,19	0,20	1,61	1,63

From Table 9 it follows that calculated according to the statistics generated $\varepsilon = 0,01$ and realization ω_1, the optimal distribution of beds has the form: $x_1^*(t) = 110, x_2^*(t) = 7, x_3^*(t) = 58, x_4^*(t) = 5, x_5^*(t) = 72$.

Table 10. Optimization problem on the generated data in 2014 with $\varepsilon = 0,01$ and ω_2.

Hospital	№ 34	№ 24	№ 14	№ 11	№ 37
$x_j^*(t)$	104	5	54	5	72
$a_j(t)$	173327,68	325128,87	407642,41	354476,88	373509,57
$q_j(t)$	12,04	8,87	14,39	8,25	13,11
$l_j(t)$	340,62	347,74	355,85	345,86	338,51
$g_j(t)$	0,20	0,20	0,22	1,73	1,63

From Table 10 it follows that calculated according to the statistics generated $\varepsilon = 0,01$ and implementation ω_2, the optimal distribution of beds has the form: $x_1^*(t) = 104, x_2^*(t) = 5, x_3^*(t) = 54, x_4^*(t) = 5, x_5^*(t) = 72$.

Table 11. Optimization problem on the generated data in 2014 with $\varepsilon = 0,01$ and ω_3.

Hospital	№ 34	№ 24	№ 14	№ 11	№ 37
$x_j^*(t)$	103	5	54	5	71
$a_j(t)$	174240,28	326840,73	409788,70	356343,26	375476,15
$q_j(t)$	12,10	8,92	14,47	8,29	13,18
$l_j(t)$	342,41	349,57	357,72	347,68	340,29
$g_j(t)$	0,20	0,20	0,22	1,74	1,64

From Table 11 it follows that calculated according to statistics generated at $\varepsilon = 0,01$ and implementation ω_3, the optimal distribution of beds has the form: $x_1^*(t) = 103, x_2^*(t) = 5, x_3^*(t) = 54, x_4^*(t) = 5, x_5^*(t) = 71$.

The values obtained by applying the method of generating new statistics according to the normal law in 2014 with $\varepsilon = 0,1$ and three different implementations of ω_1, ω_2, ω_3 are presented in Tables 12, 13 and 14.

Table 12. Optimization problem on the generated data in 2014 with $\varepsilon = 0,1$ and ω_1.

Hospital	№ 34	№ 24	№ 14	№ 11	№ 37
$x_j^*(t)$	110	7	58	5	72
$a_j(t)$	161230,94	302437,75	379192,57	329737,53	372066,96
$q_j(t)$	11,20	8,25	13,39	7,68	13,06
$l_j(t)$	316,85	323,47	331,01	321,72	337,20
$g_j(t)$	0,19	0,19	0,20	1,61	1,63

From Table 12 it follows that calculated according to statistics generated at $\varepsilon = 0,1$ and ω_1, the optimal distribution of beds has the form: $x_1^*(t) = 110, x_2^*(t) = 7, x_3^*(t) = 58, x_4^*(t) = 5, x_5^*(t) = 72$.

Table 13. Optimization problem on the generated data in 2014 with $\varepsilon = 0,1$ and ω_2.

Hospital	№ 34	№ 24	№ 14	№ 11	№ 37
$x_j^*(t)$	95	5	50	5	70
$a_j(t)$	188248,37	353117,17	442733,77	384991,56	377889,03
$q_j(t)$	13,07	9,63	15,63	8,96	13,27
$l_j(t)$	369,94	377,67	386,48	375,64	342,48
$g_j(t)$	0,22	0,22	0,24	1,88	1,65

From Table 13 it can be seen that calculated according to statistics generated at $\varepsilon = 0,1$ and implementation ω_2, the optimal distribution of beds is as follows: $x_1^*(t) = 95, x_2^*(t) = 5, x_3^*(t) = 50, x_4^*(t) = 5, x_5^*(t) = 70$.

Table 14. Optimization problem on the generated data in 2014 with $\varepsilon = 0,1$ and ω_3.

Hospital	№ 34	№ 24	№ 14	№ 11	№ 37
$x_j^*(t)$	107	5	56	5	72
$a_j(t)$	168164,99	315444,68	395500,48	343918,53	373561,20
$q_j(t)$	11,68	8,61	13,96	8,01	13,12
$l_j(t)$	330,47	337,38	345,25	335,56	338,55
$g_j(t)$	0,19	0,19	0,21	1,68	1,63

From Table 14 it can be seen that calculated according to statistics generated at $\varepsilon = 0,1$ and implementation ω_3, the optimal distribution of beds is as follows: $x_1^*(t) = 107, x_2^*(t) = 5, x_3^*(t) = 56, x_4^*(t) = 5, x_5^*(t) = 72$.

4 Conclusion

Based on the results presented in Tables 1, 2, 3, 4, 5, 6, 7, 8, 9, 10, 11, 12, 13 and 14, we can conclude that the accuracy of the ε measurement significantly affects the obtained optimal distribution of beds in the network of medical institutions. Moreover, **this influence is not linear**. For example, in 2015, with $\varepsilon = 0,01$ (Table 3), the optimal distribution of beds differed from that obtained from actual data (Table 1) for four hospitals. And for one hospital out of four ($j = 2$), the difference was significant, two times. In the case of $\varepsilon = 0,1$ (Table 7), the optimal distribution of beds differed from the actual one for all the hospitals in the network, but no such significant difference as with $\varepsilon = 0,01$ and $j = 2$ was observed.

References

1. Fedotkin M.A.: Non-traditional Problems of Mathematical Modeling of Experiments. FIZMATLIT, Moscow (2018)
2. Fedotkin, M., Proydakova, E., Edeleva, A.: Mathematical and instrumental methods for constructing a model of the economy of the hospital's functioning. Vestnik of Lobachevsky State University of Nizhni Novgorod. Ser. Soc. Sci. 4(56), 54–64 (2019)
3. Fedotkin, M., Proydakova, E., Edeleva, A.: Mathematical and instrumental methods for constructing a model of the economy of the hospital's functioning. Vestnik of Lobachevsky State University of Nizhni Novgorod. Ser. Soc. Sci. 2(58), 55–65 (2020)
4. Proydakova, E.: A system with a fixed rhythm and variable service intensity. Bull. Volga State Acad. Water Transp. **50**, 73–78 (2017)

Optimal Control Problem with State Constraint for Hyperbolic-Type PDE

A. I. Egamov[(✉)]

Lobachevsky University of Nizhny Novgorod,
Gagarin Avenue, 23, Nizhny Novgorod 603950, Russia
albert.egamov@itmm.unn.ru

Abstract. The optimal control problem with a terminal-type objective function is solved. The controlled process is described by a hyperbolic PDE with initial and boundary conditions of the second type. The special control is selected so that the state constraint is executed throughout the entire action: the integral of the spatial coordinate of the square of the solution is equal to one. The optimal control problem is considered. Note that this problem is purely theoretical, but has good practical prospects. It is proved that for such a particular equation, its solution can be represented through the solution of a standard linear initial-boundary value problem of hyperbolic type, which in turn allows us to apply the method of separated variables. The transition from the original problem to the problem of optimization by the Fourier coefficients of the solution of a linear problem is shown. The solution of the initial-boundary value problem is reduced to the system of differential equations of the second order. Further we demonstrate a transition to a finite shortened system of first-order differential equations. The solution can be obtained arbitrarily close to the solution of an infinite-dimensional system by increasing dimensionality of the shortened system. A description of the algorithm for iterating over the control coefficients for finding a solution to the optimal control problem for a shortened system and constructing a minimizing sequence are given. The value of the objective function for a shortened system by increasing dimension can be obtained arbitrarily close to the original optimal value.

Keywords: Integro-differential PDE · Hyperbolic equation · State constraint · Infinite-dimensional system of ODEs · Shortened system

1 Introduction

The various problems and difficulties encountered, we work by systems with distributed parameters are presented in the introduction of the monograph [1]. It contains a detailed bibliography related to this scientific area. The controlled processes with state constraints, which is imposed on the solution of distributed system, make the optimal control problems for such processes particularly difficult. In some cases, when there is only one state constraint, it is possible to

© Springer Nature Switzerland AG 2021
D. Balandin et al. (Eds.): MMST 2020, CCIS 1413, pp. 205–219, 2021.
https://doi.org/10.1007/978-3-030-78759-2_18

satisfy it, if feedback control is used. This assumes that the integral over the spatial domain will be included in the control [2]. The advantages of this approach compared to other methods for solving similar problems are described in the introduction [3].

There is dependence between the PDE being studied and standard linear equations. The class of integro-differential equations to which be method of this article applies is discussed, for example, in [2–4]. The resulting optimization problem with a terminal-type objective function and a state constraint is solved. The methods and algorithms for finding optimal control are similar to those used in [5] for a PDE that is reduced to a linear parabolic equation.

At present, optimal control problems for second-order hyperbolic equations are only partially investigated. Any progress in this direction allows us to take a step forward in the direction of their further study and development of the General Theory. New methods for solving a certain class of such problems or a set of known methods that were not previously considered simultaneously in relation to such an optimization problem, are of a particular interest and undoubtedly deserve close attention.

2   The Controlled Process and Supporting Statements

We consider the initial-boundary value problem. There has the set $Q = [0, l] \times [0, T]$, $l > 0$, $T > 0$ with the boundary Σ consisting of points $\{(x, t) : t = 0$ or $(l - x)x = 0\}$. The function $y(x, t) \in C^2(Q \setminus \Sigma) \cap C^1(Q)$ is called a solution to the initial-boundary value problem if $y(x, t)$ satysfies on set $Q \setminus \Sigma$ the equation:

$$y''_{tt}(x, t) = a^2 y''_{xx}(x, t) + u(x, t). \tag{1}$$

Moreover the boundary and initial conditions are valid:

$$y'_x(0, t) = y'_x(l, t) = 0, \tag{2}$$

$$y(x, 0) = \varphi(x), \quad y'_t(x, 0) = \psi(x), \tag{3}$$

where $u(x, t)$ is a continuous control function, $a \in R$. The initial functions are

$$\varphi(x) \in C^3[0, l], \quad \psi(x) \in C^2[0, l]. \tag{4}$$

In addition, the function $\varphi(x)$ and the function $\psi(x)$ satisfy conditions based on (2) for $t = 0$ and $\varphi(x)$ is a positive function.

Let the control $u(x, t)$ be a feedback control and have the form:

$$u(x, t) = \left(b(x) + \eta(t) \right) y(x, t) - 2q(t) y'_t(x, t), \tag{5}$$

where the control functions $b(x)$, $\eta(t)$ are continuous functions, and $q(t)$ is a continuously differentiable function.

Let the squares of the norm in the spaces $C[0, l]$ and $C(Q)$ be written as

$$||b(x)||_2^2 = \int_0^l b^2(x)\, dx \quad \text{and} \quad ||u(x, t)||_Q^2 = \int_0^T \int_0^l u^2(x, t)\, dx\, dt$$

respectively. Denote $\widehat{\varphi}_0 = ||\varphi'_x(x)||_2^2 + ||\psi(x)||_2^2 + ||\varphi(x)||_2^2$, it is clear that $\widehat{\varphi}_0 > 0$.

3 The Optimization Problem

There is a formulation of the optimization problem:

There are constraints on the coordinates of the solution

$$||y(x,t)||_2 = 1, \quad t \in [0,T], \tag{6}$$

and on the control function:

$$||b(x)||_2 \leq K, \tag{7}$$

where K is a positive constant. We will believe, that the function $b(x)$ is the continuously differentiable function and there are additional constraints: $b'_x(0) = b'_x(l) = 0$.

For a fixed time $T > 0$ to minimize the objective function of terminal type

$$J^{00}(T) = \int_0^l (y(x,T) - Y(x))^2 \, dx = ||y(x,T) - Y(x)||_2^2 \rightarrow \min, \tag{8}$$

where $Y(x)$ is fixed continuous function with constraint: $||Y(x)||_2 = 1$ and it satisfies conditions based on (2).

Is there any relationship between the control functions $b(x)$, $\eta(t)$ and $q(t)$? The answer to this question is given in the article [2]. The control function $\eta(t) = -R[y]$, so $R[y]$ is an integral operator of the form

$$R[y] \equiv R_y(t) = \int_0^l (b(x)y^2(x,t) - a^2 y_x'^2(x,t) + y_t'^2(x,t)) \, dx, \tag{9}$$

where the function $y(x,t)$ is the solution of the problem (1)–(5), (9).

To fulfill the condition (6), it is necessary [2], in addition to the one described above, that the continuously differentiable function $q(t)$ is a solution of the Cauchy problem for the Riccati equation

$$q_t'(t) + q^2(t) = R_y(t), \quad q(0) = 0. \tag{10}$$

Depending on the function $R_y(t)$ the function $q(t)$ can be bounded on the segment $[0,T]$ or unbounded, but for any admissible functions $y(x,t)$ and $R_y(t)$, the function $q(t)$ exists in some neighborhood of zero. In this article, we are interested in cases when the function $q(t)$ is bounded on the segment $[0,T]$. A more convenient check for the boundedness of the $q(t)$ function will be described below. From what has been said and the equality (5), it is clear that the control has the form:

$$u(x,t) = b(x)y(x,t) - 2q(t)y_t'(x,t) - R[y]y(x,t). \tag{11}$$

Thus, the condition (6) is valid. Next, instead of $J^{00}(T)$, we will write $J^{00}(T, b(x))$ thus showing the importance of the function $b(x)$ for the objective function.

4 Auxiliary Initial-Boundary Value Problem

The solution of the original problem can be written using the solution of a linear auxiliary problem. The function $z(x,t) \in C^2(Q \setminus \Sigma) \cap C^1(Q)$ is solution of a standard hyperbolic equation of the 2nd order

$$z_{tt}''(x,t) = a^2 z_{xx}''(x,t) + b(x)z(x,t) + f(x,t), \quad (x,t) \in Q \setminus \Sigma \tag{12}$$

with the second-type boundary conditions

$$z_x'(0,t) = z_x'(l,t) = 0, \tag{13}$$

and initial conditions

$$z(x,0) = \varphi(x), \quad z_t'(x,0) = \psi(x). \tag{14}$$

The unique solution of this problem exists and one can be found by the separation of variables method. Smoothness conditions (4) for initial functions are necessary for immediate differentiation of the series obtained by applying the Fourier method, see [7]. For $f(x,t) \equiv 0$, the problem is called homogeneous. For this problem (12)–(14) there is the energy inequality [7]:

$$||z(x,\tau)||_{Q_t}^2 \leq K_1(t)(||\varphi(x)||_2^2 + ||\varphi_x'(x)||_2^2 + ||\psi(x)||_2^2 + ||f(x,\tau)||_{Q_t}^2), \tag{15}$$

where $K_1(t)$ is the positive increasing function that depends only on t, a and $b(x)$, the set $Q_t = [0,l] \times [0,t]$. According to (7), the set of the different valid functions $b(x)$ is bounded and closed, then there is $\max\limits_{b(x)} K_1(T) = K_0$.

Existence. Denote by $P[w]$ is the integral operator of the form

$$P[w] = \left(\int_0^l w^2(x,t)\,dx \right)^{1/2}. \tag{16}$$

Let $p(t) \equiv P[z]$, where the function $z(x,t)$ is the solution of the problem (12)–(14) with $f(x,t) \equiv 0$.

In the article [2], we prove the theorem and its consequences on the dependence between the solution of the initial-boundary value problem described by the integro-differential equation and the solution of the linear homogeneous problem (12)–(14). Also the dependencies between the functions $p(t)$, $q(t)$, and $R_y(t)$ are derived there.

In relation to this problem statement, it is formulated as follows:

Theorem 1. *We believe for $\forall t \in [0,T]$: $P[z] \neq 0$, where $z(x,t)$ is the solution to the problem (12)–(14) with homogeneous equation (12) ($f(x,t) \equiv 0$). The equalities associated with the initial functions*

$$\int_0^l \varphi^2(x)\,dx = 1; \quad \int_0^l \varphi(x)\psi(x)\,dx = 0. \tag{17}$$

are valid. Then on the set Q there is a unique solution to the problem (1)–(3), (9)–(11), (17), is represented as

$$y(x,t) = \frac{z(x,t)}{P[z]} \equiv \frac{z(x,t)}{p(t)}. \tag{18}$$

By direct verification from (9), (16), (18), we make sure that there is the dependence between $p(t)$, $q(t)$ and $R_y(t)$ has the form

$$\frac{p''_{tt}(t)}{p(t)} = R_y(t), \quad q(t) = \frac{p'_t(t)}{p(t)}. \tag{19}$$

We assume that given the constants T, a, K and the given initial functions $\varphi(x)$, $\psi(x)$, the function $z(x,t) \neq 0$ for $t \in [0,T]$. This statement is equivalent to the inequality: $P[z] \neq 0$. The set of the different valid functions $b(x)$ is bounded and closed, then there is $\min\limits_{b(x), t\in[0,T]} p(t) = p_m > 0$, where p_m is a constant, $p(t) \equiv P[z(x,t)]$, where the function $z(x,t)$ is the solution to the homogeneous problem (12)–(14) with the control function $b(x)$ respectively.

Uniqueness. Let us assume that there are two solutions of the problem (1)–(3), (9)–(11), (17). Each of them will have an auxiliary task of the type (12)–(14) and its function $R_y(t)$ and hence its functions $q_i(t)$ and $p_i(t)$, $i = \overline{1,2}$, used in Theorem 1. But the auxiliary task is identical for them. Therefore, there are two solutions $y_1(x,t) = z(x,t)/p_1(t)$ and $y_2(x,t) = z(x,t)/p_2(t)$. From (9), (19) it follows that

$$p''_{tt} = pR_y(t) = p(t) \int_0^l (b(x)y^2(x,t) - a^2 y'^2_x(x,t) + y'^2_t(x,t))\,dx$$

$$= \frac{1}{p(t)} \int_0^l (b(x)z^2(x,t) - a^2 z'^2_x(x,t) + z'^2_t(x,t) - 2z(x,t)z_t(x,t))\frac{p'_t(t)}{p(t)}$$

$$+z^2(x,t)\frac{p'^2_t(t)}{p^2(t)})\,dx = \frac{1}{p(t)}(\gamma_0(t) + \gamma_1(t)\frac{p'_t(t)}{p(t)} + \gamma_2(t)\frac{p'^2_t(t)}{p^2(t)}) \equiv G(t,p,p'_t).$$

It is clear that the continuous by its variables function $G(t,p,p'_t)$ is continuously differentiable by variables on p and p'_t. Because the continuous function $p(t) \neq 0$ and hence $p(t) \geq p_m > 0$, it and its partial derivatives are the bounded functions. The functions $\gamma_i(t)$, $i = \overline{1,3}$, are the same as for $y_1(x,t)$ that for $y_2(x,t)$. By the generalized Cauchy-Picard theorem for the Cauchy problem of a 2-order differential equation [9] there is uniqueness solution of the differential equation $p''_{tt} = G(t,p,p'_t)$. It is follows that $p_1(t) \equiv p_2(t)$. Contradiction. Therefore the problem (1)–(3), (9)–(11), (17) has a unique solution.

5   The Expansion of the System of Cosines

The system of cosines: $v_k(x) = \cos(\lambda_k x)$, $\lambda_k = \pi k/l$, $k = \overline{0,+\infty}$, is a complete orthogonal system on the segment $[0,l]$, for which is valid the equalities

$\int_0^l v_k(x)\,dx = 0$, $k = \overline{1,+\infty}$. We decompose the system of the cosines of the function $z(x,t)$, $b(x)$, $\varphi(x)$, $\psi(x)$: $z(x,t) = \sum\limits_{k=0}^{+\infty} \xi_k(t)v_k(x)$, $b(x) = \sum\limits_{k=0}^{+\infty} b_k v_k(x)$ and $\varphi(x) = \sum\limits_{k=0}^{+\infty} \varphi_k v_k(x)$, $\psi(x) = \sum\limits_{k=0}^{+\infty} \psi_k v_k(x)$, $Y(x) = \sum\limits_{k=0}^{+\infty} Y_k v_k(x)$. By direct calculation, we make sure that the square of the norm $||v_0(x)||_2^2 = l$, $||v_j(x)||_2^2 = l/2$, $j = \overline{1,+\infty}$. Therefore

$$||z(x,t)||_2 = \left(\sum_{j=0}^{+\infty} \xi_j^2(t)||v_j||_2^2 \right)^{1/2} = \left(l\xi_0^2(t) + \frac{l}{2}\sum_{j=1}^{+\infty} \xi_j^2(t) \right)^{1/2} \equiv \Theta(\overline{\xi}(t)), \quad (20)$$

where $\overline{\xi}$ is the corresponding vector consisting of $\xi_j(t)$, $j = \overline{0,+\infty}$. It follows from Theorem 1 that

$$y(x,t) = \sum_{k=0}^{+\infty} \theta_k(t)v_k(x), \quad \text{where} \quad \theta_k(t) = \frac{\xi_k(t)}{\Theta(\overline{\xi})}.$$

Theorem 2. *Let the function $z_j(x,t)$ be the solution of the problem (12)–(14) with $f(x,t) \equiv 0$, for which $b(x) \equiv b_j(x)$, $j = \overline{1,2}$ respectively. For any $\varepsilon > 0$, there exists the positive $\delta = \varepsilon^2(K_0^2\widehat{\varphi}_0)^{-1}$, such that if the inequality $||b_2(x) - b_1(x)||_2^2 < \delta$ holds for the continuous functions $b_1(x)$ and $b_2(x)$, then $||z_2(x,t) - z_1(x,t)||_Q < \varepsilon$.*

We fix the small $\varepsilon > 0$. If we assume that the function $\zeta(x,t) = z_2(x,t) - z_1(x,t)$, then it is the solution of the problem

$$\zeta_{tt}''(x,t) = a^2\zeta_{xx}''(x,t) + b_2(x)\zeta(x,t) + (b_2(x) - b_1(x))z_1(x,t), \quad (21)$$

satisfying the boundary and initial conditions

$$\zeta_x'(0,t) = \zeta_x'(l,t) = 0, \quad \zeta(x,0) = 0, \quad \zeta_t'(x,0) = 0. \quad (22)$$

The Eq. (21) is a special case of Eq. (12), the inhomogeneous part $f(x,t) = (b_2(x) - b_1(x))z_1(x,t)$ For this task the energy inequality (15) Applying the energy inequality twice, first to the problem (21), (22), then to the problem (12)–(14), we get

$$||z_2(x,t) - z_1(x,t)||_2^2 = ||\zeta(x,t)||_2^2 \le K_0||b_2(x) - b_1(x)||_2^2 \cdot ||z_1(x,t)||_{Q_T}^2 < K_0^2\delta\widehat{\varphi}_0 = \varepsilon^2.$$

Corollary 1. *If the inequality $||z_2(x,t) - z_1(x,t)||_Q < \varepsilon$ is valid, then the inequality $|J^{00}(b_2(x),T) - J^{00}(b_1(x),T)| < \varepsilon K_2$ is valid too, K_2 is a constant.*

Let functions be note $y_1(x,t) = z_1(x,t)/p_1(t)$, $y_2(x,t) = z_2(x,t)/p_2(t)$, then

$$|J^{00}(b_2(x),T) - J^{00}(b_1(x),T)| = \int_0^l (y_2(x,T) - y_1(x,T))(y_2(x,T) + y_1(x,T)$$

$$-2Y(x))\,dx \le ||y_2(x,T) - y_1(x,T)||_2 \cdot ||y_2(x,T) + y_1(x,T) - 2Y(x)||_2 \le \frac{4\varepsilon}{p_m}.$$

Corollary 2. *If the sequence of functions* $\widehat{b}_n(x) \to b(x)$ *by the norm, then for any* $k = \overline{0, +\infty}$, *the convergence is performed* $\widehat{\xi}_{nk}(T) \to \xi_k(T)$, $\widehat{\xi}_{nk}(t)$ – *Fourier coefficients of the function* $\widehat{z}_n(x,t)$. *The function* $\widehat{z}_n(x,t)$ *is the solution of the homogeneous problem* (12)–(14), *for which* $b(x) \equiv \widehat{b}_n(x)$.

The true estimation $|\widehat{\xi}_{nk}(T) - \xi_k(T)| < \sqrt{2\varepsilon/l}$, $k = \overline{0, +\infty}$, follows from Theorem 2 and the expression (20).

6   The Infinite-Dimensional System of Differential Equations

Substituting decomposition of functions on the system of cosines in Eq. (12) and making standard transformations, described in [7], we get:

$$\sum_{k=0}^{+\infty} \xi''_{ktt}(t)v_k(x) = \sum_{k=0}^{+\infty} -a^2\lambda_k^2 \xi_k(t)v_k(x) + \sum_{k=0}^{+\infty} b_k v_k(x) \sum_{j=0}^{+\infty} \xi_j(t)v_j(x).$$

Using the formula "product of cosines": $v_k v_j = 0,5(v_{|k-j|} + v_{k+j})$, we reduce the product of series to a series

$$\sum_{k=0}^{+\infty} b_k v_k(x) \times \sum_{j=0}^{+\infty} \xi_j(t)v_j(x) = \sum_{j=0}^{+\infty} \kappa_j v_j(x)$$

where $\kappa_j = \sum\limits_{i=0}^{+\infty} c_{ji}\xi_i$, and the expressions for c_{ji} are presented below. Next, we multiply both sides by $v_k(x)$, $k = \overline{0, +\infty}$, and take the integral over x from 0 to l. The Eq. (12) is written as an infinite-dimensional system of differential equations with constant coefficients and initial conditions

$$\xi''_{tt}(t) = (C - \Lambda)\xi(t), \quad \xi_k(0) = \varphi_k, \quad \xi'_{kt}(0) = \psi_k, \quad k = \overline{0, +\infty}, \qquad (23)$$

where $\xi(t)$ is infinite-dimensional vector-function with components $\xi_i(t)$, $i = \overline{0, +\infty}$. Matrix C is stationary matrix of infinite dimensions with elements:

$$c_{00} = b_0, \; c_{0j} = \frac{1}{2}b_j, j = \overline{1, +\infty}, \; c_{ii} = b_0 + \frac{1}{2}b_{2i}, i = \overline{1, +\infty}, \qquad (24)$$

$$c_{ij} = \frac{1}{2}(b_{i+j} + b_{|i-j|}), \; i = \overline{1, +\infty}, \; j = \overline{0, +\infty}. \qquad (25)$$

Matrix $\Lambda = \mathrm{diag}(0, a^2\lambda_1^2, ..., a^2\lambda_j^2, ...)$ is infinite-dimensional stationary and diagonal matrix.

For the infinite-dimensional system of the first-order differential equations the theorem of existence has known. Therefore the infinite-dimensional system of the second-order differential equations (23) is to transform into a system of the first-order differential equations. In addition, we need to get rid of the members

$a^2\lambda_k^2\xi_k(t)$, because $\lim\limits_{k\to+\infty}\lambda_k = +\infty$. It will prevent, in the future, to use the convergence theorem for solutions of shortened problems.

Denote $\mu_k = a\lambda_k$, $k = \overline{1, +\infty}$. We will do the change of variables described in [8], thereby reducing an infinite-dimensional system of second-order differential equations to an infinite-dimensional system of first-order differential equations, Let i be an imaginary unit. For $k = \overline{1, +\infty}$, we make change of variables (writing through complex variables is more compact):

$$\xi_k(t) = w_k(t)\exp(i\mu_k t) + w_{-k}(t)\exp(-i\mu_k t), \tag{26}$$

$$\xi_k'(t) = i\mu_k w_k(t)\exp(i\mu_k t) - i\mu_k w_{-k}(t)\exp(-i\mu_k t). \tag{27}$$

The functions $w_k(t)$ and $w_{-k}(t)$ are complex conjugate functions: $w_{\pm k}(t) = A_k(t) \pm iB_k(t)$, $k = \overline{1, +\infty}$. Substituting $t = 0$ in (26), (27), we get the initial conditions:

$$A_k(0) = \frac{\varphi_k}{2}, \quad B_k(0) = -\frac{\psi_k}{2\mu_k}, \quad k = \overline{1, +\infty}. \tag{28}$$

Then from (26), (27) we have

$$w_k'(t)\exp(i\mu_k t) + w_{-k}'(t)\exp(-i\mu_k t) = 0, \quad k = \overline{1, +\infty}. \tag{29}$$

We will use here the system (23) without the zero row. Differentiating (27) and substituting (23), we get

$$i\mu_k w_k'(t)\exp(i\mu_k t) - i\mu_k w_{-k}'(t)\exp(-i\mu_k t)$$

$$= c_{k0}\xi_0(t) + \sum_{j=1}^{+\infty} c_{kj}(w_j(t)\exp(i\mu_j t) + w_{-j}(t)\exp(-i\mu_j t)), \quad k = \overline{1, +\infty}. \tag{30}$$

So from (29), (30), we obtain an infinite system of differential equations for $k = \overline{1, +\infty}$.

$$w_k'(t) = \frac{-i\exp(-i\mu_k t)}{2\mu_k}\left(c_{k0}\xi_0(t) + \sum_{j=1}^{+\infty} c_{kj}(w_j(t)\exp(i\mu_j t) + w_{-j}(t)\exp(-i\mu_j t))\right).$$

We denote

$$\omega_{2k}(t) = A_k(t), \omega_{2k+1}(t) = B_k(t), k = \overline{1, +\infty}; \ \omega_0(t) = \xi_0(t), \omega_1(t) = \xi_0'(t), \tag{31}$$

then for $k = \overline{1, +\infty}$, system (23), with (28) and (31) can be rewritten in the following way:

$$\omega_0'(t) = \omega_1(t), \ \omega_1'(t) = b_0\omega_0(t) + \sum_{j=1}^{+\infty}\frac{b_j}{2}(\omega_{2j}(t)\cos(\mu_j t) + \omega_{2j+1}(t)\sin(\mu_j t)), \tag{32}$$

$$\omega_{2k}'(t) = -\frac{c_{k0}}{2\mu_k}\sin(\mu_k t)\omega_0(t) - \sum_{j=1}^{+\infty}\gamma_{2k,j}^0(t)\omega_{2j}(t) + \sum_{j=1}^{+\infty}\gamma_{2k,j}^1(t)\omega_{2j+1}(t), \tag{33}$$

$$\omega'_{2k+1}(t) = -\frac{c_{k0}}{2\mu_k} \cos(\mu_k t)\omega_0(t) - \sum_{j=1}^{+\infty} \gamma^0_{2k+1,j}(t)\omega_{2j}(t) + \sum_{j=1}^{+\infty} \gamma^1_{2k+1,j}(t)\omega_{2j+1}(t);$$

$$(34)$$

the initial conditions are

$$\omega_0(0) = \varphi_0, \ \omega_1(0) = \psi_0, \ \omega_{2k}(0) = \frac{\varphi_k}{2}, \ \omega_{2k+1}(0) = -\frac{\psi_k}{2\mu_k}, \ k = \overline{1, +\infty}. \quad (35)$$

For $k \geq 1$ the next expressions take place

$$\gamma^0_{2k,j}(t) = \frac{c_{kj}}{\mu_k} \cos(\mu_j t) \sin(\mu_k t), \quad \gamma^1_{2k,j}(t) = \frac{c_{kj}}{\mu_k} \sin(\mu_j t) \sin(\mu_k t), \quad (36)$$

$$\gamma^0_{2k+1,j}(t) = \frac{c_{kj}}{\mu_k} \cos(\mu_j t) \cos(\mu_k t), \quad \gamma^1_{2k+1,j}(t) = \frac{c_{kj}}{\mu_k} \sin(\mu_j t) \cos(\mu_k t). \quad (37)$$

7   The Shortened System

We consider the "shortened" system, that is, by fixing N, we look for a solution to the finite shortened $(2N + 2)$-dimensional system, taking $\omega_k(t)$, $k = \overline{(2N + 2), +\infty}$, equals to zero. The initial conditions of the first $(2N + 2)$ unknowns are taken from the initial conditions (35). This shortening of the task (32)–(37) gives us information about the first $N + 1$ unknowns of the system (23)–(25), that is, about $\xi_k(t)$, $k = \overline{0, N}$. The conditions under which the solutions of a shortened problem converge at $t = T$ and $N \to +\infty$ to the solution of an infinite-dimensional system are described in [6].

Statement 1. *In the notation of this paper, the solution of a shortened system tends at $t = T$ to the value of the solution of an infinite system of differential equations*

$$\frac{d\omega_k}{dt} = f_k(t, \omega_1, \omega_2, ...), \quad k = \overline{0, +\infty};$$

it is enough that
  the following conditions were met:

a) functions $f_k(t, \omega_1, \omega_2, ...)$ are continuous over a set of variables;
b) the functions f_k satisfy with respect to the variables $\omega_1, \omega_2, ..., ...$ the enhanced Cauchy-Lipschitz condition.
c) $f_k(t, 0, 0, ...) \leq \overline{f}(t)$, where $\overline{f}(t)$ is a function that is continuous on the segment $[0, T]$.

The fulfillment of conditions a) and c), as well as b) for $k = 0$ is obvious (it is clear from (32)–(37)).

Let us check that condition b) is met, for $k \geq 1$:

$$\left| f_k(t, \omega_1, ..., \omega_m'', \omega_{m+1}', \omega_{m+2}', \omega_{m+3}'...) - f_k(t, \omega_1, ...\omega_m'', \omega_{m+1}'', \omega_{m+2}'', \omega_{m+3}''...) \right|$$

$$= \left| \sum_{j=m+1}^{+\infty} f_k(t, ..., \omega_{j-1}'', \omega_j', \omega_{j+1}', \omega_{j+2}'...) - f_k(t, ...\omega_{j-1}'', \omega_j'', \omega_{j+1}', \omega_{j+2}'...) \right|$$

$$\leq \left| \sum_{j=m+1}^{+\infty} \frac{\partial f_k}{\partial \omega_j}(t, \omega_1, ..., \omega_{j-1}'', \omega_{jcp}, \omega_{j+1}', \omega_{j+2}'...)(\omega_j' - \omega_j'') \right|$$

$$\leq \sum_{j=m+1}^{+\infty} \left| \frac{\partial f_k}{\partial \omega_j}(t, \omega_1, ..., \omega_{j-1}'', \omega_{cp}', \omega_{j+1}', \omega_{j+2}'...) \right| |\Delta\omega| \leq \varepsilon_{m1} |\Delta\omega|,$$

where $\Delta\omega = \max\limits_{j=m+1,+\infty} |\omega_j' - \omega_j''|$, and

$$\sum_{j=m+1}^{+\infty} \left| \frac{\partial f_k}{\partial \omega_j}(t, \omega_1, ..., \omega_{j-1}, \omega_{cp}', \omega_{j+1}', \omega_{j+2}'...) \right| = \sum_{j=m+1}^{+\infty} |\gamma_{k,[j/2]}^s(t)| = \varepsilon_{m1},$$

for $k = \overline{2, +\infty}$, where $s = 0$ or $s = 1$; $s = j(\mathrm{mod}\, 2)$. This expression for $k = 1$, see below when this case will be considered. To test the enhanced Cauchy condition, it remains to prove that $\lim\limits_{m \to +\infty} \varepsilon_{m1} = 0$.

Let $m' = [(m+1)/2]$, $\eta = [k/2]$, writing (24), (25) and taking into account (35), it is easy to see that

$$\sum_{j=m+1}^{+\infty} |\gamma_{k,[j/2]}^s(t)| \leq \sum_{j=m+1}^{+\infty} \frac{|c_{[k/2],[j/2]}|}{\mu_k} \leq \sum_{\theta=m'}^{+\infty} \frac{|b_{|\theta-\eta|}| + |b_{\theta+\eta}|}{2\mu_k} \leq \sum_{\theta=m'}^{+\infty} \frac{|b_{|\theta-\eta|}|}{\mu_k}. \tag{38}$$

Lemma 1. *The series of Fourier coefficients of a continuously differentiable function $b(x)$, decomposed in the system of cosines, converges absolutely.*

Note that

$$b_k = \frac{2}{l} \int_0^l b(x) v_k \, dx = \frac{2}{l} \left. b(x) \frac{l}{\pi k} \sin(\lambda_k x) \right|_0^l - \frac{l}{\pi k} \frac{2}{l} \int_0^l b_x'(x) \sin(\lambda_k x) \, dx = -\frac{l \bar{b}_k}{\pi k},$$

where $\bar{b}_k$ are the Fourier coefficients of the function $b_x'(x)$, so

$$\sum_{k=0}^{+\infty} |b_k| = \frac{l}{\pi} \sum_{k=0}^{+\infty} \frac{|\bar{b}_k|}{k} \leq \frac{l}{2\pi} \sum_{k=0}^{+\infty} \left(\frac{1}{k^2} + \bar{b}_k^2 \right).$$

Series of $\sum\limits_{k=0}^{+\infty} \bar{b}_k^2$ is converges due to the convergence of Parseval's equality for the function $b_x'(x)$ on the segment $[0, l]$, so the series of Fourier coefficients of the continuously differentiable function $b(x)$ converges absolutely.

We consider 3 cases:

1 case $k = 1$:

$$\varepsilon_{m1} = \sum_{j=m+1}^{+\infty} 0,5 \, |b_{j/2} \cos(\mu_{j/2} t)|, \text{ if } k = 1, \, j \text{ is even, or}$$

$$\varepsilon_{m1} = \sum_{j=m+1}^{+\infty} 0,5 \, |b_{(j-1)/2} \sin(\mu_{(j-1)/2} t)|, \text{ if } k = 1, \, j \text{ is odd, and the majoriz-}$$

ing sequence converges to 0: $\lim\limits_{m \to +\infty} 0,5 \sum\limits_{j=m+1}^{+\infty} |b_{[j/2]}| = 0$, since the series of Fourier coefficients of the function $b(x)$ converges absolutely.

Let K_1 be the sum of a series of modules b_j. Fix some $\varepsilon > 0$, then there is a natural M_0 such that for $j \geq [M_0/2]$, the inequality $|K_1 - S_j^{|b|}| < \varepsilon$ is satisfied, $S_j^{|b|}$ – partial sum of a series made up of $|b_i|$. Denote $[lK_1/\varepsilon a\pi k] + 1 = M_1$ and $m_0 = \max\{M_0, M_1\}$.

2 case; $k = \overline{2, m_0}$, then $\eta \leq m_0/2$, for $m_0 \to +\infty$, the right-hand side (38) can be evaluated

$$\varepsilon_{m1} \leq \frac{1}{\mu_k} \sum_{\theta=m_0}^{+\infty} |b_{|\theta-\eta|}| \leq \frac{l}{a\pi k}(K_1 - S_{[m_0/2]}^{|b|}) < \varepsilon \frac{l}{a\pi k}.$$

3 case; $k = \overline{(m_0+1), +\infty}$, then the right-hand side (38) can be evaluated by the expression

$$\varepsilon_{m1} \leq \frac{1}{\mu_k} \sum_{\theta=m_0}^{+\infty} |b_{|\theta-\eta|}| \leq \frac{lK_1}{m_0\pi} \leq \frac{lK_1}{M_1 a\pi k} < \varepsilon.$$

Thus, in all cases it can be seen that for all k and for any m it is true $\lim\limits_{m \to +\infty} \varepsilon_{m1} = 0$, thus proving that the strengthened Cauchy-Lipschitz condition – condition b) holds.

So, for a fixed T and $N \to +\infty$, the solutions of the shortened task tend the solution of an infinite-dimensional system (33) with initial conditions (35).

We denote the solution of the Cauchy problem of the shortened problem corresponding to the system (33) by $(2N + 2)$-dimensional vector function $\omega^N(t) = (\omega_0^N(t), ..., \omega_{2N+1}^N(t))$. A shortened system of ODEs, $(k = \overline{1, N})$, represented as:

$$\omega_{0t}^{N'}(t) = \omega_1^N(t), \omega_{1t}^{N'}(t) = b_0\omega_0^N(t) + \sum_{j=1}^{N} \frac{1}{2}b_j(\omega_{2j}^N(t)\cos(\mu_j t) - \omega_{2j+1}^N(t)\sin(\mu_j t)),$$

$$\omega_{2kt}^{N\,'}(t) = -\frac{c_{k0}}{2\mu_k}\sin(\mu_k t)\omega_0^N(t) - \sum_{j=1}^{N}\gamma_{2k,j}^0(t)\omega_{2j}^N(t) + \sum_{j=1}^{N}\gamma_{2k,j}^1(t)\omega_{2j+1}^N(t),$$

$$\omega_{2k+1t}^{N'}(t) = -\frac{c_{k0}}{2\mu_k}\cos(\mu_k t)\omega_0^N(t) - \sum_{j=1}^{N}\gamma_{2k+1,j}^0(t)\omega_{2j}^N(t) + \sum_{j=1}^{N}\gamma_{2k+1,j}^1(t)\omega_{2j+1}^N(t),$$

with initial conditions $w_0^N(0) = \varphi_0$, $w_0^N(0) = \psi_0$, $w_{2k}^N(0) = \varphi_k/2$, $w_{2k+1}^N(0) = -\psi_k/2\mu_k$. According to the Theorem from [6],

$$\lim_{N \to +\infty} w_j^N(T) = w_j(T), \quad j = \overline{0, 2N+1}. \tag{39}$$

Reverse substitution:

$$\xi_0(t) = w_0(t), \quad \xi_k(t) = 2w_{2k}(t)\cos(\mu_k t) - 2w_{2k+1}(t)\sin(\mu_k t), \quad k = \overline{1, N}. \tag{40}$$

If we solve the shortened system, we obtain an approximate solution of infinite system (32)–(37), and hence the solution of system (23)–(25) too. There are limits based on the limits (39)

$$\lim_{N \to +\infty} w_0^N(T) = \xi_0(T); \lim_{N \to +\infty} w_{2k}^N(T)\cos(\mu_k T) - w_{2k+1}^N(T)\sin(\mu_k T) = \frac{\xi_k(T)}{2}. \tag{41}$$

Carefully assessing the system (23)–(25), we note that for the shortened problem with a fixed N only the parameters of the b_k, $k = \overline{0, 2N}$ can affect on functions $\xi_k^N(T)$, $k = \overline{0, N}$, (and hence on functions $w_j^N(t)$, $j = \overline{0, 2N+1}$). As a result, using the function

$$\tilde{b}_N(x) = \sum_{k=0}^{2N} b_k v_k(x) \tag{42}$$

is sufficient in the shortened problem corresponding to the number N. Vector $(b_0, ..., b_{2N})^T$ uniquely sets the function $\tilde{b}_N(x)$. Note, however, that in this case of a finite vector of control parameters $(b_0, ..., b_{2N})^T$, it uniquely defines a continuously differentiable control function $\tilde{b}_N(x)$. Similarly, in a shortened problem corresponding to the number N, it is sufficient to use the functions

$$\tilde{\varphi}_N(x) = \sum_{k=0}^{N} \varphi_k v_k(x), \quad \tilde{\psi}_N(x) = \sum_{k=0}^{N} \psi_k v_k(x). \tag{43}$$

8   The Study of an Optimization Problem

We move on to the optimal control task. It is easy to see that the optimal value of the objective function (8) $J^{00}(T) = 0$ for $\theta_k^*(T) = Y_k$, $k = \overline{0, +\infty}$, and it is impossible to improve this result. It is not clear whether it is possible to achieve it, however, it can be used as a kind of reference point. We transform the objective function (8):

$$J^{00}(b(x), T) = \int_0^l (y(x, T) - Y(x))^2\, dx = 2 - l(2Y_0\theta_0(T) + \sum_{j=1}^{\infty} Y_j\theta_j(T)).$$

The objective function can be rewritten in the equivalent form (that is, the objective function reach an extremum on the same set of control parameters)

$$J(b(x), T) = 2Y_0\theta_0(T) + \sum_{j=1}^{\infty} Y_j\theta_j(T) \to \max \quad \text{or}$$

$$J(b_0, ..., b_j, ..., T) = \left(2Y_0\xi_0(T) + \sum_{j=1}^{\infty} Y_j\xi_j(T)\right) \Theta^{-1}(\overline{\xi}(T)) \to \max \qquad (44)$$

according to Theorem 1. The constraint on the control function (7) written as

$$lb_0^2 + \frac{l}{2}\sum_{i=1}^{+\infty} b_i^2 \le K^2. \qquad (45)$$

Given the remark at the end of the previous paragraph for the shortened problem and (42), the constraint (45) will take the form:

$$lb_0^2 + \sum_{i=1}^{2N} \frac{l}{2}b_i^2 \le K^2. \qquad (46)$$

Then the objective function (44) of the shortened problem is written:

$$J_N(b_0, ..., b_{2N}, T) = (2Y_0\xi_0^N(T) + \sum_{j=1}^{N} Y_j\xi_j^N(T)) \Theta^{-1}(\overline{\xi}^N(T)) \to \max, \qquad (47)$$

here $\Theta(\overline{\xi}^N(T)) = \left(l(\xi_0^N(T))^2 + \frac{l}{2}\sum_{j=1}^{N}(\xi_j^N(T))^2\right)^{1/2}$, since $\xi_j^N(t) \equiv 0$, $j \ge N+1$.

The objective function is continuous across their variables $\xi_k^N(T)$, $k = \overline{0, N}$, since $p(t) \ge p_m > 0$, and (41) holds. Then the equality

$$J_N(b_0, ..., b_{2N}, T) \to J(b_0, ..., b_{2N}, ..., T) \qquad (48)$$

is valid when $N \to +\infty$. This statement follows from the result of the convergence theorem for solutions of the shortened systems.

9   The Algorithm for Finding the Optimal Value

So, the optimization problem for a shortened system takes the form: *to maximize the objective function* (47) *under the condition of the control function* (46).

Let us take a sufficiently large N. We will use the $(2N+2)$-dimensional paralleleriped: $-\sqrt{l^{-1}}K \le b_0 \le \sqrt{l^{-1}}K$, $-\sqrt{2l^{-1}}K \le b_j \le \sqrt{2l^{-1}}K$, $j = \overline{1, 2N+1}$. We apply brute force method to find the minimizing sequence. The certain small step h_j by j-coordinate is chosen so that the series of h_j^2 converges: $\sum_{j=0}^{+\infty} h_j^2 = S_h$ (for example, $h_j = \delta_0/(j+1)$, $S_h = \pi^2\delta_0^2/6$, where δ_0 – some small enough positive constant). For each iteration point that satisfies the condition (46), the control vector $(b_0, ..., b_{2N})^T$ is constructed, and the shortened problem is solved. Then $(N+1)$-dimensional vector $(\xi_0^N(T), ..., \xi_N^N(T))^T$ and $J_N(b_0, ..., b_{2N}, T)$ are calculated by the received $(2N+2)$-dimensional solution vector $(\omega_0^N(T), ..., \omega_{2N+1}^N(T))^T$ for problem (32)–(37), whose value is checked for the maximum of the objective function (47). Further the search continues. In this process, the found maximum value of $J_N^*(\bar{b}_N^*, T)$ and the corresponding optimal control parameters of the vector $(b_0^*, ..., b_{2N}^*)^T$ and vector $(\xi_0^{N*}(T), ..., \xi_N^{N*}(T))^T$ are stored. For further investigation of the problem, we can recommend saving all possible vectors $\bar{b}_N^*$ and $\bar{\xi}_N^*(T)$ at which the maximum value of $J_N^*(\bar{b}_N^*, T)$ is reached. Next, we take the number $(N+1)$ and get optimal for this shortened problem $\bar{b}_{N+1}^*$ and $J_{N+1}^*(\bar{b}_{N+1}^*, T)$

Let stopping the algorithm be that the condition $0 < J_{N+1}^*(\bar{b}_{N+1}^*, T) - J_N^*(\bar{b}_N^*, T) < 0.5\widetilde{\varepsilon}$ where $\widetilde{\varepsilon}$ – some margin of error. It is easy to see that such a stop of the algorithm will definitely work. As a result, considering various N, we get the minimizing sequence $\bar{b}_N^*$ and optimal value $J_N^*(\bar{b}_N^*, T)$, which with increasing N can be arbitrarily close to the optimal value of the original problem.

Let $C(b(x))$ is the set of the continuously differentiable function $b(x) \in C^1[0,l]$ satisfying the condition (45). Denote $J_{00} = \sup\limits_{b(x) \in C(b)} J(b(x), T)$. For any sufficiently small ε there is the continuously differentiable function $b(x)$ such that $J_{00} - J(b(x), T) < \varepsilon$. It follows from (45), that starting from some N_1, the inequality $0,5l \cdot \sum\limits_{j=N_1+1}^{\infty} b_j^2 < \varepsilon_1$ is true.

Consider the infinite-dimensional vector that is closest to $(b_0, ..., b_{2N}, ...)^T$. A vector of its first N components always falls into the brute force search for shortened task: $\bar{b}^{bf} = (b_0^{bf}, b_1^{bf}, ..., b_{2N}^{bf}, ...)^T$, where the coefficients $b_j^{bf} = s\delta_0/(j+1)$, $s \in Z$ and $|s\delta_0/(j+1)| \leq K\sqrt{(\text{sign}\, j + 1)l^{-1}}$. It defines the function $b^{bf}(x)$. If the function $b^{bf}(x)$ is not continuously differentiable function, then for $j > N_2$ we replace $b_j^{bf} = s\delta_0/(j+1)$ with $b_j^{bf} = 0$. The new function will also be called function $b_j^{bf}(x)$. We estimate the norm for sufficiently small δ_0 and ε_1,

$$||b(x) - b^{bf}(x)||_2^2 \leq \sum_{j=0}^{N_1} l \frac{(\delta_0)^2}{(j+1)^2} + \varepsilon_1 \leq \sum_{j=0}^{\infty} l \frac{(\delta_0)^2}{(j+1)^2} + \varepsilon_1 \leq \frac{l(\delta_0\pi)^2}{6} + \varepsilon_1 < \delta.$$

Also from Theorem 2 the statement follows: $|J(b(x), T) - J(b^{bf}(x), T)| < \varepsilon$. Consider shortened systems for an infinite-dimensional system with parameters b_j^{bf}, $j = \overline{0, +\infty}$. Recall that any partial sum of a series decomposed over the system of cosines is the continuously differentiable function. Also as proved above from the theorem about the solutions of the shortened system, the statement follows: there is a sufficiently large N_2 for which it is true $|J(b^{bf}(x), T) - J_N(\widetilde{b}_N^{bf}(x), T)| < \varepsilon$. Next, for $N > \max\{N_1, N_2\}$, it follows

$$|J(b^{bf}(x), T) - J_N(\widetilde{b}_N^{bf}, T)| \leq 2\varepsilon. \tag{49}$$

According to the algorithm for finding the maximum of the shortened problem, the inequality is fulfilled $J_N^*(b_0^*, ..., b_{2N}^*, T) \geq J_N(\widetilde{b}_N^{bf}, T)$. The inequality $J_N^*(b_0^*, ..., b_{2N}^*, T) \geq J(b(x), T) - 2\varepsilon$ is valid (see (49)). Then inequality

$J_N^*(b_0^*, ..., b_{2N}^*, T) \geq J_{00} - 3\varepsilon$ is valid too. Due to the randomness of choosing the constant ε, $J_N^*(\widetilde{b}_N^*, T) \to J_{00}$ is executed, when $N \to +\infty$.

10 Conclusion

Specific feedback control plays a dual role in the optimization problem. First, the state constraint is automatically satisfied; second, there is a substitution of variables that reduces a nonlinear integro-differential equation to a linear one. This allows you to apply the Fourier method. With respect to the Fourier coefficients, an infinite-dimensional system of the second-order differential equations arises. It is further proved that the method of shortened systems can be applied to the infinite-dimensional system of ODEs obtained after some transformations. The algorithm based on the brute force method allows you to find the optimal value of the objective function with any accuracy.

References

1. Egorov, A.I., Znamenskaya, L.N.: Vvedenie v teoriyu upravleniya sistemami s raspredelennymi parametrami [Introduction to the theory of control of systems with distributed parameters]. S.Pb.: Lan' (2017) (in Russian)
2. Burago, P.N., Egamov, A.I.: O svyazi reshenij nachal'no-kraevyh zadach dlya nekotorogo klassa integro-differencial'nyh uravnenij s chastnymi proizvodnymi i linejnogo giperbolicheskogo uravneniya [On the connection between solutions of initial boundary-value problems for a some class of integro-differential PDE and a linear hyperbolic equation] // Zhurnal Srednevolzhskogo matematicheskogo obshchestva. **21**(4), 413–429 (2019) (in Russian). https://doi.org/10.15507/2079-6900.21.201904. 413-429
3. Kuzenkov, O.A., Novozhenin, A.V.: Optimal control of measure dynamics. Commun. Nonlinear Sci. Numer. Simul. **21**(1–3), 159–171 (2015). https://doi.org/10. 1016/j.cnsns.2014.08.024
4. Egamov, A.I.: The existence and uniqueness theorem for initial-boundary value problem of the same class of integro-differential PDEs. In: Bychkov, I., Kalyagin, V.A., Pardalos, P.M., Prokopyev, O. (eds.) NET 2018. SPMS, vol. 315, pp. 173–186. Springer, Cham (2020). https://doi.org/10.1007/978-3-030-37157-9_12
5. Egamov, A.I.: Construction of a minimizing sequence for the problem of cooling of the given segments of the rod with phase constraint. Uchenye Zapiski Kazanskogo Universiteta. Seriya Fiziko-Matematicheskie Nauki **162**(2), 193–210 (2020) (in Russian). https://doi.org/10.26907/2541-7746.2020.2.193-210
6. Valeyev, K.G., Zhautykov, O.A.: Beskonechnyye sistemy differentsialnykh uravneniy. [Infinite system of differential equations]. Alma-Ata: Nauka Kazakhskoy SSR (1974) (in Russian)
7. Smirnov, V.I.: Kurs vysshej matematiki, T.4. Ch.2 [Course of higher mathematics, Vol. 4. Part 2]. M.: Izd-vo "Nauka" (1974)
8. Bogolyubov, N.N., Mitropolskiy, Yu.A.: Asimptoticheskiye metody v teorii nelineynykh kolebaniy. [Asymptotic methods in the theory of nonlinear oscillations] M.: GIFML (1958) (in Russian)
9. Akhmerov, R.R., Sadovsky, B.N.: Osnovy teorii obyknovennykh differentsialnykh uravneniy [Fundamentals of the theory of ordinary differential equations]. Novosibirsk (2002) (in Russian)

Numerical Investigation and Optimization of Output Processes in Cyclic Control of Conflicting Flows

Andrey Fedotkin$^{(\boxtimes)}$ (ID) and Evgeniy Kudryavtsev (ID)

Lobachevsky State University of Nizhny Novgorod, Gagarin Avenue 23,
Nizhny Novgorod 603950, Russia
`fandr@vmk.unn.ru`

Abstract. A non-classical queuing waiting system are considered. The queuing system serves conflicting flows with control in a class of cyclic algorithms. Conflicting flows mean that they cannot be summed up and this does not allow you to reduce the problem to a simpler case with a single flow. Requirements from different conflicting flows are served at non-overlapping intervals. In addition, there are additional time intervals— readjustments, due to which the problem of conflicting flows is resolved. Such systems are adequate models of real-world systems for processing and transmitting information, technological systems, transport systems, etc.

Keywords: Conflicting flows · Homogeneous Markov sequence · Conditional distribution · Markov process

1 Introduction

Unlike most well-known works, the so-called non-local description of the requirements flow proposed in [1–10] is used to construct a mathematical model of output flows. The description of output flows includes the state of the service device and the values of queues for conflicting flows. Note that the functioning of the system under consideration for servicing non-homogeneous requirements and controlling conflict flows in continuous time is a complex non-Markov process. Therefore, researching the system characteristics and the output flows properties in continuous time are a difficult task. Using the theoretical results of works [11–20], this article substantiates the method of numerical investigation of the system by simulation methods using computer and information technologies. The results of studies of the dynamics of output processes for servicing requirements on a simulation model are interpreted on the problem of managing conflict non-homogeneous traffic flows at isolated intersections.

Supported by Lobachevsky State University of Nizhny Novgorod.

D. Balandin et al. (Eds.): MMST 2020, CCIS 1413, pp. 220–231, 2021.
https://doi.org/10.1007/978-3-030-78759-2_19

2 Problem Statement at the Content Level

The problem of cyclic control of m conflicting flows $\overline{\Pi}_1, \overline{\Pi}_2, \ldots, \overline{\Pi}_m$ non-homogeneous requirements during their maintenance by the system is considered. Input flows $\overline{\Pi}_j, j \in \{1, 2, \ldots, m\}$ are considered conflicting and independent. Conflicting input flows means that the maintenance of the flows must occur at non-overlapping intervals. Moreover, the specified intervals should be separated by time intervals during which the service of any requirements is prohibited. The flow $\overline{\Pi}_j$ requests only enter to the O_j drive with an unlimited number of waiting places. In the system without loss of requests, it is possible to record the outgoing flow N_j with an unlimited supply of pending requirements in the O_j storage and with maximum use of the resources of the service device for each $j \in \{1, 2, \ldots, m\}$.

Flows $\Pi_1, \Pi_2, \ldots, \Pi_m$ are called saturation flows. The need to introduce saturation flows arises primarily in those real queuing systems in which it is clear in advance that the service durations of different requirements can be determined by the state of the service system and, as a result, be dependent and have different distribution laws. As an example, the process of crossing the stop line by vehicles with a green traffic light allowing it can be cited. If there is a queue, the first cars only start moving and move more slowly than those who arrive at the intersection at the green light interval and continue driving at maximum speed. The service device has $2m$ states

$$\Gamma^{(1)}, \Gamma^{(2)}, \ldots, \Gamma^{(2m)}.$$

For any fixed $n \in \{1, 2, \ldots, 2m\}$, the duration of stay in the state $\Gamma^{(n)}$ is equal to T_n. For all $j \in \{1, 2, \ldots, m\}$ in state $\Gamma^{(2j-1)}$, only the requirements of flow $\overline{\Pi}_j$ are served. The maximum possible number of serviced requirements of the flow $\overline{\Pi}_j$ in the state $\Gamma^{(2j-1)}$ is determined by the saturation flow Π_j and is equal to l_j. For all $j \in \{1, 2, \ldots, m\}$ in state $\Gamma^{(2j)}$, the requirements of each of the flows are not served. The service device changes states cyclically in the following sequence:

$$\Gamma^{(1)} \rightarrow \Gamma^{(2)} \rightarrow \ldots \rightarrow \Gamma^{(2m)} \rightarrow \Gamma^{(1)} \rightarrow \ldots$$

So, the service of conflicting flows occurs in non-overlapping time intervals, which are separated by changeover intervals. An adequate mathematical model of this kind of real problems is the control systems of service with a variable structure [1].

3 Numerical Study of Output Processes

We consider input flows $\overline{\Pi}_1, \overline{\Pi}_2, \ldots, \overline{\Pi}_m$ to be non-ordinary Poisson random Gnedenko–Kovalenko processes [1]. Then the sequence of calling moments at which the requirements arrive in the system for each flow $\overline{\Pi}_j$ is a Poisson process with the parameter λ_j. Moreover, let's assume that at each of these moments, one or two applications appear, respectively, with probabilities p_j or $q_j = 1 - p_j$.

Let the random variable $\eta_j(t)$ for each $j \in \{1, 2, \ldots, m\}$ determine the number of received requests to the storage O_j along the flow $\overline{\Pi}_j$ for the time interval $[0, t)$. An important characteristic of the process of cyclic management of conflicting flows of requirements is the loading of the service system for each flow and its overall loading for all or only some pre-fixed flows. Unfortunately, determining the load of managed unconventional queuing systems is always a difficult task. In classical single-channel systems with an unlimited queue, Poisson input flow, and demand service, according to the exponential law, the probability that there will be at least one demand in the system in stationary mode $\lambda \mu^{-1} < 1$. Here, the parameters λ and μ are the intensities of receipt and, respectively, demand service. Let $\theta(t)$ is the total length of those time intervals between zero and the moment t, during which there will be at least one requirement in the system. It is well known that for any $\varepsilon > 0$ there is a limit equality

$$\lim_{t \to \infty} \mathbf{P}(|t^{-1}\theta(t) - \lambda\mu^{-1}| < \varepsilon) = 1.$$

A random variable of the form $t^{-1}\theta(t)$ determines the average relative occupancy time of the system over the interval $[0, t)$ and approximately coincides with the value $\lambda\mu^{-1}$ in the sense of convergence in probability for sufficiently large values of t. Therefore, the constant $\lambda\mu^{-1}$ is naturally called the load or occupancy measure of the system.

We will now consider the problem of managing conflicting flows of requirements in the class of cyclic algorithms in a similar way. Then the probability ρ_j that in stationary mode there will be at least one request in the service system on flow $\overline{\Pi}_j$ can be called the system loading on this flow. Let's say we want to determine the total load $\rho_{1,2}$ of the system, for example, by two flows $\overline{\Pi}_1$ and $\overline{\Pi}_2$. Then, due to the independence of the input flows, saturation flows and cyclic switching of the states of the service device, it is possible to obtain [4] that the probability of having at least one machine in at least one of the flows $\overline{\Pi}_1$ and $\overline{\Pi}_2$ is equal to $\rho_1 + \rho_2 - \rho_1\rho_2$. Therefore, we can assume that for flows $\overline{\Pi}_1$ and $\overline{\Pi}_2$ the total load is

$$\rho_{1,2} = \rho_1 + \rho_2 - \rho_1\rho_2.$$

Unfortunately, for the process of controlling conflicting traffic flows at $l_j \neq 1$, it is not possible to define a simple formula for the probability ρ_j. Only for $l_j = 1$ in [4] it was found that for the stationary mode, the probability of having at least one requirement at the intersection along the flow $\overline{\Pi}_j$ is equal to

$$\rho_j = 1 - \frac{e^{\lambda_j T}}{2m}(1 + e^{-\lambda_j T_{2j}} + e^{-\lambda_j(T_{2j}+T_{2j}+1)} + \ldots + e^{-\lambda_j(T-T_{2j}-1)})(1 - \lambda_j(1 + q_j)),$$

$T = T_1 + T_2 + \ldots + T_{2m}$. Even this formula is difficult to calculate without a computer. Therefore, there is a difficult problem of determining the estimate of $\widetilde{\rho}_j$ for loading ρ_j on the flow $\overline{\Pi}_j$ at $l_j \neq 1$.

If the necessary and sufficient conditions for the existence of a stationary mode in a managed queuing system coincide and simply depend on the intensities of receipt and maintenance of requirements, then these conditions are usually the basis for determining the load or, in extreme cases, evaluating the load. So the necessary and sufficient conditions for existence of stationary regime in the system according to the flux j according to the results of [4] can be written in the form

$$\lambda_j T(1 + q_j)/l_j < 1,$$

or the equivalent form

$$\lambda_j T(1 + q_j)/[\mu_j T_{2j-1}] < 1.$$

Here μ_j^{-1} determines the average service time requirements of input flow $\overline{\Pi}_j$ and $l_j = [\mu_j T_{2j-1}]$. Hence, as an estimate of $\widetilde{\rho}_j$ it is possible to offer a formula of the form

$$\widetilde{\rho}_j = \widetilde{\rho}_j(\lambda_j, q_j, \mu_j, T, T_{2j-1}) = \lambda_j T(1 + q_j)/[\mu_j T_{2j-1}].$$

We will consider the chosen estimate $\widetilde{\rho}_j$ acceptable, or suitable, if it satisfies the following natural requirements:

1. to estimate $\widetilde{\rho}_j$ the following inequality holds $0 < \widetilde{\rho}_j < 1$;
2. at T and T_{2j-1}, the evaluation value $\widetilde{\rho}_j(\lambda_j, q_j, \mu_j, T, T_{2j-1})$ does not decrease with increasing each of the parameters λ_j, q_j and does not increase with each parameter μ_j, T_{2j-1};
3. if some changes to the set of parameters λ_j, q_j, μ_j, T and T_{2j-1} value $\lambda_j T(1 + q_j) - [\mu_j T_{2j-1}]$ approaches zero, then the value of the estimate $\widetilde{\rho}_j$ should tend to one.

It is easy to verify that the proposed estimate of $\widetilde{\rho}_j$ for

$$\lambda_j T(1 + q_j) < [\mu_j T_{2j-1}]$$

is appropriate. In the future, for the sake of simplicity, we will call the evaluation of $\widetilde{\rho}_j$ a quasi-load on the flow of $\overline{\Pi}_j$. Now we can give the following appropriate assessment

$$\widetilde{\rho}_{1,2}(\lambda_1, q_1, \mu_1, \lambda_2, q_2, \mu_2, T, T_1, T_3) = \widetilde{\rho}_1 + \widetilde{\rho}_2 - \widetilde{\rho}_1 \widetilde{\rho}_2$$

to download $\rho_{1,2} = \rho_1 + \rho_2 - \rho_1\rho_2$ system in two flows $\overline{\Pi}_1$ and $\overline{\Pi}_2$.

Monitor various real experiments show that with increasing system load typically increases during T_{per} the transition process for its entry into the quasi-stationary mode and time computer modeling to calculate its probability and the numerical characteristics with specified degree of accuracy and reliability. Because of this, when the choice of algorithm parameters to determine when T_{per} the end of the transition process, and the job accuracy and reliability of calculations of probability and its numerical characteristics substantially overlook the importance of load estimation system. For the stationary mode of the system at $l_j \neq 1$, it is analytically impossible to obtain observable formulas for the laws of distribution of queue lengths, service waiting time, and, finally, for the laws of distribution of output flows. To obtain estimates of these distributions, some numerical characteristics for the solution of the optimization problem to the minimum weighted average waiting time and service requirements in the arbitrary flow developed a computer simulation videomodel the cyclic control m flows.

Let the symbol $v = 1, 2, \ldots$ specify the sequence number of the request when it enters the storage O_j of the service system. Denote by $\gamma^0_{j,v}$ the time (in seconds) waiting for the v-th request of flow $\overline{\Pi}_j$ to start serving in a system with zero initial queues and by $\gamma^+_{j,v}$ the time (in seconds) waiting for the v-th request of flow $\overline{\Pi}_j$ to start serving in a system with specified non-zero initial queues. At the first stage of simulation, the time T_{per} (in seconds) of the transition process or the moment when the system reaches a quasi-stationary mode is calculated. For this purpose, for each flow $\overline{\Pi}_j$, the arithmetic averages were calculated sequentially in

$$\widetilde{M}(\gamma^0_{j,v}) = v^{-1}(\gamma^0_{j,1} + \gamma^0_{j,2} + \ldots + \gamma^0_{j,v}), \quad v = 1, 2, \ldots,$$

observed waiting times in a system with zero initial queues and calculating the arithmetic mean

$$\widetilde{M}(\gamma^+_{j,v}) = v^{-1}(\gamma^+_{j,1} + \gamma^+_{j,2} + \ldots + \gamma^+_{j,v}), \quad v = 1, 2, \ldots,$$

observed waiting times in a system with a given value $x_{j,0} > 0$ of the initial queue. The time T_{per} for the quasi-stationary regime or the end of time quasiperiodic process in the system was considered when the first time occurred a multiple of k at v, the condition of the

$$|(\widetilde{M}(\gamma^0_{j,v}) - \widetilde{M}(\gamma^+_{j,v}))/\widetilde{M}(\gamma^0_{j,v})| < \delta$$

for all $j = 1, 2, \ldots, m$. Here, the natural number k and $0 < \delta < 1$ are the specified parameters of the algorithm for determining the estimate of the time T_{per} of the transition process. At the second stage, only the system with zero initial queues was simulated in order to calculate estimates for each of the m flows with a given accuracy ε for the main characteristics of the system in the quasi-stationary mode. In this case, the accuracy value ε of a particular estimate was equal to the product of a certain constant Δ by the value of this estimate. On the simulation model for the flow $\overline{\Pi}_j$ in the quasi-stationary mode with a given reliability β and a given accuracy ε, the following estimates were calculated:

1. assessment $\widetilde{M}(\gamma_j)$ and $\widetilde{D}(\gamma_j)$ for mathematical expectation and dispersion, respectively, for the time γ_j waiting the maintenance of arbitrary application $\overline{\Pi}_j$;
2. evaluation of the form

$$\widetilde{M}(\gamma) = \left(\sum_{j=1}^{m} \lambda_j(1 + q_j)\right)^{-1} \sum_{j=1}^{m} \lambda_j(1 + q_j)\widetilde{M}(\gamma_j)$$

for the expectation of a weighted average time γ of waiting to start service application arbitrary flow;

3. evaluation $\widetilde{M}(\varkappa_j)$ and $\widetilde{D}(\varkappa_j)$ for mathematical expectation and respectively variance of random length $\varkappa_j$ queue of cars flow $\overline{\Pi}_j$ at an arbitrary moment of switching service device condition $\Gamma^{(2j-1)}$;
4. evaluation $\widetilde{M}(\xi'_j)$ and $\widetilde{D}(\xi'_j)$ for the mathematical expectation and variance respectively of a random number ξ'_j requirements of flow $\overline{\Pi}_j$, leaving the drive O_j in the interval T_{2j-1};
5. the value $\widetilde{\rho}_j$ of the evaluation of kwasiborski system for each flow $\overline{\Pi}_j$, and a value of $\widetilde{\rho}_{1,2}$ assessment for the total system load, for example, for flows $\overline{\Pi}_1$ and $\overline{\Pi}_2$;
6. estimation of the distribution law and view of the relative frequency histogram for the value ξ'_j.

As an illustration of the effectiveness of the joint application of analytical methods and the method of simulation modeling, we present the solution of the following specific problems. The first task is a qualitative and numerical study of the process of cyclic control of only two of the most intensive traffic flows $\overline{\Pi}_1$ and $\overline{\Pi}_2$ at the intersection. Therefore, the traffic light (service device) has four phases or states $\Gamma^{(1)}$, $\Gamma^{(2)}$, $\Gamma^{(3)}$ and $\Gamma^{(4)}$. The second task—definition of quasi-optimal duration T_1^* green phase (state) $\Gamma^{(1)}$ for flow $\overline{\Pi}_1$ and quasi-optimal duration T_3^* green phase (state) $\Gamma^{(3)}$ for flow $\overline{\Pi}_2$ according to the condition of minimum for evaluation $\widetilde{M}(\gamma)$ the mathematical expectation of the average weighted waiting time of the machine maintenance of the arbitrary transport flow.

Software implementation of the simulation model on a computer is performed by means of Code Gear RAD Studio 2009 development in the Object Pascal language. The simulation model can work both in the mode when requests leave the system in groups, and in the mode when requests are served sequentially one by one as they arrive. The computer simulation model allows not only to calculate the main characteristics of the intersection operation with a given degree of accuracy and reliability in the counting mode, and on this basis to find quasi-optimal flow control, but also allows you to observe in video mode the entire process of servicing requirements and managing conflict flows on the example of car traffic at the intersection.

Consider the example of real intersection, for which $\lambda_1 = 0.16$ vehicle/h, $q_1 = 0.3$, $\mu_1 = 1$ vehicle/s, $\lambda_2 = 0.22$ vehicle/h, $q_2 = 0.4$, $\mu_2 = 1$ vehicle/h, $T_2 = T_4 = 4$ s. Using a simulation model with

$$k = 2, \ \delta = 0.1, \ \beta = 0.9, \ \Delta = 0.02$$

and modified method of coordinatewise descent was determined quasi-optimal values of $T_1^* = 10$ s and $T_3^* = 15$ s. the durations of the green phases of the traffic light for flow $\overline{\Pi}_1$ and, respectively, for flow $\overline{\Pi}_2$. The values T_1^* and T_3^* provide a quasi-load of $\rho_{1,2} = 0.8989$ and a value of $\widetilde{M}(\gamma)$ equal to 10.903 s. For comparison, we note that the durations $T_1 = 41$ s and $T_3 = 51$ s used in practice give a value of $\widetilde{M}(\gamma) = 20.265$ s and a value of $\widetilde{\rho}_{1,2} = 0.8049$.

Table 1 shows some of the results of counting at points on the curve of equal quasi-loads, and Table 2 shows all the results of counting on the line $T_1 + T_2 = 25$.

Table 1. Values of estimates of the main characteristics on the curve of equal quasi-loads.

T	T_1	T_3	$\widetilde{M}(\gamma_1)$	$\widetilde{M}(\gamma_2)$	$\widetilde{M}(\gamma)$	$\widetilde{M}(\varkappa_1)$	$\widetilde{M}(\varkappa_2)$	$\tilde{\rho}_1$	$\tilde{\rho}_2$	$\tilde{\rho}_{1,2}$	$\widetilde{D}(\varkappa_1)$	$\widetilde{D}(\varkappa_2)$	$\widetilde{D}(\xi_1')$
30	9	13	12.883	10.096	11.22	5.1486	6.1482	0.6933	0.711	0.9113	8.7867	11.635	5.598
31	9	14	14.322	9.2331	11.28	5.4963	5.9406	0.7164	0.682	0.9098	10.570	10.270	5.534
32	10	14	12.201	10.529	11.183	5.0690	6.6167	0.6656	0.704	0.9010	8.1432	12.443	6.499
33	10	15	13.148	9.3841	10.903	5.3938	6.1513	0.6864	0.678	0.8989	8.5311	11.086	6.616
34	10	16	14.324	9.1314	11.226	5.8074	6.1556	0.7072	0.655	0.8989	9.3487	11.067	6.409

Table 2. Values of estimates of the main characteristics on the line $T_1 + T_2 = 25$.

T	T_1	T_3	$\widetilde{M}(\gamma_1)$	$\widetilde{M}(\gamma_2)$	$\widetilde{M}(\gamma)$	$\widetilde{M}(\varkappa_1)$	$\widetilde{M}(\varkappa_2)$	$\tilde{\rho}_1$	$\tilde{\rho}_2$	$\tilde{\rho}_{1,2}$	$\widetilde{D}(\varkappa_1)$	$\widetilde{D}(\varkappa_2)$	$\widetilde{D}(\xi_1')$
33	8	17	25.566	7.3854	14.739	8.0262	5.4289	0.858	0.5979	0.9429	25.515	9.0236	3.167
33	9	16	17.434	8.2612	11.957	6.433	5.6971	0.763	0.6353	0.9134	13.925	9.8279	4.901
33	10	15	13.148	9.3841	10.903	5.3938	6.1513	0.686	0.6776	0.8989	8.5311	11.086	6.616
33	11	14	11.571	10.996	11.239	5.0696	6.7562	0.624	0.726	0.8970	8.2046	12.092	7.463
33	12	13	10.104	13.042	11.923	4.6664	7.3472	0.572	0.7819	0.9066	7.2232	16.161	8.385
33	13	12	8.8784	18.37	14.09	4.2971	9.2859	0.528	0.847	0.9278	6.3238	23.606	8.732
33	14	11	7.9915	33.408	23.223	4.1897	14.067	0.490	0.924	0.9613	5.5462	66.607	8.696

Note that the full search includes 431 points. Quasi-optimal duration $T_1^* = 10$ and $T_3^* = 15$ phases of the lights provide an overall quasijarus $\rho_{1,2} = 0.8989$ and the minimum value of $\widetilde{M}(\gamma)$ equal to 10.903, and thereby solve the problem of optimization by criterion $\widetilde{M}(\gamma)$. For comparison, note that in practice a longer duration $T_1 = 41$ and $T_3 = 51$ provide a common quasijarus $\rho_{1,2} = 0.80486$ and give value assessment $\widetilde{M}(\gamma)$ the weighted average time γ of waiting for the machine maintenance random flow at that intersection, equal to the amount of 20,265. Estimates for the mean waiting time of an arbitrary flow, as a rule, decreases with decreasing values of the period T of the cyclic control. For example, for this intersection at a fixed $T = 60$, the quasi-optimal value for T_1 is 20, which provides a total quasi-load of $\rho_{1,2} = 0,84114$ and an estimate of $\widetilde{M}(\gamma) = 13,808$. This estimate is less than the value of 20.265, which corresponds to the values of $T_1 = 41$ and $T_3 = 51$. Note that for this intersection in the region of the existence of a stationary regime, the minimum value of the total quasi-load is 0.7661, which is achieved at $T = 10000$, $T_1 = 4028$.

Qualitative and numerical studies on the simulation model allow us to draw a very important conclusion for output flows in the case of quasi-optimal control of conflicting flows in the class of cyclic algorithms. In the case of quasi-optimal control of conflicting flows in the class of cyclic algorithms, the estimate $\widetilde{D}(\xi')$ of the weighted average variance of the output flow takes relatively small values. For example, to the intersection with the settings $\lambda_1 = 0.16$, $q_1 = 0.3$, $\mu_1 = 1$, $\lambda_2 = 0.22$, $q_2 = 0.4$, $\mu_2 = 1$ and $T_2 = T_4 = T_0 = 4$, with quasi-optimal control evaluation $\widetilde{D}(\xi')$ weighted average of the variance of output is equal to 9.5337.

Table 3. Values of performance ratings.

T	T_1	T_3	$\widetilde{M}(\gamma_1)$	$\widetilde{M}(\gamma_2)$	$\widetilde{M}(\gamma)$	$\widetilde{D}(\xi_1')$
33	10	15	13.148	9.3841	10.903	9.5337
40	12	20	15.096	8.9403	11.398	12.827
60	20	32	18.223	10.854	13.808	22.235
80	27	45	23.493	12.316	16.812	30.993
100	34	58	28.722	14.096	19.979	41.793

In Table 3 for the specified period durations $T = 33$, 40, 60, 80, 100 cyclic control is given the appropriate values:

1. quasi-optimal parameters of the durations T_1 and T_3 phases of the automatic traffic lights;
2. evaluation $\widetilde{M}(\gamma_1)$ for the mean waiting time of the machine maintenance flow Π_1;
3. evaluation $\widetilde{M}(\gamma_2)$ for the mean waiting time of the machine maintenance flow Π_2;
4. evaluation $\widetilde{M}(\gamma)$ for the average weighted waiting time of the machine maintenance random flow;
5. assessment $\widetilde{D}(\xi')$ for a weighted average of the variance of the output flow.

From this table, it is easy to see that there is a sharp increase in the estimate $\widetilde{D}(\xi')$ for the weighted average variance of the output flow when deviating from the quasi-optimal values of the control parameters. It is clear that the output flows of cars from a certain intersection go to the next intersection adjacent to it and are already input flows for the next intersection. In practice, it is well known that the control algorithm at the intersection will be simpler (for example, with a fixed switching rhythm) and the more successful the smaller the variance of the input flow, i.e. the more standardized the output and input flows are. This conclusion confirms the often put forward thesis for random experiments with control that a relatively large value of the variance of some characteristic of a random experiment is the result of suboptimal control.

Let us consider another example of a qualitative-numerical study of the process of cyclic control of conflict traffic flows $\overline{\Pi}_1$ and $\overline{\Pi}_2$ at a real intersection using simulation modeling. For example, an intersection was chosen for which

$$\lambda_1 = 89/2024 \approx 0.043, \quad q_1 = 40/89 \approx 0.45,$$

$$\mu_1 = 0.9, \quad \lambda_2 = 0.1; \quad q_2 = 0.5, \quad \mu_2 = 1, \quad T_2 = T_4 = 4.$$

In this case, the statistical data of the input flow $\overline{\Pi}_1$ coincide with the observations of the real traffic flow, which are given by Bartlett in [21]. The values of the parameters μ_1, μ_2, T_2 and T_4 were also selected from the experience of a large number of observations of traffic at real intersections. The simulation was carried out at different values of the green light durations T_1 and T_3 for the flow $\overline{\Pi}_1$ and, accordingly, for flow $\overline{\Pi}_2$ from the range

$$\{(T_1, T_3): \lambda_1(1+q_1)(T_1+T_3+8)-[\mu_1 T_1] < 0, \lambda_2(1+q_2)(T_1+T_3+8)-[\mu_2 T_3] < 0\}.$$

The curve of equal loads for flows $\overline{\Pi}_1$ and $\overline{\Pi}_2$ is determined by an equation of the form

$$\lambda_2(1+q_2)/[\mu_2 T_3] = \lambda_1(1+q_1)/[\mu_1 T_1].$$

After calculating the estimate of each characteristic of the system with a given accuracy $\Delta = 0.01$ and a given reliability $\beta = 0.9$, the simulation of the process of movement at this intersection ended.

The value $\widetilde{M}(\gamma)$ of the weighted average time γ of waiting for the start of servicing of the machine of an arbitrary flow at such an intersection is 6.8966, and the value of the estimate $\widetilde{T}_{per}$ of the time T_{per} of the transient process is 12426 s.

As a result of analytical studies and a large number of experiments on the simulation model, several unexpected conclusions can be drawn. For example, from the calculations it turns out that with an increase in the estimate $\widetilde{\rho}_{1,2}$ for the load of the intersection, the estimate $\widetilde{M}(\gamma)$ of the weighted average waiting time, which is one of the main numerical characteristics of the system, increases significantly according to a nonlinear law. On the contrary, with an increase in the estimate $\widetilde{\rho}_{1,2}$ for the load of the intersection, the estimate $\widetilde{D}(\xi')$ for the weighted average variance of the output flow, which is determined by the formula

$$\widetilde{D}(\xi') = (\lambda_1(1+q_1)\widetilde{D}(\xi_1') + \lambda_2(1+q_2)\widetilde{D}(\xi_2'))(\lambda_1(1+q_1) + \lambda_2(1+q_2))^{-1}.$$

The noted non-linear nature of the dependence of the weighted average waiting time and the weighted average variance of the output flow on the load is in good agreement with observations with heavy traffic on the highway, when on average more than 1800 vehicles per hour arrive at the intersection.

Let's give an example of traffic flow control at an intersection with parameters:

$$\lambda_1 = 0.05, \ q_1 = 0.4, \ \mu_1 = 0.9, \ \lambda_2 = 0.1, \ q_2 = 0.5,$$

$$\mu_2 = 1, \ T_2 = T_4 = T_0 = 4, \ T_{min} = 18, \ T_{max} = 31.$$

From Table 4, we directly obtain that the minimum value of the estimate of the weighted average waiting time for the beginning of the crossing through the intersection of an arbitrary machine is equal to 6.849. This value is reached at $T_1^{(1)} = 5$ and $T_3^{(1)} = 9$. On the straight line $T_1 + T_3 = 14$, the estimates were determined $\widetilde{M}(\gamma_1)$, $\widetilde{M}(\gamma_2)$, $\widetilde{M}(\gamma)$, $\widetilde{M}(\varkappa_1)$, $\widetilde{M}(\varkappa_2)$, $\widetilde{\rho}_1$, $\widetilde{\rho}_2$, $\widetilde{\rho}_{1,2}$, $\widetilde{D}(\varkappa_1)$, $\widetilde{D}(\varkappa_2)$, $\widetilde{D}(\xi_1')$, $\widetilde{D}(\xi_2')$. The results of calculating these estimates are shown in Table 5.

From Table 5, we determine the quasi-optimal durations T_1^* and T_3^*, which are equal to 5 and, respectively, 9. In this case, the length of the cyclic control period is 22 and the minimum value $\widetilde{M}(\gamma)$ is 6.849. So, to determine the quasi-optimal durations T_1^* and T_3^* by the modified coordinate descent method, the required number of test pairs (T_1, T_3) is 21.

Table 4. Estimates values of the main characteristics on the curve of equal quasi-loads.

T	T_1	T_3	$\widetilde{M}(\gamma_1)$	$\widetilde{M}(\gamma_2)$	$\widetilde{M}(\gamma)$	$\widetilde{M}(\varkappa_1)$	$\widetilde{M}(\varkappa_2)$	$\tilde{\rho}_1$	$\tilde{\rho}_2$	$\tilde{\rho}_{1,2}$	$\widetilde{D}(\varkappa_1)$	$\widetilde{D}(\varkappa_2)$	$\widetilde{D}(\xi'_1)$	$\widetilde{D}(\xi'_2)$
18	4	6	8.995	6.532	7.314	1.193	2.036	0.42	0.45	0.68	1.950	3.535	1.282	3.489
19	4	7	10.17	5.924	7.276	1.329	1.983	0.44	0.40	0.66	2.267	3.349	1.338	3.943
20	4	8	10.87	5.571	7.236	1.379	1.987	0.46	0.37	0.66	2.268	3.299	1.329	4.363
21	5	8	8.665	6.162	6.961	1.269	2.147	0.36	0.39	0.61	2.028	3.686	1.790	4.669
22	5	9	9.324	5.730	**6.849**	1.324	2.100	0.38	0.36	0.61	2.099	3.498	1.820	4.930
23	5	10	10.15	5.469	6.957	1.427	2.081	0.40	0.34	0.60	2.335	3.478	1.910	5.369
24	6	10	8.958	6.030	6.969	1.362	2.270	0.33	0.36	0.57	2.195	3.712	2.293	5.547
25	6	11	9.625	5.726	6.970	1.443	2.247	0.35	0.34	0.57	2.315	3.754	2.345	5.961
26	6	12	10.09	5.550	6.987	1.474	2.240	0.36	0.32	0.57	2.346	3.690	2.331	6.129
27	7	12	9.274	5.984	7.025	1.461	2.393	0.31	0.33	0.54	2.268	3.926	2.629	6.228
28	7	13	9.817	5.814	7.079	1.540	2.386	0.32	0.32	0.54	2.391	3.952	2.709	6.748
29	7	14	10.41	5.627	7.151	1.611	2.400	0.33	0.31	0.54	2.502	3.970	2.809	6.980
30	8	14	9.972	6.092	7.323	1.625	2.521	0.30	0.32	0.52	2.494	4.193	3.048	7.168
31	8	15	10.57	5.932	7.405	1.710	2.548	0.31	0.31	0.52	2.690	4.198	3.192	7.338

Table 5. Estimates values of the main characteristics on the straight line $T_1 + T_3 = 14$.

T	T_1	T_3	$\widetilde{M}(\gamma_1)$	$\widetilde{M}(\gamma_2)$	$\widetilde{M}(\gamma)$	$\widetilde{M}(\varkappa_1)$	$\widetilde{M}(\varkappa_2)$	$\tilde{\rho}_1$	$\tilde{\rho}_2$	$\tilde{\rho}_{1,2}$	$\widetilde{D}(\varkappa_1)$	$\widetilde{D}(\varkappa_2)$	$\widetilde{D}(\xi'_1)$	$\widetilde{D}(\xi'_2)$
22	3	11	38.03	4.263	15.00	3.384	1.787	0.77	0.30	0.83	11.77	2.929	0.560	5.129
22	4	10	13.18	4.991	7.592	1.617	1.954	0.51	0.33	0.67	2.806	3.230	1.365	5.062
22	5	9	9.324	5.730	**6.849**	1.324	2.100	0.38	0.36	0.61	2.099	3.498	1.820	4.930
22	6	8	7.712	6.612	6.956	1.207	2.270	0.30	0.41	0.59	1.803	3.710	2.049	4.599
22	7	7	6.728	7.908	7.509	1.105	2.486	0.25	0.47	0.60	1.755	4.354	2.136	4.343
22	8	6	5.758	9.527	8.309	1.000	2.741	0.22	0.55	0.64	1.535	4.782	2.185	3.686
22	9	5	5.050	12.82	10.24	0.969	3.271	0.19	0.66	0.72	1.436	6.285	2.264	2.803
22	10	4	4.480	26.15	20.03	0.930	5.201	0.17	0.82	0.85	1.404	19.68	2.424	1.483

Thus, to determine the quasi-optimal durations T_1^* and T_3^*, the proposed modified coordinate descent method significantly reduces the number of test pairs (T_1, T_3) from 134 to 21.

In practice, with heavy traffic on the highway, as a rule, long durations T of the cyclic control period are used. In this case, it is possible to choose such durations T_1 and T_3 of the traffic light phases at which the point (T_1, T_3) is relatively far from the boundaries of the region of existence of the stationary mode. Therefore, with an increase in the intensity of the input flows of cars within certain limits, the stationary mode at the intersection will still be preserved. However, at large values of the durations T, T_1 and T_3, transport delays at the intersections greatly increase.

4 Conclusion

The system of cyclic conflict flows control was investigated. An analytical study of the mathematical model was carried out. Some important numerical characteristics of the system operation cannot be found analytically. The simulation method is used to solve this problem. Simulation modeling also makes it possible to find quasi-optimal values of system parameters.

References

1. Fedotkin, M.A., Fedotkin, A.M.: Analysis and optimization of output processes under cyclic control of conflicting traffic flows Gnedenko-Kovalenko. Automatics telemechanics. Russ. Acad. Sci. **12**, 92–108 (2009)
2. Fedotkin, M.A., Fedotkin, A.M., Kudryavtsev, E.V.: Construction and analysis of a mathematical model of the spatial and temporal characteristics of traffic flows. Autom. Control Comput. Sci. **48**(6), 358–367 (2014)
3. Fedotkin, M.A., Fedotkin, A.M., Kudryavtsev, E.V.: Nonlocal description of the time characteristic for input flows by means of observations. Autom. Control Comput. Sci. **49**(1), 29–36 (2015). https://doi.org/10.3103/S0146411615010034
4. Fedotkin, A.M.: Determination of the stationary regime of recurrent Markov chains by the iterative-majorant method. Bull. Lobachevsky Univ. Nizhny Novgorod **4**, 130–140 (2009)
5. Fedotkin, M.A., Fedotkin, A.M., Kudryavtsev, E.V.: Dynamic models of non-uniform traffic flow on highways. Avtomatika i telemekhanika. Russ. Acad. Sci. **8**, 149–164 (2020)
6. Fedotkin, A.M., Fedotkin, M.A.: Model for refusals of elements of a controlling system. In: Proceedings of the first French-Russian Conference on "Longevity, Aging and Degradation Models in Reliability, Public Health, Medicine and Biology, LAD 2004", vol. 2, pp. 136–151. SPU, Saint Petersburg (2004)
7. Fedotkin, A.M.: Mathematical models of traffic flows on the highway and at the intersection controlled by the cyclic algorithm. Lobachevsky University of Nizhny Novgorod, Department in VINITI 11.01.09, vol. 5–B2009, pp. 1–30 (2009)
8. Fedotkin, M.A., Fedotkin, A.A.: Output processes for cyclic control of non-ordinary flows. In: Proceedings of the International Scientific Conference "Probability Theory, Stochastic Processes, Mathematical Statistics and Applications", pp. 362–369. BGU, Minsk (2008)
9. Fedotkin, A.M.: Distributions arithmetic properties of the output process of Gnedenko-Kovalenko flows cyclic control. Lobachevsky University of Nizhny Novgorod, Department in VINITI 14.04.09, vol. 213–B2009, pp. 1–24 (2009)
10. Fedotkin, A.M.: Properties of a controllable vector Markov chain with a countable number of states, satisfying the recurrence relations. Bull. Lobachevsky Univ. Nizhny Novgorod **6**, 122–141 (2009)
11. Fedotkin, A.A., Fedotkin, A.M.: Study of the Gnedenko-Kovalenko flow properties. Bull. Lobachevsky Univ. Nizhny Novgorod **6**, 156–160 (2008)
12. Fedotkin, A.A., Fedotkin, A.M.: Bartlett's traffic flow implementation study. Bull. Lobachevsky Univ. Nizhny Novgorod **3**, 195–199 (2013)
13. Fedotkin, A.M., Golisheva, N.M.: Cyclic control of conflict Gnedenko-Kovalenko flows. Bull. Lobachevsky Univ. Nizhny Novgorod **4**, 382–388 (2014)
14. Fedotkin, M.A., Vysotsky, A.A.: An optimal control in time-sharing systems during a transient period. In: Proceedings of the International Conference "Distributed Computer Communication Networks (DCCN 1996)". University, Tel-Aviv, pp. 164–173 (1996)

15. Fedotkin, M.A., Litvak, N.V.: A conflict flows control by an information about queue lengths. In: Proceedings of the International Conference "Distributed Computer Communication Networks (DCCN 1997)". University, Tel-Aviv, pp. 68–72 (1997)
16. Fedotkin, M.A., Litvak, N.V.: Random processes of adaptive control for conflict flows. In: Proceeding of the International Conference "Prague Stochastic 1998", Prague, pp. 147–152 (1998)
17. Fedotkin, M.A.: Conflict networks of a queuing in conditions of callbacks and random medium. In: Proceedings of the International Conference "Computer Science and Information Technologies (CSIT 1999)", Yerevan, pp. 47–55 (1999)
18. Fedotkin, M.A., Zorine, A.V.: Optimal control of conflict flows with repeated service. In: Korolev, V., Andronov, A., Bocharov, P. (eds.) Proceedings of the XXIV International Seminar on Stability Problems for Stochastic Models, pp. 126–132. TTI, Riga (2004)
19. Fedotkin, M.A., Litvak, N.V.: On the class of algorithms for adoption traffic control. In: Proceedings of the International Conference "Distributed Computer Communication Networks (DCCN 1996)", pp. 73–77. University, Tel-Aviv (1996)
20. Fedotkin, M.A., Kudryavtsev, E.V.: Simulation of adaptive control system with conflict flows of non-homogeneous requests. In: Proceedings of the Twelfth International Conference "Computer Data Analysis and Modeling: Stochastics and Data Science", pp. 163–166. BSU, Minsk (2019)
21. Bartlett, M.S.: The spectral analysis of point processes. J. R. Stat. Soc. Ser. B 25(2), 264–296 (1963)

Parametric Randomization for Accelerating the Nested Global Optimization

Vladimir Grishagin[(✉)] and Victor Gergel

Lobachevsky State University, Gagarin Avenue 23, 603950 Nizhni Novgorod, Russia
{vagris,gergel}@unn.ru

Abstract. In the framework of the nested optimization scheme for reducing a multidimensional global search problem to a family of sub-problems of less dimensions an information-statistical method for solving internal subproblems of the nested scheme is considered. The method's efficiency depends essentially on parameters influencing its convergence speed and reliability. Two techniques of the parameter randomization aimed at the acceleration of the reliable global search are studied. The experimental results of efficiency evaluation for the considered methods on two multidimensional multiextremal classes of benchmarks of different dimensions widely used for testing global optimization methods are presented. The results demonstrate advantages of the randomized techniques compared to prototypical method with invariable values of parameters.

Keywords: Global optimization · Dimensionality reduction · Parameter randomization

1 Introduction

Many problems of decision making can be described as optimization models with a single objective function being multidimensional and multiextremal. Such the models are widely spread in optimal design, machine learning, forecasting problems, etc., and within these areas have important practical applications. Multiextremality causes the significant complexity of the optimization problems in the multidimensional case as the necessity of finding out the global solution leads to the exponential growth of the number of function evaluations. The complexity and diversity of the global optimization problems have been drawing attention of many scientists (see, for instance, fundamental monographs [1–7]) and served as a source of different approaches both theoretically substantiated and intuitive (nature-inspired, or meta-heuristic).

This work was supported by the Ministry of Science and Higher Education of the Russian Federation, project no. 0729-2020-0055, and by the Research and Education Mathematical Center, project no. 075-02-2020-1483/1.

D. Balandin et al. (Eds.): MMST 2020, CCIS 1413, pp. 232–246, 2021.
https://doi.org/10.1007/978-3-030-78759-2_20

One of the productive approaches to solving multidimensional optimization problems consists in transformation of the initial problem to an equivalent (in the sense of coincidence of solutions) set of the subproblems of less dimensions, as a rule, univariate ones for which there exist many efficient global optimization methods. This idea was realized, for example, in the methods which use Peano-type mappings of multidimensional domains into one-dimensional space [4,8]. Another approach proposed in [9,10] is based on the recursive reduction of a multidimensional problem to a family of univariate subproblems solved by effective algorithms of one-dimensional global optimization [4,9–16]. This approach was theoretically substantiated [4,21] and confirmed its quality in comparison with other global optimization methods [17–21].

The comparative results of the algorithms on the base of the nested scheme have shown the combination of this scheme with the information-statistical univariate algorithm of global search is very promising [4]. This algorithm has been constructed as an optimal statistical procedure in the framework of the model that considers the objective function as a realization of some stochastic process with properties close to probabilistic analogue of the Lipschitz condition.

Functioning of this algorithm depends considerably on a parameter (called reliability parameter) included in an estimation of the Lipschitz constant used by the algorithm in the course of optimization. If the parameter is small then the sufficient condition of convergence to global optimum can be violated, at the same time, too large value of the parameter leads to significant increasing the number of objective function evaluations. The well-founded choice of the parameter value depends on the information of the problem to be solved. If you know the Lipschitz constant of the optimized function (as a rule, it is not realistic) then the parameter should be chosen so that the adaptive estimation of the Lipschitz constant L is greater than $2L$. Another consideration is the more complicated objective function is optimized the greater parameter should be taken. So, with small parameter we can lose the convergence to global optimum, however, with large parameter value the convergence will be guaranteed, but the method will spent too many evaluations of the objective function. Usually, to choose a compromise fixed value of the parameter is difficult. In the paper we propose to realize such the compromise via randomization of the parameter choice when two parameters, one is small, the second is sufficiently great, are randomly chosen with a probability.

Two techniques for forming this probability are considered. The first is to use just a constant probability. The second technique provides the choice of the parameter as the result of a 2×2 zero-sum game in accordance with rules proposed in [4]. The main idea of this consideration consists, briefly, in the following. Let us have two different values of the parameter for choosing them in optimization. We can model this situation as a game with nature which can be in two states corresponding to given parameter values. The researcher, obviously, does not know the state of the nature and chooses his/her own value of the parameter. So, both the researcher and the nature have strategies consisting in assignment of parameter values. If the players (researcher and nature) choose the same parameters, the payoff is equal to zero. Otherwise, the payoff is defined according to the model in the framework of which the algorithm has been derived. Under this consideration the choice of the parameter is determined in accordance with probabilities of the game model.

For evaluation of randomization efficiency a representative experiment on hundreds of multiextremal functions of different dimensions from known test classes [15,17,23] being traditional for testing the global optimization methods has been performed on the base of building the operational characteristics [17,22] of the methods compared.

The rest of the paper is organized as follows. Section 2 contains the statement of global optimization problems and the general scheme of nested optimization. Section 3 is devoted to the consideration of the randomization techniques applied to information-statistical algorithm of global search. Section 4 describes results of experimental testing the methods considered. Section 5 concludes the paper.

2   Global Optimization and Nested Dimensionality Reduction

In the paper a minimization problem

$$F^* = F(x) \to \min, x \in P_N \subseteq R^N \tag{1}$$

is considered where the objective function $F(x)$ is multiextremal in the feasible domain

$$P_N = \{x \in R^N : a_j \leq x \leq b_j, 1 \leq j \leq N\} \tag{2}$$

and finding the global minimum F^* over the box P_N in N-dimensional Euclidean space R^N is the goal of problem solving.

The objective function is additionally supposed to satisfy over the box P_N the Lipschitz condition, i.e., for all $x', x'' \in P_N$

$$|F(x') - F(x'')| \leq L \|x' - x''\|, \tag{3}$$

where $\|\bullet\|$ denotes the Euclidean norm and the Lipschitz constant $L > 0$.

There exist many numerical methods for solving Lipschitzian problems (1)–(2), some of them are referenced in Introduction. We will deal with the algorithms elaborated on the base of ideas of dimensionality reduction connected with the nested optimization scheme that allows us to replace the search of global minimum in a multidimensional problem with solving a family of univariate problems. To explain briefly the main scheme of the nested optimization let us introduce a family of reduced functions in the following manner.

Let a function $F(x)$ satisfy in the domain (2) the Lipschitz condition (3). Setting $F^N(x) \equiv F(x)$ by definition, let us construct a family of functions

$$F^i(\xi_i) = \min\{F^{i+1}(\xi_i, x_{i+1}) : x_{i+1} \in [a_{i+1}, b_{i+1}]\}, 1 \leq i \leq N-1, \tag{4}$$

where $\xi_i = (x_1, ..., x_i), 1 \leq i \leq N$.

According to [9] and [10] the basic relation of the nested optimization scheme

$$\min_{x \in P_N} F(x) = \min_{x_1 \in [a_1,b_1]} \min_{x_2 \in [a_2,b_2]} \ldots \min_{x_N \in [a_N,b_N]} F(x) \tag{5}$$

takes place. It means that instead of solving the multidimensional problem (1)–(2) the one-dimensional problem

$$F^1(x_1) \to \min, \ x_1 \in [a_1, b_1] \subset R^1, \tag{6}$$

can be solved.

However, according to (4), each calculation of the function $F^1(x_1)$ at some fixed point $x_1 \in [a_1, b_1]$ requires solving the subproblem

$$F^2(x_1, x_2) \to \min, \ x_2 \in [a_2, b_2] \subset R^1,$$

which is a one-dimensional minimization problem with respect to x_2, since x_1 is fixed (given by the problem (6)).

In turn, each evaluation of the function $F^2(x_1, x_2)$ with fixed x_1, x_2 generates solving the one-dimensional subproblem

$$F^3(\xi_2, x_3) \to \min, \ x_3 \in [a_3, b_3] \subset R^1,$$

etc., up to solving the univariate subproblem

$$F^N(\xi_{N-1}, x_N) \equiv F(\xi_{N-1}, x_N) \to \min, \ x_N \in [a_N, b_N] \subset R^1, \tag{7}$$

where ξ_{N-1} is fixed (given in preceding subproblems).

Thus, solving the problem (1)–(2) can be reduced to solving a family of «nested» one-dimensional subproblems

$$F^i(\xi_{i-1}, x_i) \to \min, \ x_i \in [a_i, b_i] \subset R^1, \tag{8}$$

where the fixed vector $\xi_{i-1} \in P_{i-1}$ from (2).

Under Lipschitz condition (3) for the objective function $F(x)$, every reduced function $F_i(\xi_i)$ is Lipschitzian as well with the same constant L. So, we can apply for solving subproblems of the family (8) algorithms of univariate Lipschitzian optimization, for example, the wide class of characteristical methods [17] including information-statistical algorithms [4,12,24,25], etc., methods by Piyavskij [10] and Shubert [26], Bayesian algorithms [3,27–29] and many others.

One of such the algorithms will be considered in the next section. It is a modification of the basic information-statistical algorithm of global search [4], but as opposed to it applies randomization techniques for the parameter choice.

3   Randomized Choice of the Algorithm's Parameter

Let us rewrite one-dimensional subproblems (8) in a unified form

$$f(t) \to min, t \in P_1 = [a, b]. \tag{9}$$

Under general consideration, a numerical method solving a problem (9) builds in P_1 a sequence of points $\{t^k\} = \{t^1, t^2, \ldots, t^k, \ldots\}, t^i \in [a, b], i = 1, 2, \ldots,$ and computes in these points values $v^i = f(t^i)$ of the objective function $f(t)$. Hereinafter the term "trial" will be used for designation of the objective function evaluation at a point. The algorithm to be considered will be presented in the characteristical form [17].

3.1 Computational Scheme of the Algorithm

Two first trials are executed by the algorithm at the points $t^1 = a$ and $t^2 = b$ with values $v^1 = f(t^1)$ and $v^2 = f(t^2)$.

Let $k \geq 2$ trials have been executed at points $t^1, t^2, \ldots, t^k$ within P_1 and values $v^i = f(t^i), 1 \leq i \leq k$, have been obtained. In order to get the point of the next $(k+1)$-th trial it is necessary to implement the following operations.

Step 1. Define a set

$$T_k = \{t_0, t_1, \ldots, t_{k-1}\}$$

of $k - 1$ points in the domain $P_1 = [a, b]$ consisting of the coordinates of the preceding trials t^i under assumption that the set T_k is ordered (by the subscript) in the increasing order of the coordinates, i.e.,

$$a = t_0 < t_1 < \ldots < t_{k-1} = b. \tag{10}$$

and juxtapose to the points $t_i, 0 \leq i \leq k - 1$, the values $v_i = f(t_i), 0 \leq i \leq k - 1$, of the objective function $f(t)$ calculated at the preceding iterations.

Step 2. Compute

$$M = \max_{1 \leq i \leq k-1} \left| \frac{v_i - v_{i-1}}{t_i - t_{i-1}} \right| \tag{11}$$

and accept for $j = 1, 2$

$$\mu_j = \begin{cases} r_j M, & M > 0, \\ 1, & M = 0, \end{cases} \tag{12}$$

where $r_1 > 1$ and $r_2 > 1$ are parameters of the method called *reliability parameters*.

Step 3. For each interval $(t_{i-1}, t_i), 1 \leq i \leq k - 1$, calculate the values

$$C_j(i) = \mu_j(t_i - t_{i-1}) + \frac{(v_i - v_{i-1})^2}{\mu_j(t_i - t_{i-1})} - 2(v_i + v_{i-1}), j = 1, 2, \tag{13}$$

called *characteristics* of the interval.

Step 4. Find the numbers q_1 and q_2 of intervals $(t_{q_j-1}, t_{q_j}), j = 1, 2$, which the maximal characteristics

$$C_j(q_j) = \max_{1 \leq i \leq k-1} C_j(i), j = 1, 2, \tag{14}$$

correspond to.

Step 5. For a given probability $p, 0 < p < 1$, get a value γ of a random variable uniformly distributed over the interval $[0, 1]$, accept $q = q_1, \mu = \mu_1$ if $\gamma < p$ and $q = q_2, \mu = \mu_2$ if $\gamma \geq p$. In other words, choose the parameter r_1 with the probability p and r_2 with the probability $1 - p$.

Step 6. Calculate

$$t^{k+1} = \frac{t_q + t_{q-1}}{2} - \frac{v_q - v_{q-1}}{2\mu} \tag{15}$$

as the point of the next trial and calculate the value $v^{k+1} = f(t^{k+1})$.

From the general theory of characteristical algorithms [17] it is easy derived the sufficient condition of convergence to the global minimum

$$\max\{\mu_1, \mu_2\} > 2L$$

where L is the Lipschitz constant of the function $f(t)$.

As a representative of the characteristical algorithms, the described method can use the termination criterion in the form

$$t_q - t_{q-1} < \epsilon, \tag{16}$$

where $\epsilon > 0$ is a predefined accuracy of the search.

3.2   Techniques of Parameter Randomizing

In the computational scheme of the algorithm the manner of assignment for the probability p was not specified. The simplest way is to take a constant value for all iterations. The other techniques can be more flexible when the probability changes adaptively depending on the situation in the course of optimization. Under such consideration the dynamic choice of the probability as the optimal result of a zero-sum game will be used. This model has been proposed in [4] in the framework of the information-statistical approach to building optimization methods as statistical decision procedures when the optimized function is considered as a realization of a stochastic process. The algorithm described in previous subsection is an example of application of this approach. The game model for parameter choice supposes there are two possible values of the reliability parameter, r_1 and r_2, and is not known which value is better for optimization. Following [30] the unknown value of the parameter can be considered as a *state of nature*, and the problem of optimal parameter selection can be stated (under stochastic assumptions about objective function) as a *game with nature*. In this game players (researcher and nature) have strategies consisting in the choice of parameter value, either r_1 or r_2. The loss matrix of the game is

	r_1	r_2
r_1	0	$D(r_1, r_2)$
r_2	$D(r_2, r_1)$	0

Here in the case $q_1 \neq q_2$ for q_1, q_2 from (14)

$$D(r_1, r_2) = \omega_1 + \omega_2 - C_2(q_2),$$

$$D(r_2, r_1) = \omega_1 + \omega_2 - C_1(q_1),$$

where $C_1(q_1), C_2(q_2)$ from (14) and

$$\omega_j = r_j M(t_{q_j} - t_{q_{j-1}}) - (v_{q_j} - v_{q_{j-1}}), j = 1, 2.$$

If $q_1 = q_2 = q$ then

$$D(r_1, r_2) = \frac{1}{2}|v_q - v_{q-1}| \left(1 - \frac{r_2}{r_1}\right), D(r_2, r_1) = \frac{1}{2}|v_q - v_{q-1}| \left(1 - \frac{r_1}{r_2}\right).$$

In this game the optimal mixed strategy of the first player (researcher) realizes the choice of the parameter r_1 with the probability

$$p = \frac{D(r_1, r_2)}{D(r_1, r_2) + D(r_2, r_1)}$$

and the choice of r_2 with the probability $1 - p$.

After simplification we can rewrite the expression for the probability p as

$$p = \begin{cases} \frac{r_1}{r_1 + r_2}, & q_1 = q_2, \\ \frac{\omega_1 + \omega_2 - C_2(q_2)}{2\omega_1 + 2\omega_2 - C_1(q_1) - C_2(q_2)}, & q_1 \neq q_2. \end{cases} \tag{17}$$

It should be noted that if $r_1 = r_2$ then the randomized methods turn into the basic algorithm of global search.

4   Computational Experiments

For the efficiency estimation of the randomization compared to versions with constant parameter values two series of experiments have been carried out on the sets of complicated multiextremal functions from well-known test classes [15, 23] widely used for the experimental study of global optimization methods. For comparison the concept of operational characteristics [17, 22] has been used. This approach presents the results in a graphical form that allows visual comparing the effectiveness of the competitive algorithms. Later this approach has been generalized and presented as the concept of operational zones [31]. The notion of the operational characteristic consists in the following. There is a set of optimization problems and each of them is solved by an optimization method with fixed values of its parameters. After this experiment we can calculate the average number κ of trials (evaluations of objective functions in problems solved) executed by the method during optimization and the number π of problems solved successfully. Repeating optimization of the test problems with other parameters we obtain several pairs (κ, π) and the set of those is called *operational characteristic* of the method. When operational characteristics of several methods are placed on the same plane (κ, π) it is easily to compare with each other. Namely, if for given κ the operational characteristic of one method is placed higher than the operational characteristic of the other method, then the first algorithm having spent the same amount of resources (trials) has solved more problems than the second method.

The results of testing are presented for three versions of the algorithm described in Subsect. 3.1:

- Algorithm of Global Search with Constant reliability parameter r (acronym AGSC);
- Algorithm of Global Search with Fixed probabilty p at Step 5 of the computational scheme (acronym AGSF) for the random choice of the reliability parameter from two possible values r_1 and r_2;
- Algorithm of Global Search with Adaptive probability p (acronym AGSA) determined from the solution (17) of the game model of optimal choice from two parameters r_1 and r_2.

These methods were tested with different parameters r, r_1, r_2 and p (the latter parameter only for AGSF) and in each variant the points of operational characteristics were built for several values of the accuracy ϵ from the termination criterion (16).

The first series of experiments was carried out on the test class [15, 17] consisting of 100 two-dimensional multiextremal problems. Table 1 contains the average numbers of trials κ and the numbers of successfully solved problems π in dependence on the accuracy ϵ for AGSC with different values of the reliability parameter r.

Table 1. Results for invariable reliability parameter

ϵ		0.05	0.03	0.2	0.01	0.008	0.004	0.002	0.001	0.0005
$r = 1.5$	κ	101	124	168	258	284	479	899	1722	4480
	π	61	78	83	82	82	82	84	84	85
$r = 2$	κ	174	226	299	469	533	821	1432	2473	5888
	π	72	88	90	91	91	94	96	96	97
$r = 2.5$	κ	226	326	432	691	794	1255	1936	3518	6164
	π	76	94	97	98	98	99	99	99	99
$r = 3$	κ	281	424	553	913	1048	1662	2567	4079	7085
	π	74	97	99	99	99	100	100	100	100
$r = 4$	κ	419	637	910	1516	1760	2671	4018	6128	9716
	π	78	97	100	100	100	100	100	100	100
$r = 5$	κ	497	817	1226	2110	2458	3839	5829	8779	13610
	π	76	94	100	100	100	100	100	100	100

Operational characteristics based on the data from Table 1 are shown in Fig. 1. The abscissa axis corresponds to the criterion κ and is presented in the logarithmic scale and the ordinate axis reflects the criterion π.

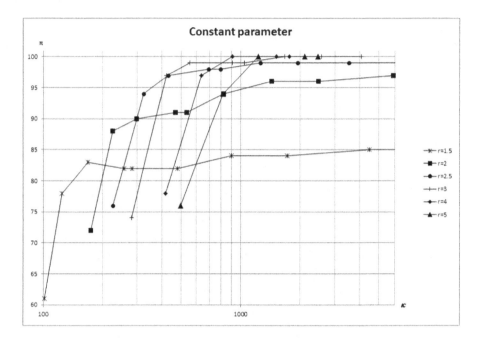

Fig. 1. Operational characteristics of AGSC

As we can see, AGSC with small parameters do not solve all the test problems even for high given accuracy because for some test problems the sufficient condition of global convergence does not meet. On the other hand, algorithm with high parameter values executes essentially more trials.

Now let us consider the simple randomization when parameters r_1 and r_2 are chosen with a fixed probability at all the iterations of the method. First of all, two parameters are taken with equal probability $p = 0.5$ and then the first parameter is chosen with the probability $p = 0.3$ and the second with $p = 0.7$. The experimental results can be found in Table 2. These results are presented as operational characteristics in Fig. 2 along with data of AGSC for comparison.

In comparison with AGSC, randomizing the parameters gives better results if the accuracy ϵ is rough, however, for high reliability when it is necessary to solve all the problems AGSC attains this aim faster than AGSF with equal probabilities of randomizing. It takes place because of influence of lesser parameter and if we diminish its probability then randomizing algorithm becomes more positive. In particular, the method with $r_1 = 2, r_2 = 5, p = 0.3$ exceeds AGSC. So, we can improve functioning the algorithm with invariable parameter by means of random mixing different values of the reliability factor, however, instead of looking for a good parameter value we have to find an efficient probability.

At last, let us look at results of randomized mixing small and high parameters in the framework of the game model (17) in two versions when AGSA works with $r_1 = 1.5, r_2 = 4$ and with $r_1 = 2, r_2 = 5$. Table 3 contains the results of this experiment.

In Table 3 the rows labeled as Speedup contain acceleration coefficients of AGSA compared to AGSC with r_2 and the same accuracy ϵ. Here the acceleration coefficient α is defined as

$$\alpha = \frac{\kappa_C}{\kappa_A}, \tag{18}$$

Table 2. Results of AGSF

ϵ		0.05	0.03	0.2	0.01	0.008	0.004	0.002	0.001	0.0005
$r_1 = 1.5, r_2 = 4$	κ	158	201	267	452	564	967	1889	3884	6927
$p = 0.5$	π	66	84	93	97	97	99	99	100	100
$r_1 = 2, r_2 = 5$	κ	211	313	438	746	892	1695	2944	5140	9440
$p = 0.5$	π	69	93	99	99	99	99	100	100	100
$r_1 = 1.5, r_2 = 4$	κ	218	315	379	703	882	1718	2930	5160	8280
$p = 0.3$	π	82	95	96	98	98	100	100	100	100
$r_1 = 2, r_2 = 5$	κ	276	417	611	1135	1418	2543	4493	7284	11638
$p = 0.3$	π	75	96	100	100	100	100	100	100	100

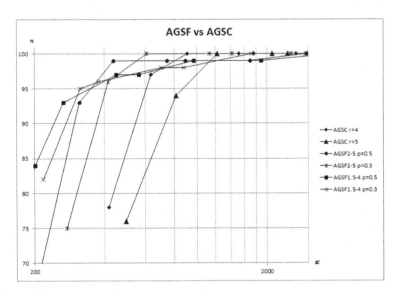

Fig. 2. Operational characteristics of AGSF and AGSC

Table 3. Results for adaptive randomization by AGSA

ϵ		0.05	0.03	0.2	0.01	0.008	0.004	0.002	0.001	0.0005
$r_1 = 1.5$	κ	220	291	389	693	854	1634	2770	4839	80073
$r_2 = 4$	π	82	96	96	100	100	100	100	100	100
Speedup		1.90	2.19	2.34	2.19	2.06	1.64	1.45	1.27	1.20
$r_1 = 2$	κ	295	417	603	1085	1336	2367	4150	6678	11343
$r_2 = 5$	π	78	99	100	100	100	100	100	100	100
Speedup		1.68	1.96	2.03	1.94	1.84	1.62	1.40	1.31	1.20

where κ_C is number of trials spent by AGSC and κ_A is number of trials executed by AGSA.

Operational characteristics of AGSA and AGSC with $r = 4$ and $r = 5$ are presented in Fig. 3.

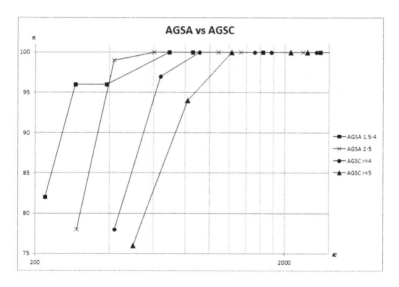

Fig. 3. Operational characteristics of AGSA and AGSC

As it follows from Table 3 and Fig. 3 the randomized algorithm has signifi-cant advantage over opponent with invariable parameter. AGSA accelerates the search in all the cases and can be faster up to two times. Moreover, as opposed to AGSF it is free from the necessity to guess appropriate probability of ran-domizing.

The second experiment with algorithms mentioned above was conducted for 100 multiextremal 3-dimensional functions from the test class GKLS [23] with tuned complexity. The test functions defined in the box P_3 with $a_i = -1, b_i = 1, 1 \leq i \leq 3$, were generated with the following parameters of the class:

- number of local minima - 50;
- radius of the attraction region of the global minimizer - 0.2;
- distance from the global minimizer to the vertex of the paraboloid - 0.5.

Some of the most interesting results can be found in Table 4.
The line Speedup contains the acceleration coefficients (18).
The top parts of operational characteristics built according to data from Table 4 are shown in Fig. 4.

Table 4. GKLS experiments

ϵ		0.05	0.03	0.2	0.01	0.008
AGSC	κ	10334	20797	35939	75971	89820
$r = 4$	π	40	89	100	100	100
AGSC	κ	14264	32912	60497	129524	153115
$r = 5$	π	36	99	100	100	100
AGSF $p = 0.3$	κ	7994	15990	29692	77506	97785
$r_1 = 2, r_2 = 5$	π	44	96	99	100	100
AGSA	κ	5026	9721	16516	40097	54046
$r_1 = 1.5, r_2 = 4$	π	64	92	99	100	100
AGSA	κ	7797	14748	27973	68842	92112
$r_1 = 2, r_2 = 5$	π	65	92	100	100	100
Speedup		1.83	2.23	2.16	1.88	1.66

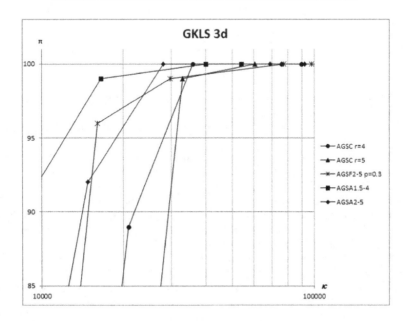

Fig. 4. GKLS experiment

As well as in the 2-dimensional case, the algorithm AGSA with randomization on the base of the game theory model demonstrates the best results. It accelerates the global optimum search in comparison with the method AGSC with invariable parameter up to 2.3 times for different accuracies and achieves the global solution sought for all the test functions faster than its rivals.

5   Conclusion

In this paper the nested optimization scheme in combination with the core information-statistical algorithm of global search is considered. For accelerating the optimization two techniques of algorithm's parameter randomization are studied. These techniques allow balancing rapid convergence and reliability of the method which depend contradictorily on its parameter. Efficiency of these techniques are estimated in a representative experiment on two sets of complicated multiextremal functions of different dimensions taken from well-known test classes widely used for testing global optimization algorithms. The results of the experiment demonstrate the randomization techniques allow achieving significant acceleration in comparison with the core algorithm with invariable parameter. The most efficient technique is the random choice of the parameter realized as an optimal mixed strategy in a game theory model.

As a further way to continue investigations in this direction, the comparison and combination with acceleration techniques on the base of other approaches, for example, local tuning [15, 24, 32], monotonous transformations [19, 20] and multiple Lipschitz constants [33], could be interesting. Moreover, the development of parallel versions of the considered algorithms can be promising for solving optimization problems of high dimensions.

References

1. Horst, R., Pardalos, P.M.: Handbook of Global Optimization. Kluwer Academic Publishers, Dordrecht (1995)
2. Pintér, J.D.: Global Optimization in Action. Kluwer Academic Publishers, Dordrecht (1996)
3. Mockus, J., Eddy, W., Mockus, A., Mockus, L., Reklaitis, G.: Bayesian Heuristic Approach to Discrete and Global Optimization. Kluwer Academic Publishers, Dordrecht (1996)
4. Strongin, R.G., Sergeyev, Y.D.: Global Optimization with Non-convex Constraints: Sequential and Parallel Algorithms. Kluwer Academic Publishers, Dordrecht (2000)
5. Zhigljavsky, A.A., Žilinskas, A.: Stochastic Global Optimization. Springer, New York (2008). https://doi.org/10.1007/978-0-387-74740-8
6. Paulavicius, R., Žilinskas, J.: Simplicial Global Optimization. Springer, New York (2014). https://doi.org/10.1007/978-1-4614-9093-7
7. Sergeyev, Y.D., Kvasov, D.E.: Deterministic Global Optimization: An Introduction to the Diagonal Approach. Springer, New York (2017). https://doi.org/10.1007/978-1-4939-7199-2

8. Sergeyev, Y.D., Strongin, R.G., Lera, D.: Introduction to Global Optimization Exploiting Space-Filling Curves. Springer, New York (2013). https://doi.org/10.1007/978-1-4614-8042-6

9. Carr, C.R., Howe, C.W.: Quantitative Decision Procedures in Management and Economic: Deterministic Theory and Applications. McGraw-Hill, New York (1964)

10. Piyavskij, S.A.: An algorithm for finding the absolute extremum of a function. Comput. Math. Math. Phys. **12**, 57–67 (1972)

11. Evtushenko, Yu.G.: Numerical Optimization Techniques. Translation Series in Mathematics and Engineering. Optimization Software Inc., Publication Division, New York (1985)

12. Grishagin, V.A., Strongin, R.G.: Optimization of multiextremal functions subject to monotonically unimodal constraints. Eng. Cybern. **22**, 117–122 (1984)

13. Shi, L., Ólafsson, S.: Nested partitions method for global optimization. Oper. Res. **48**, 390–407 (2000)

14. van Dam, E.R., Husslage, B., Hertog, D.: One-dimensional nested maximin designs. J. Glob. Opt. **46**, 287–306 (2010)

15. Gergel, V.P., Grishagin, V.A., Israfilov, R.A.: Local tuning in nested scheme of global optimization. Procedia Comput. Sci. **51**, 865–874 (2015)

16. Gergel, V., Goryachih, A.: Multidimensional global optimization using numerical estimates of objective function derivatives. Optim. Methods Softw. 1–21 (2019)

17. Grishagin, V.A., Sergeyev, Y.D., Strongin, R.G.: Parallel characteristical algorithms for solving problems of global optimization. J. Global Optim. **10**(2), 185–206 (1997)

18. Grishagin, V.A., Israfilov, R.A.: Multidimensional constrained global optimization in domains with computable boundaries. In: CEUR Workshop Proceedings, vol. 1513, pp. 75–84 (2015)

19. Grishagin, V.A., Israfilov, R.A.: Global search acceleration in the nested optimization scheme. In: AIP Conference Proceedings, vol. 1738, p. 400010 (2016)

20. Grishagin, V., Israfilov, R., Sergeyev, Y.: Comparative efficiency of dimensionality reduction schemes in global optimization. In: AIP Conference Proceedings, vol. 1776, p. 060011 (2016)

21. Grishagin, V., Israfilov, R., Sergeyev, Y.: Convergence conditions and numerical comparison of global optimization methods based on dimensionality reduction schemes. Appl. Math. Comput. **318**, 270–280 (2018)

22. Grishagin, V.A.: Operating characteristics of some global search algorithms. Probl. Stat. Optim. **7**, 198–206 (1978). Zinatne, Riga. (in Russian)

23. Gaviano, M., Kvasov, D.E., Lera, D., Sergeyev, Y.D.: Algorithm 829: software for generation of classes of test functions with known local and global minima for global optimization. ACM Trans. Math. Softw. **29**(4), 469–480 (2003)

24. Sergeyev, Y.D., Mukhametzhanov, M.S., Kvasov, D.E., Lera, D.: Derivative-free local tuning and local improvement techniques embedded in the univariate global optimization. J. Optim. Theory Appl. **171**(1), 186–208 (2016)

25. Strongin, R.G., Markin, D.L.: Minimization of multiextremal functions with non-convex constraints. Cybernetics **22**, 486–493 (1986)

26. Shubert, B.O.: A sequential method seeking the global maximum of a function. SIAM J. Numer. Anal. **9**(3), 379–388 (1972)

27. Kushner, H.J.: A new method of locating the maximum point of an arbitrary multipeak curve in the presence of noise. Trans. ASME, Ser. D. J. Basic Eng. **86**, 97–106 (1964)

28. Locatelli, M.: Bayesian algorithms for one-dimensional global optimization. J. Global Optim. **1**, 57–76 (1997)

29. Žilinskas, A.: Axiomatic characterization of a global optimization algorithm and investigation of its search strategy. Oper. Res. Lett. **4**, 35–39 (1985)
30. De Groot, M.: Optimal Statistical Decisions. McGraw-Hill, New York (1970)
31. Sergeyev, Y.D., Kvasov, D.E., Mukhametzhanov, M.S.: On the efficiency of nature-inspired metaheuristics in expensive global optimization with limited budget. Sci. Rep. **8**(453), 1–9 (2018)
32. Sergeyev, Y.D., Nasso, M.-C., Mukhametzhanov, M.S., Kvasov, D.E.: Novel local tuning techniques for speeding up one-dimensional algorithms in expensive global optimization using Lipschitz derivatives. J. Comput. Appl. Math. **383**, 113134 (2021)
33. Paulavičius, R., Sergeyev, Y.D., Kvasov, D.E., Žilinskas, J.: Globally-biased BIRECT algorithm with local accelerators for expensive global optimization. Expert Syst. Appl. **144**, 113052 (2020)

Modeling Vertical Migrations of Zooplankton Based on Maximizing Fitness

Oleg Kuzenkov, Elena Ryabova$^{(\boxtimes)}$, Amparo Garcia,
and Anton Degtyarev

Lobachevsky State University, Gagarin Avenue 23, Nizhny Novgorod 603950, Russia
{oleg.kuzenkov,elena.ryabova}@itmm.unn.ru

Abstract. The purpose of the work is to calculate the evolutionarily stable strategy of zooplankton diel vertical migrations from known data of the environment using principles of evolutionary optimality and selection.

At the first stage of the research, the fitness function is identified using artificial neural network technologies. The training sample is formed based on empirical observations. It includes pairwise comparison results of the selective advantages of a certain set of species. Key parameters of each strategy are calculated: energy gain from ingested food, metabolic losses, energy costs on movement, population losses from predation and unfavorable living conditions. The problem of finding coefficients of the fitness function is reduced to a classification problem. The single-layer neural network is built to solve this problem. The use of this technology allows one to construct the fitness function in the form of a linear convolution of key parameters with identified coefficients.

At the second stage, an evolutionarily stable strategy of the zooplankton behavior is found by maximizing the identified fitness function. The maximization problem is solved using optimal control methods. A feature of this work is the use of piecewise linear approximations of environmental factors: the distribution of food and predator depending on the depth.

As a result of the study, mathematical and software tools have been created for modeling and analyzing the hereditary behavior of living organisms in an aquatic ecosystem. Mathematical modeling of diel vertical migrations of zooplankton in Saanich Bay has been carried out.

Keywords: Diel vertical migrations of zooplankton · Fitness function · Ranking order · Machine-learned ranking · Pattern recognition · Optimal control

1 Introduction

The phenomena of daily recurring vertical migrations of zooplankton were discovered more than two hundred years ago [1]. The study of the marine zooplankton's behavior is of great importance due to zooplancton is a key link in the food

© Springer Nature Switzerland AG 2021
D. Balandin et al. (Eds.): MMST 2020, CCIS 1413, pp. 247–259, 2021.
https://doi.org/10.1007/978-3-030-78759-2_21

chain. It plays a decisive role in the aquatic ecosystem; its diel migrations represent one of the most significant synchronous movements of biomass on earth. As a result, they affect carbon exchange and the climate of the planet [2–5]. In this regard, the problem of mathematical modeling of zooplankton's diel vertical migrations is of great importance [6–12].

Currently, Darwin's idea "survival of the fittest" is effectively used for modeling biological processes [13,14]. It is possible to predict the results of evolution and to study the direction of changes in ecological systems comparing fitness of different biological species. Maximizing the fitness function provides the possibility to identify evolutionarily stable hereditary behavioral strategies (i.e. strategies that persist in the community against the appearance of possible mutations [15]). In particular, the use of the fitness concept for modeling diel migrations of zooplankton provides the opportunity to explain the quantitative characteristics of the behavior and its dependence on the age of an individual [16–18]. In this case, the main difficulty is the identification of the fitness function and its parameters.

There is a general approach to solving this problem based on studying the dynamics of a population distribution over the space of hereditary elements. This approach was proposed in [19] and was further developed in a series of works [20–22]. It was shown that on the set of hereditary elements it is possible to introduce a partial ranking order reflecting selective advantages by analyzing the long-term dynamics of the corresponding numbers of individuals [23]. The fitness function is introduced as a comparison function expressing the given ranking order. Then the problem of identifying the fitness function is reduced to expressing this function through the known hereditary features of elements.

In [24], the methodology for deriving the mathematical expression of the fitness function was developed for wide classes of population models, taking into account age heterogeneity. However, the parameters and coefficients of the model cannot quite often be measured empirically, and by themselves presuppose identification making the restoration of the fitness function much more difficult. Therefore, it seems interesting to construct the fitness function directly on the basis of the known population dynamics. In this case, the problem of restoring the fitness function is a special case of the well-known ranking problem [25]. For its solution, there is a wide arsenal of computer methods, in particular, machine learning methods (learning-to-rank) [26–32]. In [33,34], the problem of ranking hereditary elements and identifying the corresponding fitness function was reduced to the problem of classification - dividing ordered pairs of elements into two classes: "the first element is better than the second" and "the second element is better than the first".

In this work, this technique is used to identify the fitness function of diel vertical migrations of zooplankton. Parameters of the fitness function are identified on the basis of empirical observations. A feature of this work is the use of piecewise linear approximations of the distribution of food and predator depending on the depth of immersion. An evolutionarily stable behavior strategy is found by maximizing the identified fitness function using optimal control methods. As a result, mathematical modeling of diel vertical migrations of zooplankton in Saanich Bay is carried out.

2   Materials and Methods

The present study is based on the following methodology for comparing the selective advantages of hereditary elements (behavior strategies) [24]. Let some compact metric space V of hereditary elements v be given. For example, such elements v can be continuous functions. Each element v at each moment of time t is assigned a certain number $\rho(v, t)$ (indicator of presence), which numerically characterizes the presence of v in the community at time t. The indicator of presence satisfies the following requirements: it is zero when the element is not present in the community; it is strictly larger than zero when the element is presented to the community; the indicator is continuously dependent on time; its tendency to zero corresponds to the loss (extinction, disappearance) of this element in the community. This indicator can be the number, biomass of the subpopulation with a given hereditary element, the density of distribution of the population in the space of hereditary elements, etc.

Using the introduced indicator of presence, the selective advantages of various hereditary elements are compared with each other, namely, it is considered that the element v is better than the element w if

$$\lim_{t \to \infty} \frac{\rho(w, t)}{\rho(v, t)} = 0. \tag{1}$$

In the case when the presence indicator is uniformly above bounded (the community size is uniformly above bounded), the limit (1) means that the element v displaces the element w from the community over time. Thus, a partial order of selective advantages is given on the set V.

It is assumed that the introduced order can be expressed using the comparison functional $J(v)$, that is, there is a functional that satisfies the condition $J(v) > J(w)$ if and only if v is better than w. Then the functional J is a fitness function reflecting the selective advantages of hereditary elements.

If the change of the presence indicator in time is uniquely determined by a finite set $M(v) = (M_1(v), \ldots, M_n(v))$ of key hereditary parameters (features) of the element v, then the functional J will be a function of these parameters: $J(v) = J(M(v))$. If this function is sufficiently smooth, then it is expedient to use Taylor's expansions for its approximation. The simplest approximation is a linear convolution of key parameters

$$J(M) = \sum_{i=1}^{n} \lambda_i M_i.$$

Here, the weights λ_i reflect the impact of each key parameter on overall fitness. The problem of identifying the fitness function is reduced to determining the values of the convolution coefficients.

It is obvious that the representation of the function J in the form of a linear combination of key parameters is not always possible. In the case when the linear approximation problem is unsolvable, it is necessary to use a higher order approximation - second, third, etc. But in this case, too, the problem is reduced to finding the coefficients of the corresponding Taylor approximation [25, 34]. Here we consider the simplest case of linear approximation, but the developed approach can be successfully applied to higher order approximations [25, 34].

If it is known that the element v is better than the element w (from the analysis of the dynamics of the presence indicator), then the inequality $J(M(v)) > J(M(w))$ should be fulfilled, respectively, the coefficients λ_i should satisfy the inequality

$$\sum_{i=1}^{n} \lambda_i M_i(v) > \sum_{i=1}^{n} \lambda_i M_i(w).$$

Knowing the results of comparing hereditary elements from a certain finite set, one can build a system of linear inequalities with respect to the convolution coefficients, which can be solved using linear programming methods [17].

Nevertheless, identification of these coefficients is also possible based on classification methods [33, 34]. Let us associate an ordered pair of elements (v, w) with a point $M(v) - M(w)$, a pair (w, v) with a point $M(w) - M(v)$ in the n-dimensional space of key parameters. Then the hyperplane

$$\sum_{i=1}^{n} \lambda_i M_i = 0$$

should separate these points from each other. A certain set of pairs of hereditary elements with known comparison results defines in a n-dimensional space two sets of points that must lie on opposite sides of this hyperplane. Thus, the problem of finding the convolution coefficients is reduced to finding the components of the normal to the separating hyperplane. This is the classification problem, for the solution of which there is a sufficient arsenal of well-proven methods [26]. For example, the separating hyperplane can be constructed using the Fisher determinant [35]. The classification problem is solved quite simply by the nearest neighbors method, but this method has limited application here, since it does not always allow one to find the coefficients of the separating hyperplane. One of the promising methods for solving this problem is the construction of a learning neural network [36, 37].

The formulated problem is also a special case of the pattern recognition problem [33, 34]. But in contrast to classical problems of this type, here it is not a simple assignment of an element to one of the two classes, but a comparison of elements according to the principle "better or worse". Such a comparison is equivalent to recognizing the belonging of ordered pairs of elements "first, second" to one of two classes: "the first is better than the second" or "the first is worse than the second."

The use of machine learning methods is more preferable than the use of traditional linear programming methods [17]. Classical methods are extremely sensitive to the accuracy of the values of key parameters for a set of elements from the training set. A small inaccuracy in the values of key parameters can lead to the incompatibility of the system of linear inequalities. To obtain an acceptable estimate of the convolution coefficients, it is necessary to use a sufficiently large training set (hundreds of elements), and this leads to technical difficulties in resolving a large system of inequalities. Adding new elements to the training set may result in the need for software updates. These problems are absent in the case of using machine learning methods.

As the experience of using various methods shows [34], the greatest effect can be obtained by using neural networks to solve the set problem. The neural networks technology provides greater flexibility of the algorithm with regard to expanding the training set, adding new experimental results of pair comparison. The use of neural networks provides a lower error rate compared to the nearest neighbors method.

It is known from the results of numerous studies that the main environmental factors affecting the behavior of zooplankton are: the degree of saturation of the water layer (with a vertical coordinate x) with food (phytoplankton) $E_0(x)$, metabolic costs $E_2(x)$ for maintaining viability in the water layer x (depends on the temperature of the layer), the number of predators (fish) $S_x(x)$ in the water layer x, the predator activity $S_t(t)$ depending on the time of day t, the presence of unfavorable factors $G(x)$ in the water layer, such as temperature, hydrogen sulfide concentration, etc. [1,6]. All of these factors are mathematically represented as functions of vertical coordinate or time.

Let us introduce a coordinate system so that $x = 0$ coincides with the water surface; $x = -D$ is the level of the lethal hydrogen sulfide concentration (maximum immersion depth); $x = -C$ is the level, below which there are neither predators feeding on zooplankton, nor phytoplankton, which feeds on zooplankton (D, C – positive constants, $C < D$). Let t be a time of day ranging from 0 to 1, with 0 being noon, $1/2$ – midnight, 1 – next noon.

We take the following approximations of external factors:

$$E_0 = \begin{cases} \sigma_1(x+C), & x > -C, \\ 0, & x < -C; \end{cases}$$

$$S_x = \begin{cases} \sigma_2(x+C), & x > -C, \\ 0, & x < -C; \end{cases}$$

$S_t = \cos 2\pi t + \epsilon \cos 6\pi t + 1; \; S = S_x \cdot S_t; \; G = \delta(x + D/2)^2; \; E_2 = \sigma_3(x + D)$.

In addition, it is assumed that the metabolic costs of zooplankton vertical migrations are proportional to the kinetic energy of movement, which in turn is proportional to the square of the speed: $E_1 = \dot{x}^2$.

On the one hand, the introduced functions E_0, E_1, E_2, S, G represent a good approximation to the actually observed data, on the other hand, their relative simplicity allows us to investigate and solve the optimization problem analytically.

Figures 1, 2, 3 and 4 show the graphs of the functions $E_0(x), G(x), S_x(x), S_t(t)$ at $D = 120$, $C = 50$, $\epsilon = -0.013$, $\sigma_1 = 0.018$, $\sigma_2 = 1.8$, $\sigma_3 = 1$, (which corresponds to the data of empirical observations $[1,6,18]$).

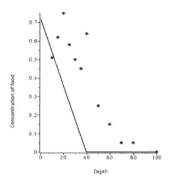

Fig. 1. Amount of food $(E_0(x))$. Dots show the empirical data.

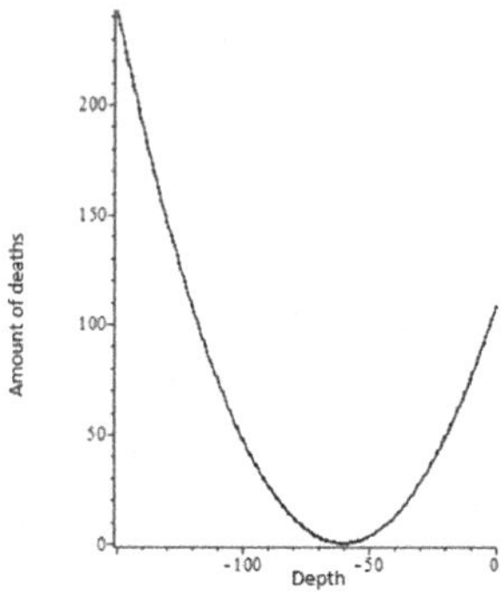

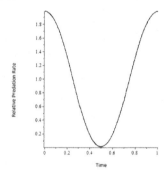

Fig. 2. Additional mortality caused by approaching habitat boundaries $(G(x))$.

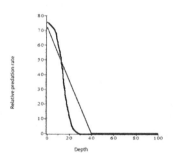

Fig. 3. Mortality due to predation $S_x(x)$. The curve line shows the given observations and the line segments represent its approximation.

Fig. 4. Number of attacks in time $(S_t(t))$.

3   Results

3.1   Fitness Identification

The linear approximations of the fitness function were built using a neural network.

Let $x(t)$ be the hereditary strategy of the zooplankton behavior, the depth of immersion depending on the time of day. It is obvious that the function $x(t)$ must be continuous periodic with a period $T = 1$ (one day). This implies the condition $x(0) = x(1)$. It is also assumed that this function is smooth.

It is possible to calculate the key parameters of the behavioral strategy v on the base of known functions of external factors

$$M_1(v) = \int_0^1 E_0(x(t))\, dt, \quad M_2(v) = -\int_0^1 S_x(x(t))S_t(t)\, dt,$$

$$M_3(v) = -\int_0^1 E_1(x(t))\, dt = -\int_0^1 \dot{x}^2\, dt,$$

$$M_4(v) = -\int_0^1 G(x(t))\, dt, \quad M_5(v) = -\int_0^1 E_2(x(t))\, dt$$

and the corresponding vector $M(v) = (M_1(v), M_2(v), M_3(v), M_4(v), M_5(v))$.

It is assumed that the fitness function depends on these parameters linearly as follows

$$J(v) = \alpha M_1(v) + \gamma M_2(v) + \beta M_3(v) + \delta M_4(v) + \xi M_5(v)$$

or

$$J(v) = \int_0^1 (\alpha E_0 - \beta E_1 - \gamma S - \delta G - \xi E_2)dt. \qquad (2)$$

Weighting coefficients $\alpha, \gamma, \beta, \delta, \xi$ determine the impact of each factor on overall fitness. The problem is to find the values of these coefficients.

The linear form of the fitness function corresponds to the energy balance equation discussed in [38]. This equation assumes that the population reproductive effect consists of the energy gain from food, minus the energy costs for vertical movements, population losses as a result of predation, and losses due to unfavorable living conditions.

To solve this problem, it is necessary to use information about known strategies of behavior. We can compare strategies v and w with each other, if we know the long-term dynamics of corresponding indicators $\rho(v, t)$ and $\rho(w, t)$. Then we can use the described above technology to estimate the coefficients $\alpha, \gamma, \beta, \delta, \xi$ on the base of comparison results for a certain set of pairs.

To solve this problem, a single-layer neural network was built, which allows us to recognize pairs of hereditary strategies by their belonging to two classes - "the first strategy is better than the second" or "the second strategy is better than the first".

This mathematically corresponds to constructing a hyperplane in a five-dimensional space separating two sets of points.

The coordinates of the normal of the constructed hyperplane correspond to the values of the required coefficients α, γ, β, δ, ξ.

For the computer solution using neural network technologies, the following standard free software was used: Scikit-learn machine learning library for the Python programming language, Pandas software library in Python for data processing and analysis.

The training sample was built taking into account the empirical results of observing the behavior of zooplankton [18,39]. It contains comparing results for 202 strategies or 2031 pairs.

The training sample was divided at a percentage of 70% for training by 30% for testing using the train_test_split module from the sklearn.model_selection library. The quality of training was assessed using the Logloss metric. The learning error in this metric is 9.99e−16. The second check method was also used, using the cross_val_score function from the sklearn.model_selection library. The recognition is performed with an accuracy of 96.3%.

Figure 5 shows a visualization of the solution to the corresponding classification problem.

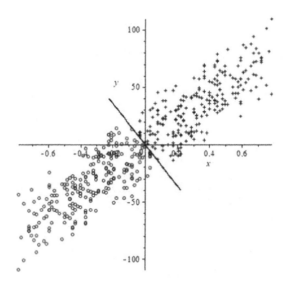

Fig. 5. Solution of the linear classification problem for two classes of pairs of strategies.

Here the projections of the points of the training sample are shown. They correspond to different pairs of strategies onto the plane of two key parameters that have the meaning of food consumed per day – M_1 and daily losses from predators – M_2. The projections have coordinates $(M_1(v) − M_1(w), M_2(v) − M_2(w))$. The crosses mark the points corresponding to the pairs (v, w) for which v is better than w; the circles mark the points for which v is worse than w. The straight line corresponds to the intersection of the separating hyperplane and the plane of the parameters M_1 and M_2. The graph shows that the hyperplane accurately separates two classes of points from each other.

Found values of fitness coefficients are $\alpha = 344.444$, $\beta = 3.25 \cdot 10^{-5}$, $\gamma = 1.461$, $\delta = 0.03$, $\xi = 2.24$.

3.2   Optimization Problem Solution

The problem of constructing the evolutionarily stable strategy for zooplankton was solved as an optimal control problem [16,22,40–42] by maximizing the fitness function (2).

Let us introduce the notation

$$u(t) = \dot{x}(t)$$

then the function u can be regarded as a control.

The conjugate system and transversality conditions have the following form [40]

$$\dot{\psi} = \begin{cases} (\alpha\sigma_1 - \gamma\sigma_2(\cos 2\pi t + \epsilon \cos 6\pi t + 1) - 2\delta(x + D/2) - \xi\sigma_3, & x > -C, \\ -2\delta(x + D/2) - \xi\sigma_3, & x < -C; \end{cases}$$

$$\psi(0) = \psi(1).$$

According to the minimum principle [40], the Hamilton function

$$H_\tau[u] = \psi u + \beta u^2$$

attains its minimum at the optimal control $u(\tau)$ for almost all times τ. Hence it follows that the optimal strategy $x(t)$ of zooplankton behavior should satisfy the following conditions

$$\ddot{x} - \tfrac{\delta}{\beta}x = \tfrac{1}{2\beta}(\gamma\sigma_2(\cos 2\pi t + \epsilon\cos 6\pi t + 1) - \alpha\sigma_1 + \delta D + \xi\sigma_3), \ x > -C;$$
$$\ddot{x} - \tfrac{\delta}{\beta}x = \tfrac{1}{2\beta}(\delta D + \xi\sigma_3), \ x < -C.$$

Note that the functional (2) is symmetric with respect to the replacement of the variable t by $\tau = 1 - t$. Therefore, the solution in the interval $0 \le t \le 1$ must be symmetric with respect to the time instant $t = 1/2$ and satisfy the condition $\dot{x}(1) = -\dot{x}(0)$. Taking into account the periodicity of the solution, we conclude that $\dot{x}(1) = \dot{x}(0) = 0$.

Then the optimal solution $x(t)$ is a continuous connection of functions

$$x = C_1 \cosh\left(\sqrt{\tfrac{\delta}{\beta}}(t - \tfrac{1}{2})\right) - \tfrac{\gamma\sigma_2}{2}\left(\tfrac{\cos 2\pi t}{4\pi^2\beta+\delta} + \tfrac{\epsilon\cos 6\pi t}{32\pi^2\beta+\delta}\right) - \tfrac{\gamma\sigma_2-\alpha\sigma_1+\delta D+\xi\sigma_3}{2\delta}, \ x > -C;$$

$$x = C_2 \cosh\left(\sqrt{\tfrac{\delta}{\beta}}t\right) - \tfrac{\delta D+\xi\sigma_3}{2\delta}, \ x < -C, t < 1/2;$$

$$x = C_2 \cosh\left(\sqrt{\tfrac{\delta}{\beta}}(1 - t)\right) - \tfrac{\delta D+\xi\sigma_3}{2\delta}, \ x < -C, t > 1/2.$$

The constants C_1 and C_2 were calculated numerically to ensure a continuous connection. One can calculate the value of the constant C_2 with a fixed arbitrary constant C_1, at which two functions are continuously connected. Then one can choose such C_1, at which functional (2) reaches its maximum. The standard Maple 17 package was used to solve the problem numerically.

Figure 6 shows the calculated trajectory of zooplankton movement in comparison with the empirically observed strategy of vertical movement of zooplankton in Saanich Bay [39]. Found constants are $C_1 = -0.003$, $C_2 = 0.054$.

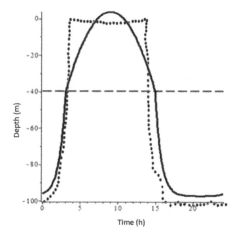

Fig. 6. Comparison with experimental data obtained on 01.04.2010 from Saanich. The dotted line indicates the path most likely followed by zooplankton, and the continuous line is the line obtained by our model with $D=120$, $C=50$, $\alpha=344.444$, $\beta=3.25\cdot10^{-5}$, $\gamma=1.461$, $\delta=0.03$, $\xi=2.24$, $\epsilon=-0.013$, $\sigma_1=0.018$, $\sigma_2=1.8$, $\sigma_3=1$ and constants $C_1=-0.003$, $C_2=0.054$.

4   Summary

This study continues a series of works by the authors devoted to modeling the behavior of a zooplankton population using the principles of evolutionary optimality and selection. It is shown how artificial neural networks can be used to identify the fitness function of living organisms. The fitness function is built on the basis of pairwise comparison of the selective advantages of a certain set of species. The problem of finding the coefficients of the fitness function is reduced to the problem of classification. The parameters of the fitness function are identified on the basis of empirical observations.

Mathematical and software tools have been created for modeling and analyzing the hereditary behavior of living organisms in an aquatic ecosystem, determining their evolutionarily stable strategy and predicting changes in the system.

A feature of this work is the use of piecewise linear approximations of the distribution of food and predator depending on the depth of immersion. An evolutionarily stable strategy of zooplankton behavior is found by maximizing the identified fitness function by optimal control methods. As a result, mathematical modeling of diel vertical migrations of zooplankton in Saanich Bay is carried out.

It should be noted that results of work were implemented in the educational process of Lobachevsky State University of Nizhny Novgorod. The results are used within studying of the discipline "Mathematical modeling of selection processes" [43,44]. They are used for the providing final qualification works of bachelors and masters. It provides the close connection of science and education and corresponds to the modern trends of the education modernization [45].

References

1. Clark, C., Mangel, M.: Dynamic State Variable Models in Ecology: Methods and Applications. Oxford University Press, Oxford (2000)
2. Kaiser, M.J., et al.: Marine Ecology: Processes, Systems, and Impacts. Oxford University Press, Oxford (2005)
3. Buesseler, K., et al.: Revisiting carbon flux through the ocean's twilight zone. Science **316**(5824), 567–570 (2007). https://doi.org/10.1126/science.1137959
4. Isla, A., Scharek, R., Latasa, M.: Zooplankton Diel vertical migration and contribution to deep active carbon flux in the NW Mediterranean. J. Mar. Syst. **143**, 86–97 (2015). https://doi.org/10.1016/j.jmarsys.2014.10.017
5. Archibald, K.M., Siegel, D.A., Doney, S.C.: Modeling the impact of zooplankton diel vertical migration on the carbon export flux of the biological pump. Global Biogeochem. Cycles **33**, 181–199 (2019). https://doi.org/10.1029/2018GB005983
6. Fiksen, O., Giske, J.: Vertical distribution and population dynamics of copepods by dynamic optimization. ICESJ Mar Sci. **52**, 483–503 (1995). https://doi.org/10.1016/1054-3139(95)80062-X
7. Ringelberg, J.: Diel Vertical Migration of Zooplankton in Lakes and Oceans. Springer, Dotrecht (2010). https://doi.org/10.1007/978-90-481-3093-1
8. Morozov, A., Arashkevich, E.: Towards a correct description of zooplankton feeding in models: taking into account food-mediated unsynchronized vertical migration. J. Theor. Biol. **262**(2), 346–360 (2009). https://doi.org/10.1016/j.jtbi.2009.09.023
9. Morozov, A., Arashkevich, E., Nikishina, A., Solovyev, K.: Nutrient-rich plankton communities stabilized via predator-prey interactions: revisiting the role of vertical heterogeneity. Math. Med. Biol. **28**(2), 185–215 (2011). https://doi.org/10.1093/imammb/dqq010
10. Arcifa, M.S., Perticarrari, A., Bunioto, T.C., Domingos, A.R., Minto, W.J.: Microcrustaceans and predators: diel migration in a tropical lake and comparison with shallow warm lakes. Limnetica. **35**(2), 281–296 (2016). https://doi.org/10.23818/limn.35.23
11. Hafker, N.S., Meyer, B., Last, K.S., Pond, D.W., Huppe, L., Teschke, M.: Circadian clock involvement in zooplankton diel vertical migration. Curr. Biol. **27**(14), 2194–2201.e3 (2017). https://doi.org/10.1016/j.cub.2017.06.025
12. Guerra, D., Schroeder, K., Borghini, M., et al.: Zooplankton diel vertical migration in the Corsica Channel (North-Western Mediterranean Sea) Detected by a Moored Acoustic Doppler Current Profiler. Ocean Sci. **15**(3), 631–649 (2019). https://doi.org/10.5194/os-15-631-2019
13. Birch, J.: Natural selection and the maximization of fitness. Biol. Rev. **91**(3), 712–727 (2016). https://doi.org/10.1111/brv.12190
14. Gavrilets, S.: Fitness Landscapes and the Origin of Species (MPB-41). Princeton University Press, Princeton (2004)
15. Gabriel, W., Thomas, B.: Vertical migration of zooplankton as an evolutionarily stable strategy. Am. Nat. **132**(2), 199–216 (1988). https://www.jstor.org/stable/2461866
16. Kuzenkov, O., Morozov, A.: Towards the construction of a mathematically rigorous framework for the modelling of evolutionary fitness. Bull. Math. Biol. **81**(11), 4675–4700 (2019). https://doi.org/10.1007/s11538-019-00602-3
17. Sandhu, S.K., Morozov, A., Kuzenkov, O.: Revealing evolutionarily optimal strategies in self-reproducing systems via a new computational approach. Bull. Math. Biol. **81**(11), 4701–4725 (2019). https://doi.org/10.1007/s11538-019-00663-4

18. Morozov, A., Kuzenkov, O., Arashkevich, E.: Modelling optimal behavioral strategies in structured populations using a novel theoretical framework. Sci. Rep. **9**, 15020 (2019). https://doi.org/10.1038/s41598-019-51310-w

19. Gorban, A.: Selection theorem for systems with inheritance. Math. Model. Nat. Phenom. **2**(4), 1–45 (2007). https://doi.org/10.1051/mmnp:2008024

20. Kuzenkov, O., Kuzenkova, G.: Optimal control of self-reproduction systems. J. Comput. Syst. Sci. **51**(4), 500–511 (2012). https://doi.org/10.1134/S1064230712020074

21. Kuzenkov, O., Novozhenin, A.: Optimal control of measure dynamics. Commun. Nonlinear Sci. Numer. Simul. **21**(1–3), 159–171 (2015). https://doi.org/10.1016/j.cnsns.2014.08.024

22. Kuzenkov, O., Ryabova, E.: Variational principle for self-replicating systems. Math. Model. Nat. Phenom. **10**(2), 115–129 (2015). https://doi.org/10.1051/mmnp/201510208

23. Kuzenkov, O.A., Ryabova, E.A.: Limit possibilities of solution of a hereditary control system. Differ. Equ. **51**(4), 523–532 (2015). https://doi.org/10.1134/S0012266115040096

24. Kuzenkov, O., Morozov, A.: Towards the construction of a mathematically rigorous framework for the modelling of evolutionary fitness. Bull. Math. Biol. **81**(11), 4675–4700 (2019). https://doi.org/10.1007/s11538-019-00602-3

25. Kuzenkov, O., Morozov, A., Kuzenkova, G.: Machine learning evaluating evolutionary fitness in complex biological systems. In: 2020 International Joint Conference on Neural Networks (IJCNN) (2020). https://doi.org/10.1109/IJCNN48605.2020.9206653

26. Mohri, M., Rostamizadeh, A., Talwalkar, A.: Foundations of Machine Learning. The MIT Press, Cambridge (2012)

27. Tax, N., Bockting, S., Hiemstra, D.: A cross-benchmark comparison of learning to rank methods. Inf. Process. Manage. **51**(6), 757–772 (2015). https://doi.org/10.1016/j.ipm.2015.07.002

28. Liu, T.-Y.: Learning to rank for information retrieval. Found. Trends Inf. Retr. **3**(6), 225–331 (2009). https://doi.org/10.1007/978-3-642-14267-3

29. Rahangdale, A., Raut, S.: Machine learning methods for ranking. Int. J. Software Eng. Knowl. Eng. **29**(06), 729–761 (2019). https://doi.org/10.1142/S021819401930001X

30. Basalin, P.D., Bezruk, K.V.: Hybrid intellectual decision making support system architecture. Neurocomputers **8**, 26–35 (2012). https://www.elibrary.ru/item.asp?id=17997728. (in Russian)

31. Basalin, P.D., Timofeev, A.E.: Fuzzy models for the functioning of the rule-based hybrid intelligent learning environment. Int. J. Open Inf. Technol. **7**(2), 49–55 (2019). http://injoit.org/index.php/j1/article/view/693

32. Basalin, P.D., Kumagina, E.A., Nejmark, E.A., Timofeev, A.E., Fomina, I.A., Chernyshova, N.N.: Rule-based hybrid intelligent learning environment implementation. Modern Inf. Technol. IT Educ. **14**(1), 256–267 (2018). http://sitito.cs.msu.ru/index.php/SITITO/article/view/360

33. Kuzenkov, O., Kuzenkova, G.: Identification of the fitness function using neural networks. Procedia Comput. Sci. **169**, 692 (2020). https://doi.org/10.1016/j.procs.2020.02.179

34. Kuzenkov, O., Morozov, A., Kuzenkova, G.: Exploring evolutionary fitness in biological systems using machine learning methods. Entropy **23**(1), 35 (2021). https://doi.org/10.3390/e23010035

35. Fisher, R.A.: The use of multiple measurements in taxonomic problems. Ann. Eugen. **7**, 179–188 (1936). https://doi.org/10.1111/j.1469-1809.1936.tb02137.x
36. Chatterjee, C., Roychowdhury, V.P: On self-organizing algorithms and networks for class-separability features. IEEE Trans. Neural Netw. **8**(3), 663–678 (1997). https://doi.org/10.1109/72.572105
37. Demir, G.K., Ozmehmet, K.: Online local learning algorithms for linear discriminant analysis. Pattern Recogn. Lett. **26**(4), 421–431 (2005). https://doi.org/10.1016/j.patrec.2004.08.005
38. Kuzenkov, O., Ryabova E., Garcia Garcia A.: Optimal control with state constraints in the problem of zooplankton's oscillations modeling. In: Proceedings of 2020 International Conference "Stability and Oscillations of Nonlinear Control Systems" (Pyatnitskiy's Conference), 3–5 June 2020, pp. 242–245. V. A. Trapeznikov Institute of Control Sciences of Russian Academy of Sciences, Moscow (2020). https://www.elibrary.ru/item.asp?id=44162370. (in Russian)
39. Ocean Networks Canada, ZAP Data from Saanich Inlet. https://www.oceannetworks.ca/zap-data-saanich-inlet. Accessed 5 Sept 2020
40. Pontryagin, L.S.: Mathematical Theory of Optimal Processes. CRC Press, Boca Raton (2000)
41. Kuzenkov, O., Ryabova, E., Sokolov, M.: Search for robust-optimal periodic migrations of aquatic organisms based on the variational selection principle. In: Proceedings of 2016 International Conference "Stability and Oscillations of Nonlinear Control Systems" (Pyatnitskiy's Conference), 1–3 June 2016) pp. 226–228. V. A. Trapeznikov Institute of Control Sciences of Russian Academy of Sciences, Moscow (2016). https://www.elibrary.ru/item.asp?id=28299619. (in Russian)
42. Kuzenkov, O.: Information technologies of evolutionarily stable behavior recognition. Modern Inf. Technol. IT Educ. **1201**, 250–257 (2020). https://doi.org/10.1007/978-3-030-46895-8_20
43. Kuzenkov, O., Kuzenkova, G., Kiseleva, T.: The use of electronic teaching tools in the modernization of the course "mathematical modeling of selection processes". Educ. Technol. Soc. **21**(1), 435–448 (2018). https://www.elibrary.ru/item.asp?id=32253185. (in Russian)
44. Kuzenkov, O., Kuzenkova, G., Kiseleva, T.: Computer support of training and research projects in the field of mathematical modeling of selection processes. Educ. Technol. Soc. **22**(1), 152–163 (2019). https://www.elibrary.ru/item.asp?id=37037790. (in Russian)
45. Kuzenkov, O., Zakharova, I.: Mathematical programs modernization based on Russian and international standards. Modern Inf. Technol. IT Educ. **14**(1), 233–244 (2018). https://doi.org/10.25559/SITITO.14.201801.233-244

Localization of Pareto-Optimal Set in Multi-objective Minimax Problems

Dmitry Balandin[1]([✉]) [ID], Ruslan Biryukov[1,2] [ID], and Mark Kogan[2] [ID]

[1] Institute of Informational Technologies, Mathematics and Mechanics,
Mathematical Center, Lobachevsky State University of Nizhny Novgorod,
Gagarin Avenue, 23, Nizhny Novgorod 603022, Russia
[2] Department of Mathematics, Architecture and Civil Engineering State University,
Il'yinskaya Street, 65, Nizhny Novgorod 603950, Russia
mkogan@nngasu.ru

Abstract. In this paper, we consider multi-objective minimax problems with criteria being maxima of functionals. We determine a domain in the criteria space containing Pareto optimal points. The upper boundary of this domain corresponds to Pareto suboptimal solutions minimizing maxima of weighted sums of these functionals, while the lower one is computed using the same Pareto suboptimal solutions. This domain allows to evaluate a "proximity" of any solutions of the multi-objective problem to Pareto optimal solutions, which minimize weighted sums of the criteria. The proposed approach is applied to multi-objective control designs for continuous and discrete LTV systems and LTI systems over finite and infinite time horizons, respectively. The criteria used are H_∞ norms with transients for several controlled outputs. Pareto suboptimal controls in such problems turn out to be H_∞ controls with transients for combined outputs. State feedback gains of these controllers are computed in terms of solutions to differential or difference LMIs. Numerical example illustrates the theoretical results.

Keywords: Multi-objective control · Pareto set · H_∞ norm with transients · Differential/Difference linear matrix inequalities

1 Introduction

Real control problems are multi-objective, as a rule. Finding the Pareto set, and hence the Pareto optimal solutions, i.e. unimprovable for all criteria simultaneously, is a complex problem. Traditionally, it is reduced to the single-criterion minimization of the so-called optimal cost function in the form of a certain convolution of the selected criteria. The most convenient convolution is a weighted sum of criteria. Multi-objective minimax problems with maxima of certain functionals as criteria are especially difficult to solve in view of the optimal cost

The work was supported by the Scientific and Education Mathematical Center "Mathematics for Future Technologies" (Project No. 075-02-2020-1483/1).

function to be a weighted sum of the maxima of different functionals. There are only a few multi-objective control problems for which Pareto optimal solutions have been found: linear-quadratic Gaussian controls in [1] and H_2 optimal controls in [2] based on the Q-parametrization of stabilizing controllers for LTI systems on the infinite time interval, as well as generalized H_2 optimal controls in [3,4] for LTV systems on a finite horizon and LTI systems on the infinite horizon. In [5,6], Pareto suboptimal controls were derived for multi-objective problems with N criteria in the form of H_∞ and γ_0 norms, whose relative losses, in comparison with the Pareto optimal ones, do not exceed $1 - \sqrt{N}/N$.

Multi-objective control problems with criteria involving H_∞ norms or H_∞ norms with transients, taking initial conditions into account explicitly, are beyond the scope of more standard design techniques as, for example, those based on Riccati equations. In order to cope with such problems the concept of the mixed H_2/H_∞ norm and the Lyapunov shaping paradigm were introduced in [7–12] to force all Lyapunov matrices used in several constraints to be the same. This is the more important source of conservatism related to the multi-objective control design. This conservatism was demonstrated by [13,14] with using genetic algorithms. The technical restriction of using a single Lyapunov function was to some extent ruled out in the approach based on extended or dilated LMI characterizations for design specifications; see [15,16]. This is obtained at the expense of imposing conservative constraints on the extra instrumental variables. In [17], a certain approach was used, which involved obtaining finite-dimensional Q-approximations of Pareto optimal controllers for the synthesis of a two-objective control. In all these studies, the question remains unanswered to what extent the values of the individual criteria in the closed-loop systems with multi-objective controls, synthesized taking into account the additional constraints or on the basis of approximations, exceed the corresponding values of the criteria for Pareto optimal controls.

To answer this key question it is required either to know Pareto set itself or locate this set inside a domain with certain boundaries. However, there have been no theoretical results and corresponding technique to date that would have allowed to do it in multi-objective control problems with H_∞ criteria. This is exactly the question our paper addresses.

The contribution of the paper is twofold. First, an universal technique for multi-objective minimax problems including those with H_∞ criteria is proposed to compute two-sided boundaries of a domain containing Pareto optimal points. We derive Pareto suboptimal solutions corresponding to the upper boundary of this domain in the criteria space and evaluate quantitatively a "proximity" of these solutions to Pareto optimal ones. Second, in order to implement this technique, we propose a new characterization of H_∞ norm with transients for LTV continuous and discrete time systems over finite horizon in terms of differential and difference LMIs, respectively.

The paper is organized as follows. In Sect. 2, we consider multi-objective minimax problems and demonstrate how to assess two-sided boundaries of the domain in which the Pareto optimal points being the minima of the weighted sum of criteria are located. These boundaries for two-objective minimax problems are upper and lower curves, between which the specified domain is located. Pareto suboptimal solutions are defined to correspond to the upper boundary of the specified domain and an assessment of their suboptimality index is given. In Sect. 3, we revisit the characterizations of H_∞ norm with transients, presented in [18] in terms of differential matrix Riccati equations for LTV continuous time systems on finite horizon and in [19,20] in terms of LMIs for LTI continuous and discrete time systems on the infinite horizon. We provide new characterizations of H_∞ norm with transients in terms of differential/difference LMIs for continuous/discrete LTV systems on finite horizon. In Sect. 4, LMI based Pareto suboptimal controls are synthesized for multi-objective problems involving H_∞ norm with transients design specifications. Illustrative example for two-objective problems is given in Sect. 5. The final conclusions are made in Sect. 6.

2   Two-Sided Boundaries of a Domain Containing Pareto Optimal Points

The problem is to find Pareto optimal solutions in a multi-objective problem with criteria $J_i(\Theta)$, $i = 1, \ldots, N$, each of which is a maximum of some nonnegative function $F_i(\Theta, \omega) \geq 0$ with respect to some variables $\omega \in \Omega$, i.e.

$$J_i(\Theta) = \sup_{\omega \in \Omega} F_i(\Theta, \omega), \quad i = 1, \ldots, N. \tag{1}$$

We remind that solution Θ_P is said to be Pareto optimal if there doesn't exist such a solution Θ that inequalities $J_i(\Theta) \leq J_i(\Theta_P)$, $i = 1, \ldots, N$ hold, with at least one inequality being strict (see, for example [21]). Pareto set is the point set in the N-dimension criterion space corresponding to all Pareto optimal solutions

$$\mathcal{P} = \{J(\Theta_P) = (J_1(\Theta_P), \ldots, J_N(\Theta_P))\}.$$

The most convenient technique of finding Pareto optimal solutions is the weighted sum scalarization, i.e. by solving a single objective problem for the so-called optimal cost function in the form of the weighted sum of criteria

$$J_\alpha(\Theta) = \sum_{i=1}^{N} \alpha_i J_i(\Theta) \quad \forall \alpha \in \mathcal{S},$$

$$\mathcal{S} = \{(\alpha_1, \ldots, \alpha_N) : \alpha_i > 0, \sum_{i=1}^{N} \alpha_i = 1\}.$$

As is well-known, parameters Θ_α minimizing the optimal cost function

$$\min_{\Theta} J_\alpha(\Theta) = J_\alpha(\Theta_\alpha) = \mu(\alpha)$$

are Pareto optimal solutions of the multi-objective problem [21]. Denote the point set in the criteria space corresponding to Θ_α for all $\alpha \in \mathcal{S}$ as follows

$$\mathcal{P}_\mathcal{L} = \{J(\Theta_\alpha) = (J_1(\Theta_\alpha), \ldots, J_N(\Theta_\alpha)) \quad \forall \alpha \in \mathcal{S}\}.$$

Generally speaking, it may not exhaust the whole Pareto set, i.e. $\mathcal{P}_{\mathcal{L}} \subseteq \mathcal{P}$.

It is rather difficult to immediately find the solutions Θ_α of the multi-objective minimax problems, since the optimal cost functions for such problems turn out to be weighted sums of the maxima of different functions. To overcome this difficulty let us estimate the optimal cost function from below, replacing the sum of the weighted maxima by the maximum of the weighted sum

$$
J_\alpha(\Theta) = \sum_{i=1}^{N} \alpha_i \sup_{\omega \in \Omega} F_i(\Theta, \omega) \geq
$$
$$
\sup_{\omega \in \Omega} \sum_{i=1}^{N} \alpha_i F_i(\Theta, \omega) = \sup_{\omega \in \Omega} F_\alpha(\Theta, \omega) = \widehat{J}_\alpha(\Theta).
$$
(2)

Let us call

$$
\widehat{J}_\alpha(\Theta) = \sup_{\omega \in \Omega} \sum_{i=1}^{N} \alpha_i F_i(\Theta, \omega)
$$

the suboptimal cost function, and

$$
\widehat{\Theta}_\alpha = \arg\min_\Theta \widehat{J}_\alpha(\Theta), \widehat{J}_\alpha(\widehat{\Theta}_\alpha) = \mu_-(\alpha) \quad \forall \alpha \in \mathcal{S}
$$
(3)

the Pareto suboptimal solutions of the multi-objective problem. We will show that one can specify the boundaries of the domain in the criteria space, which contains the subset $\mathcal{P}_L$, and thus estimate the suboptimality index of the solutions $\widehat{\Theta}_\alpha$.

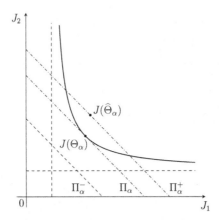

Fig. 1. Pareto optimal point $J(\Theta_\alpha)$ located between hyperplanes

The equality

$$
J_\alpha(\Theta_\alpha) = \sum_{i=1}^{N} \alpha_i J_i(\Theta_\alpha) = \mu(\alpha)
$$

means that in the criteria space, the point $J(\Theta_\alpha)$ belongs to hyperplane Π_α (see Fig. 1) with equation

$$n_\alpha^T J = \mu(\alpha), \quad n_\alpha^T = (\alpha_1, \ldots, \alpha_N).$$

The distance of this hyperplane to the origin is equal to $d_\alpha = |n_\alpha|^{-1}\mu(\alpha)$. Since

$$\mu_+(\alpha) = \sum_{i=1}^{N} \alpha_i J_i(\widehat{\Theta}_\alpha) \geq \sum_{i=1}^{N} \alpha_i J_i(\Theta_\alpha) = \mu(\alpha), \tag{4}$$

Pareto suboptimal solution $\widehat{\Theta}_\alpha$ corresponds to the point $J(\widehat{\Theta}_\alpha)$, which belongs to the hyperplane Π_α^+ with equation $n_\alpha^T J = \mu_+(\alpha)$. This hyperplane is at the distance $d_\alpha^+ = |n_\alpha|^{-1}\mu_+(\alpha) \geq d_\alpha$ from the origin.

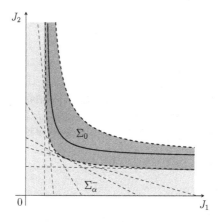

Fig. 2. Domain Σ_0 containing Pareto optimal points

Since

$$\mu(\alpha) = \sum_{i=1}^{N} \alpha_i J_i(\Theta_\alpha) \geq \widehat{J}_\alpha(\Theta_\alpha) \geq \widehat{J}_\alpha(\widehat{\Theta}_\alpha) = \mu_-(\alpha), \tag{5}$$

the distance from point $J(\Theta_\alpha)$ to the origin is not less than $d_\alpha^- = |n_\alpha|^{-1}\mu_-(\alpha)$, which is the distance from the hyperplane Π_α^- with equation $n_\alpha^T J = \mu_-(\alpha)$ to the origin, i.e. $d_\alpha \geq d_\alpha^-$. Thus, Pareto optimal point $J(\Theta_\alpha) \in \Pi_\alpha$ is located between two parallel hyperplanes Π_α^- and Π_α^+.

Let us define the sets

$$\Sigma_\alpha^- = \left\{ (J_1, \ldots, J_N) : \sum_{i=1}^{N} \alpha_i J_i < \mu_-(\alpha), J_i \geq 0 \right\},$$

$$\Sigma_\alpha^+ = \left\{ (J_1, \ldots, J_N) : \sum_{i=1}^{N} \alpha_i J_i \leq \mu_+(\alpha), J_i \geq 0 \right\}, \tag{6}$$

$$\Sigma_- = \bigcup_{\alpha \in \mathcal{S}} \Sigma_\alpha^-, \quad \Sigma_+ = \bigcup_{\alpha \in \mathcal{S}} \Sigma_\alpha^+, \quad \Sigma_0 = \Sigma_+ \backslash \Sigma_-$$

and prove that Pareto optimal points $J(\Theta_\alpha) \in \mathcal{P}_L$ belongs to the set Σ_0 (see Fig. 2).

Since

$$\sum_{i=1}^{N} \alpha_i J_i(\Theta_\alpha) \le \sum_{i=1}^{N} \alpha_i J_i(\widehat{\Theta}_\alpha) = \mu_+(\alpha),$$

we get $J(\Theta_\alpha) \in \Sigma_+$. Now we will show that, for any fixed $\widehat{\alpha} \in \mathcal{S}$, one has $J(\Theta_{\widehat{\alpha}}) \notin \Sigma_-$, i.e. $J(\Theta_{\widehat{\alpha}}) \notin \Sigma_\alpha^-$ for all $\alpha \in \mathcal{S}$. It follows from (5) that $J(\Theta_{\widehat{\alpha}}) \notin \Sigma_{\widehat{\alpha}}^-$.

Let there exist $\alpha \ne \widehat{\alpha}$ such that $J(\Theta_{\widehat{\alpha}}) \in \Sigma_\alpha^-$, i.e. $\sum_{i=1}^{N} \alpha_i J_i(\Theta_{\widehat{\alpha}}) < \mu_-(\alpha)$. Since

$$\mu_-(\alpha) = \widehat{J}_\alpha(\widehat{\Theta}_\alpha) \le \widehat{J}_\alpha(\Theta_\alpha) \le J_\alpha(\Theta_\alpha) = \mu(\alpha),$$

we have

$$\sum_{i=1}^{N} \alpha_i J_i(\Theta_{\widehat{\alpha}}) < \sum_{i=1}^{N} \alpha_i J_i(\Theta_\alpha),$$

i.e. $J_\alpha(\Theta_{\widehat{\alpha}}) < J_\alpha(\Theta_\alpha)$. However, this is contrary to the fact that Θ_α provides the minimum of the optimal cost function $J_\alpha(\Theta)$. Thus, $\Theta_\alpha \in \Sigma_0$ and we arrive at the following statement.

Theorem 2.1. *The set $\mathcal{P}_L$ corresponding to Pareto optimal solutions Θ_α minimizing optimal cost functions $J_\alpha(\Theta) = \sum_{i=1}^{N} \alpha_i \sup_{\omega \in \Omega} F_i(\Theta, \omega)$ for all $\alpha \in \mathcal{S}$ is a subset of Σ_0 defined in (6), (4), (5).*

For two-objective problems the lower and upper boundaries of the domain Σ_0 in the criteria space (J_1, J_2) are envelopes of families of straight lines

$$\alpha J_1 + (1 - \alpha)J_2 = \widehat{J}_\alpha(\widehat{\Theta}_\alpha),$$
$$\alpha J_1 + (1 - \alpha)J_2 = \alpha J_1(\widehat{\Theta}_\alpha) + (1 - \alpha)J_2(\widehat{\Theta}_\alpha).$$

Note that the straight lines $J_1 = \min_\Theta J_1(\Theta)$ and $J_2 = \min_\Theta J_2(\Theta)$ correspond to $\alpha = 1$ and $\alpha = 0$, respectively.

Now it is possible to assess the quality of Pareto suboptimal solutions $\widehat{\Theta}_\alpha$ in relation to Pareto optimal solutions Θ_α. For a quantitative estimate of the proximity between Pareto suboptimal and optimal solutions we introduce the suboptimality index

$$\eta = \max_{\alpha \in \mathcal{S}} \frac{d_\alpha^+ - d_\alpha^-}{d_\alpha^+} = \max_{\alpha \in \mathcal{S}} \frac{\mu_+(\alpha) - \mu_-(\alpha)}{\mu_+(\alpha)}.$$

The suboptimality index is determined by the relative value of the maximum "distance" between the boundaries of the set Σ_0. The closer η to zero, the more accurate the estimate of the Pareto set and the closer to each other the values of the corresponding criteria for Pareto suboptimal and optimal solutions.

3 H_∞ Norm with Transients

In this section, we consider the system performances, which will be chosen as criteria in multi-objective control problems. Let an LTV system be governed by the equation

$$\partial x = A(t)x(t) + B(t)v(t), \quad x(t_0) = x_0$$
$$z(t) = C(t)x(t) + D(t)v(t), \quad t \in [t_0, t_f], \tag{7}$$

where ∂ denotes differential operator for continuous-time systems or shift operator, i.e. $\partial x(t) = x(t+1)$, for discrete-time systems, $x \in \mathbf{R}^{n_x}$ is the state, $v \in \mathbf{R}^{n_v}$ is the disturbance and $z \in \mathbf{R}^{n_z}$ is the controlled output. H_∞ norm with transients of the system (7) on finite horizon $[t_0, t_f]$ from input v to output z under an uncertain initial state for given weighting matrices of initial state $R = R^T > 0$ and terminal state $S^T = S \geq 0$ is defined as

$$\gamma_{\infty,0} = \sup_{x_0, v} \left(\frac{\|z\|^2_{[t_0, t_f]} + x^T(t_f) S x(t_f)}{x_0^T R^{-1} x_0 + \|v\|^2_{[t_0, t_f]}} \right)^{1/2}, \tag{8}$$

where supremum is taken over all initial states $x(t_0) = x_0$ and all disturbances $v \in L_2$ or $v \in l_2$ which are not vanished simultaneously. The following notations

$$\|\xi\|^2_{[t_0, t_f]} = \int_{t_0}^{t_f} |\xi(t)|^2 dt, \quad \|\xi\|^2_{[t_0, t_f]} = \sum_{t=t_0}^{t_f - 1} |\xi(t)|^2$$

are used for continuous- or discrete-time systems, respectively. If the initial state is zero, H_∞ norm with transients becomes the standard H_∞ norm, and if the disturbance is absent, i.e. $v(t) \equiv 0$, H_∞ norm with transients becomes γ_0 norm (see [20] for details) corresponding to the maximal disturbance attenuation level caused by uncertain initial states. When $S = 0$, the terminal state is not taken into account in H_∞ norm with transients.

In [18], it was established that H_∞ norm with transients for LTV continuous-time systems on finite horizon is computed by solving the Riccati differential equation with initial and terminal conditions. The following theorem shows that H_∞ norm with transients for continuous- or discrete-time systems on finite horizon can be calculated by minimizing a linear function under differential or difference LMI constraints.

Theorem 3.1. *Let the inequality*

$$\gamma^2 I - D^T(t) D(t) > 0 \quad \forall t \in [t_0, t_f] \tag{9}$$

be fulfilled for a given γ. H_∞ norm with transients of system (7) satisfies inequality $\gamma_{\infty,0} < \gamma$ if and only if differential LMI

$$\begin{pmatrix} -\dot{Y}(t) + Y(t)A^T(t) + A(t)Y(t) & * & * \\ B^T(t) & -I & * \\ C(t)Y(t) & D(t) & -\gamma^2 I \end{pmatrix} \leq 0 \tag{10}$$

for continuous time $t \in [t_0, t_f]$ or difference LMI

$$\begin{pmatrix} -Y(t+1) & * & * & * \\ Y(t)A^T(t) & -Y(t) & * & * \\ B^T(t) & 0 & -I & * \\ 0 & C(t)Y(t) & D(t) & -\gamma^2 I \end{pmatrix} \leq 0 \tag{11}$$

for discrete time $t = t_0, \ldots, t_f - 1$, equality

$$Y(t_0) = R \tag{12}$$

and LMI

$$\begin{pmatrix} Y(t_f) & * \\ S^{1/2}Y(t_f) & \gamma^2 I \end{pmatrix} > 0 \tag{13}$$

are feasible with respect to $Y(t) > 0$ and $\gamma^2 > 0$.

Proof. We will give a sketch of proof for continuous-time case. The full proof including the discrete-time case is presented in [22]. Firstly, we will show that the inequality $\gamma_{\infty,0} < \gamma$ implies inequalities (10) or (11), (13), and equality (12). Let us define functional

$$\bar{J}(v) = \gamma^2 \left[\|v\|^2_{[t_0, t_f]} + x_0^T R^{-1} x_0 \right] - \|z\|^2_{[t_0, t_f]} \tag{14}$$

on trajectories of the system (7). The inequality $\gamma_{\infty,0} < \gamma$ is equivalent to

$$\bar{J}(v) > x^T(t_f) S x(t_f)$$
$$\forall x_0 \in \mathbf{R}^{n_x}, \ \forall v \in L_2 : x_0^T R^{-1} x_0 + \|v\|^2_{[t_0, t_f]} \neq 0. \tag{15}$$

Consider the minimization problem of the functional (14) with respect to $v(t)$, which is feasible due to (15). Let us introduce Bellman function

$$V(t, x) = \min_v \left\{ \gamma^2 \left[\|v\|^2_{[t_0, t]} + x_0^T R^{-1} x_0 \right] - \|z\|^2_{[t_0, t]} \right\},$$

where $x = x(t)$ is the system state at the moment t. The corresponding Bellman equation has the form

$$\min_{v(t)} \left(-\dot{V} - |z|^2 + \gamma^2 |v|^2 \right) = 0,$$
$$V(t_0, x) = \gamma^2 x^T R^{-1} x. \tag{16}$$

It is not difficult to verify that the solution of this minimization problem for the continuous-time system is a quadratic form $V(t, x) = x^T Q^{-1}(t) x$, where the matrix $Q(t)$ satisfies the Riccati equation

$$\dot{Q} = AQ + QA^T + QC^T CQ + (B^T + D^T CQ)^T \times$$
$$(\gamma^2 I - D^T D)^{-1}(B^T + D^T CQ), \ Q(t_0) = \gamma^{-2} R. \tag{17}$$

From Eq. (16) it follows that inequality

$$\dot{V} + |z|^2 - \gamma^2 |v|^2 \leq 0 \tag{18}$$

is fulfilled for $V(t, x) = x^T Q^{-1}(t) x$ along any trajectory of the system (7) under any disturbances $v(t)$. The last inequality can be rewritten in the form of quadratic inequality $\xi^T M \xi \leq 0$, $\xi^T = (x^T \ v^T)$, where matrix M is negatively semi-definite. Some simple manipulations with inequality $M \leq 0$ and substitution $Q = \gamma^{-2}Y$ imply (10). Note that according to (17) we have $Y(t_0) = R$. Finally, from (15) it follows

$$\bar{J}(v) \geq V(t_f, x(t_f))$$
$$= \gamma^2 x^{\mathrm{T}}(t_f)Y^{-1}(t_f)x(t_f) > x^{\mathrm{T}}(t_f)Sx(t_f)$$

that is equivalent to $\gamma^2 Y^{-1}(t_f) > S$. Subsequent application of Schur's lemma results in LMI (13). Therefore, the necessity part of the theorem is proven.

Now, let LMIs (10) or (11), (13), and equality (12) be feasible with respect to $Y(t) > 0$ and $\gamma^2 > 0$. It is easy to verify that function $V(t, x) = \gamma^2 x^{\mathrm{T}}Y^{-1}(t)x$ satisfies (18) along trajectories of system (7). Integrating or summing these inequalities over interval $[t_0, t_f]$ and taking into account (13), we get

$$\|z\|^2_{[t_0, t_f]} + x^{\mathrm{T}}(t_f)Sx(t_f) < \gamma^2 \left[x_0^{\mathrm{T}}R^{-1}x_0 + \|v\|^2_{[t_0, t_f]} \right]$$

for any disturbance v and initial state x_0. Therefore, $\gamma_{\infty, 0} < \gamma$, which concludes the proof.

For internally stable LTI system (7) with $A(t) \equiv A$, $B(t) \equiv B$, $C(t) \equiv C$, $D(t) \equiv D$, H_∞ norm with transients, standard H_∞ norm, i.e. for zero initial state, and γ_0 norm over infinite horizon are defined as

$$\gamma^s_{\infty, 0} = \sup_{x_0, v} \frac{\|z\|_{[0, \infty)}}{\left(x_0^{\mathrm{T}}R^{-1}x_0 + \|v\|^2_{[0, \infty)} \right)^{1/2}},$$

$$\gamma^s_{\infty} = \sup_{v \neq 0} \frac{\|z\|_{[0, \infty)}}{\|v\|_{[0, \infty)}}, \quad \gamma^s_0 = \sup_{x_0 \neq 0} \frac{\|z\|_{[0, \infty)}}{\left(x_0^{\mathrm{T}}R^{-1}x_0 \right)^{1/2}},$$

where superscript s corresponds to a stationary system.

Theorem 3.2. H_∞ *norm with transients for internally stable LTI system (7) over infinite horizon satisfies inequality $\gamma^s_{\infty, 0} < \gamma$ if and only if LMIs*

$$\begin{pmatrix} YA^{\mathrm{T}} + AY & * & * \\ B^{\mathrm{T}} & -I & * \\ CY & D & -\gamma^2 I \end{pmatrix} < 0, \quad Y > R \tag{19}$$

for the continuous-time system or

$$\begin{pmatrix} -Y & * & * & * \\ YA^{\mathrm{T}} & -Y & * & * \\ B^{\mathrm{T}} & 0 & -I & * \\ 0 & CY & D & -\gamma^2 I \end{pmatrix} < 0, \quad Y > R \tag{20}$$

for the discrete-time system are feasible with respect to Y and γ^2.

The proof of this theorem follows immediately from the papers [19] and [20] for continuous- and discrete-time systems, respectively.

4 Synthesizing Pareto Suboptimal H_∞ Controllers with Transients

The control problem we wish to address is that of designing a linear time-varying state-feedback controller $u = \Theta(t)x$ that reduces all controlled outputs z_i, $i = 1, \ldots, N$ of an LTV system governed by the equation

$$\partial x = A(t)x(t) + B_v(t)v(t) + B_u(t)u(t),$$
$$z_i(t) = C_i(t)x(t) + D_{v\,i}(t)v(t) + D_{u\,i}(t)u(t). \tag{21}$$

More specifically, consider a multi-objective control problem with squared H_∞ norms with transients as criteria

$$J_i(\Theta) = \sup_{x_0,\,v} \frac{\|z_i\|^2_{[t_0,\,t_f]} + x^{\mathrm{T}}(t_f)S_i x(t_f)}{x_0^{\mathrm{T}}R^{-1}x_0 + \|v\|^2_{[t_0,\,t_f]}}, \quad i = 1, \ldots, N.$$

According to (2) the suboptimal cost function for this problem can be written as

$$\widehat{J}_\alpha(\Theta) = \sup_{x_0,\,v} \frac{\|z_\alpha\|^2_{[t_0,\,t_f]} + x^{\mathrm{T}}(t_f)S_\alpha x(t_f)}{x_0^{\mathrm{T}}R^{-1}x_0 + \|v\|^2_{[t_0,\,t_f]}},$$

where

$$z_\alpha(t) = [C_\alpha(t) + D_{u\,\alpha}(t)\Theta(t)]x(t) + D_{v\,\alpha}(t)v(t), \tag{22}$$

$$C_\alpha(t) = \begin{pmatrix} \alpha_1^{1/2}C_1(t) \\ \cdots \\ \alpha_N^{1/2}C_N(t) \end{pmatrix}, D_{u\,\alpha}(t) = \begin{pmatrix} \alpha_1^{1/2}D_{u\,1}(t) \\ \cdots \\ \alpha_N^{1/2}D_{u\,N}(t) \end{pmatrix},$$

$$D_{v\,\alpha}(t) = \begin{pmatrix} \alpha_1^{1/2}D_{v\,1}(t) \\ \cdots \\ \alpha_N^{1/2}D_{v\,N}(t) \end{pmatrix}, \quad S_\alpha = \sum_{i=1}^{N} \alpha_i S_i.$$

This means that $\widehat{J}_\alpha(\Theta)$ is the squared H_∞ norm with transients of system (21) with combined output z_α and terminal state matrix S_α. Therefore, Pareto suboptimal solutions to this multi-objective problem are the corresponding optimal H_∞ controls with transients for all $\alpha \in \mathcal{S}$. To compute state-feedback gains of these control laws one should replace matrices $A(t)$, $B(t)$, $C(t)$ and $D(t)$ in LMI (10) for the continuous-time case or LMI (11) for the discrete-time case by the matrices $A(t) + B_u(t)\Theta(t)$, $B_v(t)$, $C_\alpha(t) + D_{u\,\alpha}(t)\Theta(t)$, $D_{v\,\alpha}(t)$, respectively, and introduce variables $Z(t) = \Theta(t)Y(t)$. Solving the corresponding LMIs with $Y(t_0) = R$ and terminal state matrix S_α, we get $\widehat{\Theta}_\alpha(t) = Z(t)Y^{-1}(t)$. Note that to calculate the state-feedback gain in the continuous-time case a discretization of the differential LMIs is needed. Pareto suboptimal solutions to multi-objective problems with standard H_∞ or γ_0 norms as criteria are computed in similar way under $Y(t_0) = 0$ or $B(t) \equiv 0$, respectively.

Let us consider multi-objective control problems for LTI systems described by the Eq. (21), $u = \Theta x$ with stationary matrices and criteria being squared H_∞ norms with transients

$$J_i(\Theta) = \sup_{x_0,\, v} \frac{\|z_i\|_{[0,\,\infty)}^2}{x_0^\mathrm{T} R^{-1} x_0 + \|v\|_{[0,\,\infty)}^2}, \quad i = 1, \ldots, N.$$

In this case, the suboptimal cost function will be as follows

$$\widehat{J}_\alpha(\Theta) = \sup_{x_0,\, v} \frac{\|z_\alpha\|_{[0,\,\infty)}^2}{x_0^\mathrm{T} R^{-1} x_0 + \|v\|_{[0,\,\infty)}^2}$$

for all $\alpha \in S$, where combined output $z_\alpha(t)$ is defined as in (22) with all stationary matrices. This is H_∞ norm with transients for the combined output $z_\alpha(t)$. Gain matrices $\widehat{\Theta}_\alpha$ of Pareto suboptimal state-feedbacks are computed as $\widehat{\Theta}_\alpha = ZY^{-1}$ by solving LMIs (19) or (20), in which matrices A, B, C and D are replaced by matrices $A + B_u \Theta$, B_v, $C_\alpha + D_{u\,\alpha}\Theta$ and $D_{v\,\alpha}$, respectively, and $Z = \Theta Y$.

5   Illustrative Example: Vibration Isolation

Consider a mechanical two-degree-of-freedom system consisting of an elastic body modeled by elastically linked material points connected to a moving base by means of an isolator. This system is described by the equations

$$\ddot{x}_1 = -2\beta\dot{x}_1 + \beta\dot{x}_2 - 2x_1 + x_2 + v + u,$$
$$\ddot{x}_2 = -\beta(\dot{x}_2 - \dot{x}_1) - x_2 + x_1 + v$$

with nonzero initial states, where x_1 and x_2 are coordinates of the material points with respect to the moving base, u is the active component of the isolator, v is the external disturbance coinciding up to a sign with acceleration of the moving base, $\beta = 0.1$ is the damping parameter. The vibration isolation problem is to find a stationary state-feedback control $u = \theta_1 x_1 + \theta_2 x_2 + \theta_3 \dot{x}_1 + \theta_4 \dot{x}_2$ minimizing in Pareto sense both the deformation of the mechanical system and the force generated by the isolator. Let the controlled outputs be as follows

$$z_1 = (x_1,\, x_2 - x_1)^\mathrm{T}, \qquad z_2 = -x_1 - \beta\dot{x}_1 + u,$$

then the corresponding H_∞ norms with transients over infinite horizon can be chosen as the performance measures of the system.

By using inequalities (19) for $R = I$, we find domain Σ_0 and suboptimality index $\eta = 0.2768$ (Fig. 3). The point A with coordinates $(4.256, 5.582)$, obtained for $\alpha = 0.64$, belongs to the upper boundary of domain Σ_0 and the corresponding state-feedback gain of the Pareto suboptimal controller is $\Theta = (-1.7651,\, 0.7337,\, -2.7555,\, -2.6619)^\mathrm{T}$.

For comparizon we synthesized multi-objective controls by using LMI formulations for each H_∞ norm with transients in the framework of the Lyapunov shaping paradigm. More precisely, state-feedback gains of these controllers were determined when solving the problem inf $J_2(\Theta)$ provided that $J_1(\Theta) < \gamma^2$ with γ

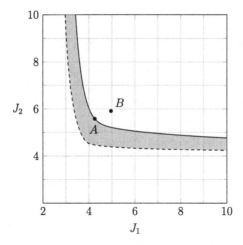

Fig. 3. Localization of Pareto set in the vibration isolation problem

being a parameter. They are computed as $\widetilde{\Theta}_\gamma = Z_\gamma Y_\gamma^{-1}$, where Y_γ and Z_γ are the solutions to the problem inf γ_2^2 subject to two pairs LMIs of the form (19), in one of which matrices A, B, C, D are replaced by matrices $A + B_u\Theta$, B_v, $C_1 + D_{u1}\Theta$, D_{v1}, respectively, and in the other by matrices $A + B_u\Theta$, B_v, $C_2 + D_{u2}\Theta$, D_{v2}, $\gamma = \gamma_2$, and $Z = \Theta Y$. The point B with coordinates $(4.959, 5.913)$ in Fig. 3 corresponds to state-feedback gain $\Theta = (-0.472, 0.252, -1.745, -1.385)^{\mathrm{T}}$ of one from these controllers.

6   Conclusion

The paper deals with multi-objective problems with maxima of several functionals as criteria. The maxima of the weighted sums of these functionals are defined as the suboptimal cost functions and their minima as Pareto suboptimal solutions. We pick the domain in criteria space that contains Pareto optimal points corresponding to minima of weighted sums of criteria. The upper and lower boundaries of this domain are computed by means of the Pareto suboptimal solutions. This allows to evaluate a "proximity" for any suboptimal solutions to the optimal ones. This approach is applied to multi-objective control problems with criteria being H_∞ norms with transients for several controlled outputs. It turned out that Pareto suboptimal controls in these problems are H_∞ controls with transients for combined outputs. The state-feedback gains of these controllers for continuous and discrete LTV systems over finite horizon are characterized in terms of differential or difference LMIs, whereas for LTI systems over infinite horizon, in terms of LMIs.

References

1. Mäkila, P.: On multiple criteria stationary linear quadratic control. IEEE Trans. Autom. Control **34**, 1311–1313 (1989)
2. Khargonekar, P., Rotea, M.: Multiple objective optimal control of linear systems: the quadratic norm case. IEEE Trans. Autom. Control **36**, 14–24 (1991)
3. Balandin, D.V., Kogan, M.M.: Multi-objective generalized H_2 control. Automatica **99**(1), 317–322 (2019)
4. Balandin, D.V., Biryukov, R.S., Kogan, M.M.: Finite-horizon multi-objective generalized H_2 control with transients. Automatica **106**(8), 27–34 (2019)
5. Balandin, D.V., Kogan, M.M.: Pareto suboptimal solutions in control and filtering problems under multiple deterministic and stochastic disturbances. In: Proceedings European Control Conference, Aalborg, pp. 2263–2268 (2016)
6. Balandin, D.V., Kogan, M.M.: Pareto suboptimal controllers in multi-objective disturbance attenuation problems. Automatica **84**(10), 56–61 (2017)
7. Bernstein, D., Haddad, W.: LQG control with an H_∞ performance bound: a Riccati equation approach. IEEE Trans. Autom. Control **34**, 293–305 (1989)
8. Khargonekar, P., Rotea, M.: Mixed H_2/H_∞ control: a convex optimization approach. IEEE Trans. Autom. Control **36**, 821–834 (1991)
9. Zhou, K., Glover, K., Bodenheimer, B., Doyle, J.: Mixed H_2 and H_∞ performance objectives I: robust performance analysis. IEEE Trans. Autom. Control **39**, 1564–1574 (1994)
10. Doyle, J., Zhou, K., Glover, K., Bodenheimer, B.: Mixed H_2 and H_∞ performance objectives II: optimal control. IEEE Trans. Autom. Control **39**, 1575–1587 (1994)
11. Scherer, C., Gahinet, P., Chilali, M.: Multiobjective output-feedback control via LMI optimization. IEEE Trans. Autom. Control. **42**, 896–911 (1997)
12. Chen, X., Zhou, K.: Multi-objective H_2/H_∞ control design. SIAM J. Control Optim. **40**(2), 628–660 (2001)
13. Takahashi, R., Palhares, R., Dutra, D., Goncalves, L.: Estimation of Pareto sets in the mixed H_2/H_∞ control problem. Int. J. Syst. Sci. **35**(1), 55–67 (2004)
14. Molina-Cristóbal, A., Griffin, I., Fleming, P., Owens, D.: Linear matrix inequalities and evolutionary optimization in multiobjective control. Int. J. Syst. Sci. **37**(8), 513–522 (2006)
15. Oliveira, M., Geromel, J., Bernussou, J.: An LMI optimisation approach to multi-objective controller design for discrete-time systems. In: Proceedings IEEE CDC, Arizona, pp. 3611–3616 (1999)
16. Ebihara, Y., Hagiwara, T.: Characterisations for continuous-time control multi-objective controller synthesis. Automatica **40**(8), 2003–2009 (2004)
17. Hindi, H., Hassibi, B., Boyd, S.: Multiobjective H_2/H_∞-optimal control via finite dimensional Q- parametrization and linear matrix inequalities. In: Proceedings 1998 American Control Conference, Philadelphia, pp. 3244–3249 (1998)
18. Khargonekar, P., Nagpal, K., Poolla, K.: H_∞ control with transients. SIAM J. Control Optim. **29**(6), 1373–1393 (1991)
19. Balandin, D.V., Kogan, M.M.: LMI based H_∞- optimal control with transients. Int. J. Control **83**(8), 1664–1673 (2010)
20. Balandin, D.V., Kogan, M.M., Krivdina, L.N., Fedyukov, A.A.: Design of generalized discrete-time H_∞-optimal control over finite and infinite intervals. Autom. Remote Control **75**(1), 1–17 (2014)
21. Ehrgott, M.: Multicriteria Optimization. Springer, Berlin - Heidelberg (2005). https://doi.org/10.1007/3-540-27659-9
22. Balandin, D.V., Biryukov, R.S., Kogan, M.M.: Multi-objective H_∞ controls with transients. Automatica. submitted for publication

Estimating a Set of the States in the Case of an Error in the Measured Output for Controlled System

Alexander A. Fedyukov$^{(\boxtimes)}$

Lobachevsky State University of Nizhni Novgorod, 603950 Nizhny Novgorod, Russia

Abstract. In the problem of state stabilization under constraints on state and control variables, it is assumed that the state of the system is measurable. However, in real situations, the state of the system is measured, as a rule, with an error. Therefore, the question of the possibility of using the obtained controller in this situation remains open. In this article, we study the problem of estimating the set of admissible initial states for a dynamic system, in which the controller obtained in the state feedback control synthesis problem under constraints imposed on state and control variables, will provide stabilization even in the case when the system state is measured with an error. The sufficient conditions are derived in terms of linear matrix inequalities to estimate the set of admissible initial states of a dynamical system. The solution is based on the application of the method of Lyapunov functions and technique of linear matrix inequalities. The key point in the proof of the theorem is the application of the S-procedure being non-defective under two constraints. As an example, the problem of stabilization of an inverted pendulum is considered. Numerical experiments have confirmed the theoretical results.

Keywords: Stabilization · Linear matrix inequalities · State feedback control

1 Introduction

There are different ways of constructing controllers [1–6], including a method based on the use of the technique of linear matrix inequalities [1]. In the problem of state stabilization, it is assumed that the state of the system is measurable and control is constructed in the form of linear state feedback. With the help of modern software (for example, software for engineering calculations MATLAB [7]), we can get the parameters of such a controller. At the same time, a situation is possible when the obtained solution cannot be physically implemented. This is due to the fact that the synthesis of linear control laws based on the linear model of the controlled object can be effectively applied only where the linear

The reported study was funded by RFBR, project number 19-31-90086.

D. Balandin et al. (Eds.): MMST 2020, CCIS 1413, pp. 273–285, 2021.
https://doi.org/10.1007/978-3-030-78759-2_23

model more or less adequately describes the real object, i.e. in a limited region of phase space. Note also that in real operating conditions the system must be in the area of its permissible states. In this regard, it becomes necessary to take into account the limitation on the phase variables of the object and control in the model. The problem of control synthesis under given constraints is complex and relevant at the present time [2,3,8].

In [2,3], the problem of synthesis of state control is considered and solved, which provides stabilization of a dynamic object under constraints on state and control variables. In the phase space, the set of admissible initial states of the system is obtained, at which the controller stabilizes the system. However, in real situations the state of the system is measured, as a rule, with an error. Therefore, the question of the possibility of using the controller obtained in [2,3] remains open in this situation.

In this article, we study the problem of estimating the set of admissible initial states for a dynamic system, in which the controller obtained in the state feedback control synthesis problem under constraints imposed on state and control variables, will also provide stabilization in the case when the state of the system is measured with an error. The sufficient conditions are derived in terms of linear matrix inequalities to estimate the set of admissible initial states of a dynamical system. The solution is based on the application of the method of Lyapunov functions and technique of linear matrix inequalities. The key point in the proof of the theorem is the application of the S-procedure being non-defective under two constraints [9]. As an example, the problem of stabilization of an inverted pendulum is considered. Numerical experiments have confirmed the theoretical results.

2   Preliminary Information

Consider a controlled object

$$\dot{x} = Ax + Bu, \quad x(0) = x_0, \tag{1}$$

$$z_i = C_i x + D_i u, \quad i = 1, 2, ..., N, \tag{2}$$

where $x \in R^n$—state of the system, $u \in R^l$ control, $z_i \in R^{m_i}$—controlled system outputs; A, B, C_i and D_i—given matrices of appropriate sizes.

The problem of stabilizing the object (1) using control in the form of linear state feedback

$$u = Kx, \tag{3}$$

which ensures the asymptotic stability of the closed-loop system (1), (2), (3) and its fulfillment for given values γ_i of the constraints

$$\max_{t \geq 0} |z_i(t)| \leq \gamma_i, \quad i = 1, 2, ..., N, \tag{4}$$

was discussed in [2,3]. Using the technique of linear matrix inequalities and the non-degradation of the S-procedure for quadratic inequalities [10], conditions were formulated on the set of initial states, starting from which the phase trajectories of system (1), closed by control (3), asymptotically approached the zero state and did not go beyond boundaries of the set defined by constraints (4). To solve the control synthesis problem in [3], a linear system with a constraint is analyzed. Consider the asymptotically stable linear system

$$\dot{x} = Ax, \tag{5}$$

$$z = Cx,$$

where the matrix A is Hurwitz, i.e. all eigenvalues of this matrix have strictly negative real parts. The problem is posed of finding a set of initial states $x(0) = x_0$, starting from which the phase trajectory does not go beyond the set defined by the constraint

$$\max_{t \geq 0} |z(t)| \leq \gamma, \tag{6}$$

for a given value $\gamma > 0$.

Note that if a function $V(x) = x^T Y^{-1} x$ with a matrix $Y = Y^T > 0$ is a quadratic Lyapunov function of system (5), then all trajectories of this system outgoing from a set $E(Y) = \{x : x^T Y^{-1} x \leq 1\}$, bounded by an ellipsoid $x^T Y^{-1} x = 1$, inscribed in the region of the phase space specified by the inequality $|z(t)| \leq \gamma$, satisfy constraint (6). In the matrix inequality $Y > 0$, the sign ">" means the positive definiteness of the matrix Y, i.e. $u^T Y u > 0$, $\forall u \in R^n$, $u \neq 0$. It is shown in this paper that the region of the phase space, defined by the union of all such sets $E(Y)$ for all possible Lyapunov functions of the indicated form, can be distinguished in terms of linear matrix inequalities.

Theorem 1. *If the matrix $Y = Y^T > 0$ satisfies the system of linear matrix inequalities*

$$YA^T + AY < 0,$$
$$\begin{pmatrix} Y & YC^T \\ CY & \gamma^2 I \end{pmatrix} \geq 0, \tag{7}$$

then all trajectories of system (5) with initial conditions $x(0) \in E(Y)$ satisfy constraint (6).

This theorem was formulated and proved in [2,3].

Note that there are a lot of matrices Y, satisfying the system of matrix inequalities (7). This, in turn, means that there are many sets of initial states determined by the corresponding ellipsoids. Therefore, there is a desire to find a set that is maximum in accordance with some criterion. In particular, maximization of the trace of the matrix Y under the constraints specified by linear matrix inequalities (7), or maximization of the volume of the corresponding ellipsoid can serve as criteria for searching for a set possessing, in a sense, "maximum" size.

In the case of analyzing an asymptotically stable linear system with several constraints, we define the set of initial states of the "largest" size as the set obtained by the intersection of ellipsoids with "maximal" sizes corresponding to each of these constraints.

The key point in solving the problem of stabilizing the plant (1) in the class of linear state feedbacks (3) under constraints (4) is the choice of a single Lyapunov function of the closed-loop system subject to constraints and the application of the S-procedure being non-defective under one constraint [10]. This allows us to represent sufficient conditions for finding the matrix of parameters of the controller (3) in terms of linear matrix inequalities. An S-procedure under one constraint is a trick that allows us to replace two inequalities for quadratic forms with their equivalent single inequality. It is as follows. Let there be an inequality

$$F(x) < 0, \quad x \neq 0, \tag{8}$$

for all $x \in R^n$, satisfying the inequality

$$G(x) \leq 0, \tag{9}$$

where $F(x)$ and $G(x)$ are quadratic forms. Then we can compose a quadratic form $S(x) = F(x) - \lambda G(x)$ and consider the inequality

$$S(x) < 0, \quad x \neq 0, \tag{10}$$

for some $\lambda \geq 0$. Replacing inequalities (8) and (9) by inequality (10) is called an S-procedure.

It is obvious that the fulfillment of (10) implies the fulfillment of (8) under condition (9). But the converse is also true. Provided that exists x_0 for which $G(x_0) < 0$, the fulfillment of inequality (8) under condition (9) implies the existence $\lambda > 0$, for which holds the inequality

$$F(x) - \lambda G(x) < 0, \quad x \neq 0.$$

In this case, it is said that the S-procedure being non-defective for one restriction. The authors use this technique in [2,3] for everyone i, which allows us to reduce the process of finding a single Lyapunov function of a closed-loop system to solving a system of linear matrix inequalities.

If a function $V(x) = x^T Y^{-1} x$ with a matrix $Y = Y^T > 0$ is a single quadratic Lyapunov function of system (1), closed by control (3), then all trajectories of this system outgoing from a set $E(Y) = \{x : x^T Y^{-1} x \leq 1\}$, bounded by an ellipsoid $x^T Y^{-1} x = 1$, inscribed in the phase space region defined by inequalities $|z_i(t)| \leq \gamma_i, i = 1, 2, ..., N$, satisfy constraints (4). In [2,3] was formulated and proved the following theorem.

Theorem 2. *If matrices $Y = Y^T > 0$, Z and values $\gamma_i > 0, i = 1, 2, ..., N$, satisfy the system of linear matrix inequalities*

$$YA^T + AY + Z^T B^T + BZ < 0,$$
$$\begin{pmatrix} Y & YC_i^T + Z^T D_i^T \\ C_i Y + D_i Z & \gamma_i^2 I \end{pmatrix} \geq 0, \quad i = 1, 2, ..., N, \tag{11}$$

then all trajectories of system (1) closed by control (3) with initial conditions $x(0) \in E(Y)$ satisfy constraints (4). The matrix of parameters of the control law (3) for a dynamical system with constraints is calculated as

$$K = ZY^{-1}. \tag{12}$$

Note that if the matrix of parameters of the control law (12) is found, then for all initial states $x(0) \in \bigcap\limits_{i=1}^{N} E(Y_i)$ the phase trajectories of system (1) closed by control (3) will asymptotically approach the zero state and not go beyond the boundaries of the set defined by constraints (4). Here the sets $E(Y_i) = \{x : x^T Y_i^{-1} x \leq 1\}$ are obtained as sets of initial states $x(0) = x_0$ for an asymptotically stable linear system for which the phase trajectory does not go beyond the limits of the set defined by the constraint $\max\limits_{t\geq 0} |z_i(t)| \leq \gamma_i$. In this case, it is desirable to choose sets $E(Y_i)$ that have, in a certain sense, "maximum" size (for example, in the sense of maximizing the trace of the matrix Y_i, or maximizing the volume of the corresponding ellipsoid).

As noted above, the key point in solving the problem of stabilization of the plant (1) in the class of linear state feedbacks (3) under constraints (4) is the choice of a single Lyapunov function of the closed-loop system taking into account the constraints. This is due to the fact that otherwise, choosing our own Lyapunov function for each constraint $\max\limits_{t\geq 0} |z_i(t)| \leq \gamma_i$, we arrive at the system of bilinear matrix inequalities

$$A^T X_i + K^T B^T X_i + X_i A + X_i B K < 0,$$
$$\begin{pmatrix} \gamma_i^2 X_i & C_i^T + K^T D_i^T \\ C_i + D_i K & I \end{pmatrix} \geq 0, \quad i = 1, 2, ..., N,$$

relatively unknown matrices $X_i = X_i^T > 0$, $i = 1, 2, ..., N$ and K. At present, there are no computationally efficient numerical methods for solving this class of problems.

Note that the result obtained in [2,3] does not allow us to indicate the "complete" set of initial states, the phase trajectories of the system from which do not violate the constraints. As an example, consider a controlled inverted pendulum

$$\ddot{\varphi} - \varphi = u, \tag{13}$$

with restrictions on φ—the angle of deviation of the pendulum link from the vertical and u—control:

$$\max\limits_{t\geq 0} |\varphi(t)| \leq 0.1, \quad \max\limits_{t\geq 0} |u(t)| \leq 1. \tag{14}$$

We represent the equation and restrictions in the form (1), (2), where

$$A = \begin{pmatrix} 0 & 1 \\ 1 & 0 \end{pmatrix}, \quad B = \begin{pmatrix} 0 \\ 1 \end{pmatrix}, \quad C_1 = \begin{pmatrix} 1 & 0 \end{pmatrix}, \quad D_1 = 0, \quad C_2 = \begin{pmatrix} 0 & 0 \end{pmatrix}, \quad D_2 = 1.$$

Control found for object (13)

$$u = -11.1888\varphi - 3.5402\dot{\varphi}, \tag{15}$$

which ensures the asymptotic stability of the closed-loop system (13), (15) and the fulfillment of constraints (14). Control (15) was obtained as a result of searching for a matrix Y with a maximum trace and satisfying the system of linear matrix inequalities (11).

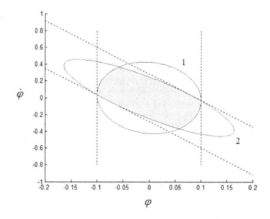

Fig. 1. Estimation of the set of admissible initial states obtained by the intersection of ellipsoids in the stabilization problem for an inverted pendulum under constraints on the angle and control

In Fig. 1 and Fig. 2 in the phase plane the dashed lines mark the restrictions

$$|\varphi(t)| \leq 0.1, \quad |u(t)| \leq 1. \tag{16}$$

In Fig. 1, ellipse 1 limits the estimate of the set of initial states, at the choice of which control (15) provides stabilization of the inverted pendulum under the first constraint, i.e. by the angle φ of deflection of the pendulum. Ellipse 2 limits the estimate of the set of initial states, when chosen, the control provides stabilization under the second constraint, i.e. with control restrictions. At the intersection of ellipses, we obtain an estimate for the region of admissible initial states for which the control stabilizes the object under two constraints. In Fig. 1 and Fig. 2 this area is marked in light gray. A phase portrait of a closed system can be constructed and analyzed. In Fig. 2, the set of admissible initial states is marked in gray, starting from which the phase trajectories of system (13), closed by control (15), asymptotically approach the zero state and do not go beyond the boundaries of the set specified by constraints (14). As an example, trajectory 1 is given for the initial state $\varphi = -0.09$, $\dot{\varphi} = 0.36$. In dark color in Fig. 2, the set of initial states is marked, at the choice of which the phase trajectories of the system will go beyond the boundaries of the region (16). As an example, trajectory 2 is given for the initial state $\varphi = -0.095$, $\dot{\varphi} = 0.56$.

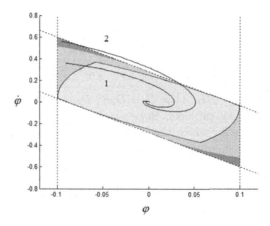

Fig. 2. The set of admissible initial states and its estimate obtained by the intersection of ellipsoids in the stabilization problem for an inverted pendulum under constraints on the angle and control

3   Formulation of the Problem

Suppose that for the object (1), (2) the stabilization problem under constraints on state and control variables is solved and the state control law (3) is found. In a real situation, the state of a dynamic system is measured with some error. In this regard, we introduce the measured output of the system

$$y = (I + \Delta(t))x, \tag{17}$$

where I—identity matrix of size $n \times n$, and the matrix $\Delta(t)$ determines the relative measurement errors of the phase variables, and satisfies the condition $\Delta^T \Delta \leq \delta^2 I$, $\delta > 0$—given parameter. Consider the problem of stabilizing system (1), (2) by the controller

$$u = Ky, \tag{18}$$

with restrictions on state and control variables (4). The question arises about the influence of errors in measuring phase variables on the fulfillment of constraints (4). In other words, the question arises of how the set of initial states of the system will change, for which controller (18) provides stabilization under constraints (4) and in the case of an error in the measured output (17).

4   Estimation of the Set of Admissible Initial States

Let us represent the measured output of system (17) as

$$y = x + w, \tag{19}$$

where $w = \Delta(t)x$. Since the uncertainty matrix $\Delta(t)$ satisfies the condition $\Delta^T \Delta \leq \delta^2 I$, then

$$w^T w \leq \delta^2 x^T x. \tag{20}$$

We write the closed-loop system (1), (2), (18), (19) in the form

$$\dot{x} = \overline{A}x + \overline{B}w, \tag{21}$$

$$z_i = \overline{C}_i x + \overline{D}_i w, \quad i = 1, 2, ..., N,$$

where $\overline{A} = A + BK$, $\overline{B} = BK$, $\overline{C}_i = C_i + D_i K$, $\overline{D}_i = D_i K$.

Consider an auxiliary problem. Suppose it is required to find the sets of admissible initial states for which the control (18) provides system stabilization (21) for each i with one constraint $\max_{t \geq 0} |z_i(t)| \leq \gamma_i$. The following theorem is true.

Theorem 3. *Let the matrix* $X_i = X_i^T > 0$ *and values* $\mu_1 > 0$, $\mu_2 > 0$, $\delta > 0$, $\gamma_i > 0$ *satisfy the system of matrix inequalities*

$$
\begin{pmatrix} \overline{A}^T X_i + X_i \overline{A} + \mu_1 \delta^2 I & X_i \overline{B} \\ \overline{B}^T X_i & -\mu_1 I \end{pmatrix} < 0,
$$
$$
\begin{pmatrix} \overline{C}_i^T \overline{C}_i + \mu_2 \delta^2 I - \gamma_i^2 X_i & \overline{C}_i^T \overline{D}_i \\ \overline{D}_i^T \overline{C}_i & \overline{D}_i^T \overline{D}_i - \mu_2 I \end{pmatrix} \leq 0. \tag{22}
$$

Then all trajectories of the closed-loop system (21) with the initial conditions $x(0) \in E(X_i)$, $E(X_i) = \{x : x^T X_i x \leq 1\}$, *satisfy the constraint*

$$\max_{t \geq 0} |z_i(t)| \leq \gamma_i.$$

Proof. In the region of phase space given by the inequality $|z_i(t)| \leq \gamma_i$, we inscribe the ellipsoid $x^T X_i x = 1$. Let us show that the fulfillment of the first inequality of system (22) ensures the fulfillment of the condition that a quadratic function $V(x) = x^T X_i x$ with a matrix $X_i = X_i^T > 0$ is a Lyapunov function for a closed system. On any trajectory of the closed-loop system (21), the condition

$$\dot{V}(x) = (\overline{A}x + \overline{B}w)^T X_i x + x^T X_i (\overline{A}x + \overline{B}w) < 0. \tag{23}$$

According to the fact that the S-procedure is not defective under one constraint, inequality (23) holds for all x, w such that $|x|^2 + |w|^2 \neq 0$, satisfying inequality (20) if and only if for some number $\mu_1 > 0$ and for all x, w performed the inequality

$$(\overline{A}x + \overline{B}w)^T X_i x + x^T X_i (\overline{A}x + \overline{B}w) - \mu_1 (w^T w - \delta^2 x^T x) < 0.$$

We write it in the form

$$
\begin{pmatrix} x \\ w \end{pmatrix}^T \begin{pmatrix} \overline{A}^T X_i + X_i \overline{A} + \mu_1 \delta^2 I & X_i \overline{B} \\ \overline{B}^T X_i & -\mu_1 I \end{pmatrix} \begin{pmatrix} x \\ w \end{pmatrix} < 0.
$$

This inequality is equivalent to the first inequality of system (22).

Let us show that the fulfillment of the second inequality of system (22) ensures the fulfillment of the condition $|z_i(t)| \leq \gamma_i$. For quadratic forms, the

S - procedure is valid under two constraints [9]. The theorem states the following. Let there be given quadratic forms $F(x) = x^T A_0 x$, $G_1(x) = x^T A_1 x$, $G_2(x) = x^T A_2 x$, where $x \in R^n$, $A_i = A_i^T \in R^{n \times n}$, $i = 0, 1, 2$ and numbers a_0, a_1, a_2. Let's make a quadratic form $S(x) = F(x) - \tau_1 G_1(x) - \tau_2 G_2(x)$ and consider the system of inequalities

$$S(x) \leq 0, \quad a_0 \geq \tau_1 a_1 + \tau_2 a_2, \tag{24}$$

with some $\tau_1 \geq 0$, $\tau_2 \geq 0$. Consider the inequality

$$F(x) \leq a_0, \tag{25}$$

which, for all $x \in R^n$, satisfies the system of inequalities

$$G_1(x) \leq a_1, \quad G_2(x) \leq a_2. \tag{26}$$

Then the fulfillment of inequalities (24) implies the fulfillment of inequality (25) under the condition (26).

Conversely, in case, if $n \geq 3$, there are numbers τ_3, τ_4 and a vector $x^0 \in R^n$ such that

$$\tau_3 A_1 + \tau_4 A_2 > 0, \quad G_1(x^0) < a_1, \quad G_2(x^0) < a_2,$$

then inequality (25) under condition (26) implies the existence of numbers $\tau_1 \geq 0$, $\tau_2 \geq 0$ for which condition (24) is satisfied.

Let's apply the statement to solve the problem. Since the S-procedure being non-defective under two constraints, the inequality $\max_{t \geq 0} |z_i(t)| \leq \gamma_i$ subject to condition (20) and condition $x^T X_i x \leq 1$ for all x, w such that $|x|^2 + |w|^2 \neq 0$, is equivalent to the existence of numbers $\mu_2 \geq 0$, $\mu_3 \geq 0$ for which performed the inequality

$$|z_i(t)|^2 - \gamma_i^2 - \mu_2(w^T w - \delta^2 x^T x) - \mu_3(x^T X_i x - 1) \leq 0. \tag{27}$$

In this case, there must be numbers μ_4, μ_5 and a vector $\begin{pmatrix} x^0 \\ w^0 \end{pmatrix}$, such that

$$\mu_4 \begin{pmatrix} X_i & 0 \\ 0 & 0 \end{pmatrix} + \mu_5 \begin{pmatrix} -\delta^2 I & 0 \\ 0 & 1 \end{pmatrix} > 0 \tag{28}$$

and

$$\begin{pmatrix} x^0 \\ w^0 \end{pmatrix}^T \begin{pmatrix} X_i & 0 \\ 0 & 0 \end{pmatrix} \begin{pmatrix} x^0 \\ w^0 \end{pmatrix} < 1, \quad \begin{pmatrix} x^0 \\ w^0 \end{pmatrix}^T \begin{pmatrix} -\delta^2 I & 0 \\ 0 & 1 \end{pmatrix} \begin{pmatrix} x^0 \\ w^0 \end{pmatrix} < 0. \tag{29}$$

We write inequality (27) in the form

$$(\overline{C_i} x + \overline{D_i} w)^T (\overline{C_i} x + \overline{D_i} w) - \gamma_i^2 - \mu_2(w^T w - \delta^2 x^T x) - \mu_3(x^T X_i x - 1) \leq 0. \tag{30}$$

Inequality (30) is true for all x, w. Means

$$\mu_3^2 \leq \gamma_i^2, (\overline{C_i} x + \overline{D_i} w)^T (\overline{C_i} x + \overline{D_i} w) - \mu_2(w^T w - \delta^2 x^T x) \leq \mu_3(x^T X_i x).$$

Hence

$$(\overline{C}_i x + \overline{D}_i w)^T (\overline{C}_i x + \overline{D}_i w) - \mu_2 (w^T w - \delta^2 x^T x) \le \gamma_i^2 (x^T X_i x). \qquad (31)$$

We write inequality (31) in the form

$$\begin{pmatrix} x \\ w \end{pmatrix}^T \begin{pmatrix} \overline{C}_i^T \overline{C}_i + \mu_2 \delta^2 I - \gamma_i^2 X_i & \overline{C}_i^T \overline{D}_i \\ \overline{D}_i^T \overline{C}_i & \overline{D}_i^T \overline{D}_i - \mu_2 I \end{pmatrix} \begin{pmatrix} x \\ w \end{pmatrix} \le 0. \qquad (32)$$

Matrix inequality (32) is equivalent to the second matrix inequality in system (22).

Find numbers μ_4, μ_5, satisfying inequality (28). Let us rewrite condition (28) as $\mu_5 > 0$, $\mu_4 X_i - \mu_5 \delta^2 I > 0$. Hence, given that $X_i = X_i^T > 0$, to satisfy these inequalities, it suffices to choose $\mu_4 = \frac{2\delta^2}{min|eig(X_i)|}$, $\mu_5 = 1$.

Let us find a vector $\begin{pmatrix} x^0 \\ w^0 \end{pmatrix}$, satisfying inequalities (29). Since $V(x) = x^T X_i x$ is the Lyapunov function, for all $x \in E(X_i)$ inequality holds $x^T X_i x \le 1$. Therefore, the first inequality (29) is satisfied if the point x^0 lies inside the ellipsoid $E(X_i)$. By virtue of inequality (20), for the second inequality (29) to hold, we choose $w^0 = \frac{\delta}{2} x^0$. The statement is proven.

Let us denote by Ξ_i the set of all matrices X_i, satisfying inequalities (22). Let us choose the trace criterion as a criterion for minimality X_i. The maximum over all $X_i \in \Xi_i$ region $E(X_i^*)$, is found by minimizing the trace of the matrix X_i. This operation is standard in the MATLAB software package for engineering calculations [7] using the CVX application.

Let us formulate sufficient conditions for finding the region of admissible initial states of a dynamic system under which control (18), with the matrix of controller parameters K obtained in the problem of control synthesis under constraints imposed on state and control variables, will provide stabilization also in the case when the state of the system is measured with an error. Let the matrices X_i^*, $i = 1, 2, ..., N$, have a minimal trace and are solutions of system (22) for values γ_i, respectively. Then all trajectories of the closed-loop system (21) with the initial conditions $x(0) \in E(X_i^*)$, $E(X_i^*) = \{x : x^T X_i^* x \le 1\}$ will satisfy the constraint $\max_{t \ge 0} |z_i(t)| \le \gamma_i$. Therefore, for all initial states $x(0) \in \bigcap_{i=1}^{N} E(X_i^*)$, the control with a given matrix of controller parameters K, stabilizes the closed-loop system under constraints (4).

5   Numerical Simulation Results

Consider a controlled inverted pendulum

$$\ddot{\varphi} - \varphi = u, \qquad (33)$$

with restrictions on φ—the angle of deviation of the pendulum link from the vertical and u—control:

$$\max_{t \geq 0} |\varphi(t)| \leq 0.1, \quad \max_{t \geq 0} |u(t)| \leq 1. \tag{34}$$

Numerical solution obtained in MATLAB package. For object (33), a number of problems have been solved. For the stabilization problem, in the absence of an error in the measurement of the state, the control is obtained

$$u = -11.1888\varphi - 3.5402\dot{\varphi}, \tag{35}$$

which ensures the asymptotic stability of the closed-loop system (33), (35) and the fulfillment of constraints (34).

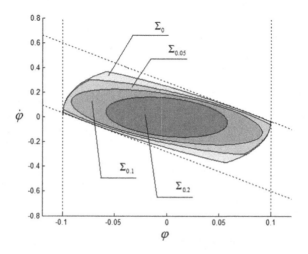

Fig. 3. Intersection of areas Σ_0, $\Sigma_{0.05}$, $\Sigma_{0.1}$ and $\Sigma_{0.2}$

In Fig. 3 in the phase plane, the dotted line marks the restrictions

$$|\varphi(t)| \leq 0.1, \quad |u(t)| \leq 1.$$

Let us estimate the change in the estimate of the region of admissible initial states of the dynamic system, at which control (35) will provide stabilization even in the case when the state of the system is measured with an error. We denote Σ_δ an estimate for the set of admissible initial states for which the control stabilizes the system for the value δ.

The algorithm for constructing area Σ_δ is as follows. Consider two ellipses. The first ellipse $E(X_1^*) = \{x : x^T X_1^* x \leq 1\}$ limits the estimate of the set of initial states, at the choice of which control (35) provides stabilization of the inverted pendulum under the first constraint, i.e. by the angle φ of deflection of the pendulum. Here, the matrix X_1^* is a matrix with a minimum trace that satisfies the system of linear matrix inequalities (22) for given values δ and γ_1.

The second ellipse $E(X_2^*) = \{x : x^T X_2^* x \leq 1\}$ limits the estimate of the set of initial states, when chosen, control (35) provides stabilization under the second constraint, i.e. with control restrictions. Here, the matrix X_2^* is a matrix with a minimum trace that satisfies the system of linear matrix inequalities (22) for given values δ and γ_2. At the intersection of ellipses, we obtain the desired estimate for the set of admissible initial states for which the control stabilizes the object under two constraints for a given value δ.

In Fig. 3 shows the areas Σ_0, $\Sigma_{0.05}$, $\Sigma_{0.1}$, $\Sigma_{0.2}$ corresponding to the values $\delta = 0$, $\delta = 0.05$, $\delta = 0.1$, $\delta = 0.2$, and shows the intersection of these areas. It follows from the figure that the area $\Sigma_{0.2}$ lies inside the area $\Sigma_{0.1}$, which in turn lies inside the area $\Sigma_{0.05}$, and the area $\Sigma_{0.05}$ lies inside Σ_0.

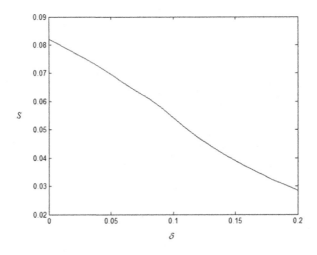

Fig. 4. The graph the dependence of the area of the region Σ_δ

Let us calculate the dependence of the area S of the region Σ_δ on the value δ, which determines the magnitude of the error in the measured output of the system. In Fig. 4 shows a graph of this dependence. In particular, the values $S(0) = 0.0819$, $S(0.05) = 0.0696$, $S(0.1) = 0.0541$, $S(0.2) = 0.0285$.

The performed calculations showed that the ellipses responsible for the constraints on the deflection angle of the pendulum link at the values $\delta = 0$, $\delta = 0.05$, $\delta = 0.1$ and $\delta = 0.2$ are close to each other. Consequently, the size of the region Σ_δ of admissible initial states is mainly influenced by both the value of the parameter value δ, and the presence of a control constraint in the problem.

6   Conclusions

The problem is posed and solved to estimate the region of admissible initial states of a dynamic system, at which the controller obtained in the problem of synthesis of state control under constraints imposed on state and control variables will also provide stabilization in the case when the system state is measured with an error. In terms of linear matrix inequalities, conditions are obtained that make it possible to estimate the set of admissible initial states of a dynamical system. The problem of stabilization of an inverted pendulum is considered as an example. Numerical experiments confirm the theoretical results.

Note that when solving practical problems of controlling real physical objects, complete information about the state of the system is usually inaccessible to measurement. In this regard, a nontrivial problem arises of stabilizing dynamic objects by the measured system output. In the future, it is planned to consider the situation when part of the phase variables or their linear combination is measured. It is supposed to solve the stabilization problem using a static controller under constraints on the phase and control variables, and also to estimate the region of admissible initial states for the obtained controller in the presence of an error in the measurements of the output variables.

References

1. Balandin, D.V., Kogan, M.M.: Sintez zakonov upravleniya na osnove lineinykh matrichnykh neravenstv (Design of Control Laws on the Basis of Matrix Inequalities). Fizmatlit, Moscow (2007)
2. Balandin, D.V., Kogan, M.M.: Linear control design under phase constraints. Autom. Remote. Control. **70**(6), 958–966 (2009). https://doi.org/10.1134/S0005117909060046
3. Balandin, D.V., Kogan, M.M.: Lyapunov function method for control law synthesis under one integral and several phase constraints. Differ. Equ. **45**(5), 670–679 (2009). https://doi.org/10.1134/S001226610905005X
4. Balandin, D.V., Kogan, M.M.: LMI based multi-objective control under multiple integral and output constraints. Int. J. Control **83**(2), 227–232 (2010). https://doi.org/10.1080/00207170903134130
5. Hu, T., Lin, Z.: Control Systems with Actuator Saturation. Birkhauser, Norwell (2001)
6. Polyak, B.T., Shcherbakov, P.S.: Robastnaya ustoichivost i upravlenie (Robust Stability and Control). Nauka, Moscow (2002)
7. Gahinet, P., Nemirovski, A., Laub, A.J., Chilali, M.: The LMI Control Toolbox. For Use with Matlab. User's Guide. MathWorks, Natick (1995)
8. Fedyukov, A.A.: Sintez stabiliziruyushchikh regulyatorov po vykhodu dlya dinamicheskikh sistem s ogranicheniyami na fazovyye peremennyye (Synthesis of dynamic regulators stabilizing systems with phase constraints). Vestnik of Lobachevsky State Univ. Nizhni Novgorod **2**(1), 152–159 (2013)
9. Polyak, B.T., Khlebnikov, M.V., Rapoport, L.B.: Matematicheskaya teoriya avtomaticheskogo upravleniya (Mathematical Theory of Automatic Control). LENAND, Moscow (2019)
10. Gelig, A.Kh., Leonov, G.A., Yakubovich, V.A.: Ustoichivost' nelineinykh sistem s needinstvennym sostoyaniem ravnovesiya (Stability of Nonlinear Systems with Nonunique Equilibrium Position). Nauka, Moscow (1978)

Multicriteria Problem of Wireless System Design Using Tricriteria Approach with Qualitative Information on the Decision Maker's Preference

Dmitrii Shaposhnikov$^{(\boxtimes)}$ ⓘ and Julia Makarova ⓘ

Lobachevsky State University of Nizhny Novgorod, Nizhny Novgorod,
Russian Federation
dsh@unn.ru

Abstract. This paper addresses the multicriteria problem of choosing a
rational solution when designing complex devices on the example of integrated radar and communication system (IRCS) operating in a changing
environment. The mathematical model is built as a multicriteria system for making optimal decisions, with particular criteria being divided
into three subsets characterizing "costs", "efficiency" and "reliability".
A "design-action" approach is proposed, in which weighting coefficients
of particular criteria relative importance are assigned within each group
independently at the design stage of the device. Further, in the process
of functioning (operation) of the technical system, based on the results of
the environment analysis, relative importance between the three gropus
of particular criteria is automatically determined and optimal functioning parameters are calculated. Relative importance weights, both at the
design stage and at the action stage, can be assigned as exact values
or calculated on the principle of a guaranteed result using the user's
qualitative preferences.

Keywords: Multicriteria optimization · Wireless system · IRCS ·
Effective solution set · Reliability

1 Introduction

Recently, wireless systems and devices have become more complex, often being
multi-modular and multifunctional. Their design is a difficult problem associated
both with the choice of parameters during design and with their configuration
during operation to adapt to changes in the external environment. These parameters are, in particular, frequency ranges and signal structure.

It is obvious that the problem of optimization of technical system parameters
is multicriteria. One of the well-known and effective in practice approaches to
its solution is to reduce the problem to bicriteria one by dividing the criteria
into two groups "cost" and "efficiency". This approach is essentially a search for

a compromise between device performance and device manufacturing, maintenance and operating costs.

In this paper it is proposed to allocate the separate group of criteria representing "reliability" which is a crucial property of technical systems. Reliability is the third fundamental aspect of any technical system along with its performance and costs. Reliability shows likelihood that a system will perform its function over a specified period of time, or will operate without failure under certain conditions. By analogy with bicriteria this method is called tricriteria. The reliability criterion is formed either as a generalized criterion from a set of particular criteria, or can be expressed as a single indicator.

In the considered three-criterion problem, it is assumed to use a generalized optimality criterion with weighting coefficients of the relative importance of particular criteria in two stages. At the design stage of the device, the decision maker formulates quantitatively or qualitatively preferences within three groups of criteria. At the stage of device operation, the system automatically prioritizes between groups of particular optimality criteria and makes the final decision on the choice of parameters.

2   Model for Making a Rational Decision on the Choice of Wireless System Parameters

Since in the presence of several particular criteria of optimality, the optimal solution (the best according to all criteria) most often does not exist and a feasible solution is chosen as rational, which can be reasonably explained to other specialists [1].

Signal Parameters of the Integrated Radar and Communication System as Variable Parameters. The problem of multicriteria search for optimal parameters of a wireless system is considered on the example of the integrated wireless system, which simultaneously performs the functions of radar and communication. These systems are used in airplanes, drones, unmanned vehicles, vehicles with automatic control support, et al. The design and management of such multifunctional systems are complex technical tasks, which are essentially multicriteria optimization problems.

For both radar and communication, an OFDM (Orthogonal frequency-division multiplexing) signal consisting of closely spaced orthogonal subcarriers is used. The total signal energy can be distributed over subchannels, and depending on this distribution, the functioning of the system changes and, accordingly, the values of the criteria describing the system [2].

Let us assume that in the investigated mathematical model, from the point of view of a given subject area, the object of choice is a vector $x = (x_1, \ldots, x_n)$ of n parameters that are used in a given technical device.

The parameter vector x must belong to a certain region of feasible solutions D, which is given a priori and is determined at the design stage. In special cases, the area D can be a discrete range of values of a sufficiently high power, a hypercube, and suggests the possible presence of complex nonlinear restrictions on the parameters [3].

Variable parameters in the system include, first, the distribution of energy over the subchannels, specified in the form of a normalized vector $x = (x_1, \ldots, x_N)$. The value x_i is the power transmitted over the i-th subchannel. The distribution can be changed directly during operation. In addition, the configurable parameters are the number of subchannels, the width of the subchannel, the center frequency, the number of pulses, the pulse repetition interval, the number of OFDM symbols, the duration of the OFDM symbol, but they are usually specified during design before commissioning. In our work, we consider the energy distribution over subchannels as variable parameters and investigate the behavior of the criteria when they change. Consider four particular criteria for assessing the performance of the IRCS.

Q_1: The communication efficiency.

Q_2: The accuracy of determining the distance to the target.

Q_3: The accuracy of determining the target speed.

Q_4: The accuracy of determining the target reflection coefficients.

It is proposed to use the channel capacity as a characteristic of the communication efficiency. According to the Shannon-Hartley theorem, the bandwidth of a channel, measured in bits per second, can be found as

$$C = B \cdot log_2(1 + \frac{S}{N_p}), \tag{1}$$

where S is the average power of the received signal, N_p is the average noise power, B is the channel bandwidth measured in hertz [4,5].

For a channel with frequency selective attenuation, which is divided into N subchannels of width Δf, there is a generalization of the bandwidth formula

$$C = \sum_{m=1}^{N} \Delta f \cdot log_2(1 + \frac{|x_m|^2 |h_m|^2}{\sigma_2^2}), \tag{2}$$

where h_m is the transfer function of the m-th subchannel, σ_m is the noise power, which is assumed to be even throughout the channel [2].

To assess the performance of the radar, we use the value of the lower bound from the Cramer-Rao inequality. The Cramer-Rao inequality is used to obtain the boundaries of the minimum measurement errors [6,7]. The Cramer-Rao boundary was chosen as the accuracy estimate because the procedure for finding it requires less computational work compared to others, such as the Battachari boundary, for which calculating high-order partial derivatives is needed, or the Barankin boundary, for which finding the maximum of the function is needed. The latter generally give a slightly more accurate lower bound than the Cramer-Rao inequality, and in the case of an unbiased estimate, the Cramer-Rao bound is an exact result [8]. Thus, the Cramer-Rao boundary is optimal meaning accuracy and computational complexity.

In the case when $\hat{\theta}(x)$ is an unbiased estimate of the parameter θ and the regularity conditions are satisfied, the Cramer-Rao inequality is written as follows:

$$D_\theta \hat{\theta}(x) \geq \frac{1}{I_n(\theta)}, \tag{3}$$

$I_n(\theta)$ is Fisher's information, $D_\theta\hat{\theta}(x)$ is the variance, for an unbiased estimate equal to $D_\theta\hat{\theta}(x) = M(\hat{\theta} - \theta_t)^2$, where θ_t is the true value of θ, M is the mathematical expectation [9]. The Cramer-Rao bounds for time delay, velocity and reflection coefficients are obtained from the Cramer-Rao matrix equal to the inverse Fisher information matrix [6,7]:

$$\begin{bmatrix} CRB(\tau) & CRB(\tau,\nu) & \mathbf{CRB}(\tau,\boldsymbol{\eta}) \\ CRB(\nu,\tau) & CRB(\nu) & \mathbf{CRB}(\nu,\boldsymbol{\eta}) \\ \mathbf{CRB}(\boldsymbol{\eta},\tau) & \mathbf{CRB}(\boldsymbol{\eta},\nu) & \mathbf{CRB}(\boldsymbol{\eta}) \end{bmatrix} = \begin{bmatrix} J_{\tau\tau} & J_{\tau\nu} & J_{\tau\eta} \\ J_{\nu\tau} & J_{\nu\nu} & J_{\nu\eta} \\ J_{\eta,\tau} & J_{\eta,\nu} & J_{\eta)} \end{bmatrix}^{-1}. \tag{4}$$

The Cramer-Rao boundary for target distance can be found through the boundary for the time delay. To improve radar accuracy, the aforementioned radar criteria should be minimized provided the total signal power is limited:

$$\mathbf{x}_{Q2} = \mathrm{argmin} CRB(R), \mathbf{x} \in \mathbb{C}^N, \mathbf{x}^H\mathbf{x} = 1; \tag{5}$$

$$\mathbf{x}_{Q3} = \mathrm{argmin} CRB(\nu), \mathbf{x} \in \mathbb{C}^N, \mathbf{x}^H\mathbf{x} = 1; \tag{6}$$

$$\mathbf{x}_{Q4} = \mathrm{argmin}(\mathrm{tr}[\mathbf{CRB}(\boldsymbol{\eta})]), \mathbf{x} \in \mathbb{C}^N, \mathbf{x}^H\mathbf{x} = 1. \tag{7}$$

The communication efficiency criterion $Q1$ is defined as the reciprocal of the channel capacity and is to be minimized.

$$\mathbf{x}_{Q1} = \mathrm{argmin} \frac{1}{\sum_{m=1}^N \Delta f \cdot log_2(1 + \frac{|x_m|^2|h_m|^2}{\sigma_2^2})}, \mathbf{x} \in \mathbb{C}^N, \mathbf{x}^H\mathbf{x} = 1. \tag{8}$$

Figure 1 shows, as an example, the values of the communication efficiency criterion for different energy distributions (different colors) for different noise levels in the channel. The calculation was carried out for the case of 8 subcarriers, the subchannel width is 0.25 MHz, and the center frequency is 5 GHz.

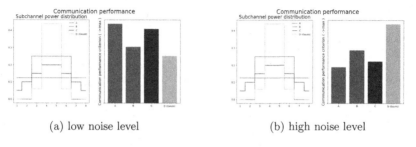

(a) low noise level   (b) high noise level

Fig. 1. Dependence of the communication efficiency criterion of IRCS on the power distribution over subchannels in the case of 8 subcarriers with different noise levels.

These pictures demonstrate that optimal solution depends on the environment which is likely to change significantly for wireless system operating situations.

3 Objects and Criteria

Objects and Criteria as Characteristics of the Selected Solution Quality

"Cost" Criteria. Let us assume that from the point of view of this subject area, a vector M of numerical characteristics (particular cost criteria) with respect to option x is used as a "cost": $P(x) = (P_1(x), ..., P_M(x))$. The generalized "cost" criterion $G(x) = G(P_1(x), ..., G_M(x))$ is constructed for final decision making using the vector of the weighting coefficients $y = (y_1, \dots, y_M)$ by various methods described in [10].

"Efficiency" Criteria. The efficiency of the system with the selected parameters x is another group of particular criteria. By analogy with the "cost" criteria, generalized efficiency criterion $F(x) = F(Q(x))$ is constructed from K particular quality criteria $Q(x) = (Q_1(x), ..., Q_K(x))$ using the vector of the weighting coefficients $w = (w_1, \dots, w_K)$.

These two groups of criteria are discussed in more detail in [11, 12].

"Reliability" Criteria. Criteria of the third group represent "reliability". Characteristics of this type are described by a vector $R(x)$ consisting of L particular reliability criteria $R(x) = (R_1(x), ..., R_L(x))$. Using the weighting coefficients $v = (v_1, \dots, v_L)$ the generalized criterion $H(x) = H(R(x))$ is constructed.

We will assume that all particular criteria above are reduced to the direction of minimization.

The weighting coefficients y, w, v reflect the relative importance of particular criteria [13]. It is assumed here that the assigned values of the importance weighting coefficients w remain unchanged for the entire region of feasible solutions D. The sets of permissible values of the weight coefficients are determined by the following relations:

$$y \in D_1 = \{z \in \mathbb{R}^M | z_i \geq 0, i = 1, ..., M; \sum_{i=1}^{M} z_i = 1\}. \tag{9}$$

$$w \in D_2 = \{z \in \mathbb{R}^K | z_i \geq 0, i = 1, ..., K; \sum_{i=1}^{K} z_i = 1\}. \tag{10}$$

$$v \in D_3 = \{z \in \mathbb{R}^L | z_i \geq 0, i = 1, ..., L; \sum_{i=1}^{L} z_i = 1\}. \tag{11}$$

The generalized criteria $G(x)$, $F(x)$ and $H(x)$ can be used in several forms, e.g. the generalized criterion "reliability" is formed as the additive criterion

$$H(x) = \sum_{k=1}^{L} v_k R_k(x), \tag{12}$$

or a generalized logical criterion (criterion of maximum caution)

$$H(x) = \max_{1 \le k \le L} (v_k R_k(x)). \tag{13}$$

4 Using Designer Preferences During the Development Phase

4.1 Three-Criteria Approach

Each configuration of the varied parameters is characterized by three criteria $G(x)$, $F(x)$ and $H(x)$ that form the area of compromises and the corresponding area of Pareto optimal solutions.

In particular, the final solution can be obtained using the additive generalized criterion

$$(G(x) + F(x) + H(x)) \to min, \tag{14}$$

here $x \in D$, values of weighting coefficients is defined by user according to (9)–(11), and all particular and generalized criteria are reduced to the direction of minimization.

In this approach, the weights are assigned by the developer at design stage independently within each group of particular criteria. In accordance with the decision-making process in the organization, preferences and, therefore, importance weight factors are set separately by different groups of specialists. This approach involves the assignment of precise importance factors, which can be obtained in various ways, e.g. by an examination.

4.2 Single Criteria Approach

This approach involves using a generalized optimal criterion (14) and weighting factors of the form (9)–(11) subject to:

$$\sum_{i=1}^{M} y_i + \sum_{i=1}^{K} w_i + \sum_{i=1}^{L} v_i = 1. \tag{15}$$

The weights y, w, v also reflect the relative importance of particular criteria, but are assigned simultaneously by the decision maker:

$$Q_i \succ Q_j \Leftrightarrow w_i \ge w_j; P_i \succ P_j \Leftrightarrow y_i \ge y_j; R_i \succ R_j \Leftrightarrow v_i \ge v_j. \tag{16}$$

Let us consider the previously proposed approach, in which the dependence of weighting coefficients on the particular criteria values at each point of feasible solution region D is analyzed. Assuming that the decision maker cannot accurately determine the numerical values of the weighting coefficients y, w, v, one can consider them as uncontrollable factors and, applying the principle of guaranteed result, go to the next decision-making model:

$$\min_{x \in D} \left\{ \max_{y,w,v} \left(G(x) + F(x) + H(x)\right)\right\}. \tag{17}$$

Here weighting coefficients y, w, v are defined according to (15). Assuming that the structure of coefficient admissible values region remains unchanged while searching for the optimal solution, then the weight coefficients are functions of the parameters x.

Assume that the decision maker has formulated additional qualitative information that establishes, for some L_q pairs of particular criteria (not necessarily for all C_n^2 admissible pairs), the preference of the i-th criterion over the j-th on the entire set D of feasible solutions:

$$e_l^1 = \{P_i \succcurlyeq P_j\} \Leftrightarrow y_i \geqslant y_j, l = 1, ..., L_1 \leqslant M(M-1)/2$$
$$e_l^2 = \{Q_i \succcurlyeq Q_j\} \Leftrightarrow w_i \geqslant w_j, l = 1, ..., L_2 \leqslant K(K-1)/2 \tag{18}$$
$$e_l^3 = \{R_i \succcurlyeq R_j\} \Leftrightarrow v_i \geqslant v_j, l = 1, ..., L_3 \leqslant L(L-1)/2$$

Thus, importance weight coefficients are calculated automatically with the decision maker's qualitative preferences in each criteria group separately. Moreover, if the ranges of particular criteria admissible values are not empty, then the qualitative information about preferences is consistent.

The static solution of the optimal parameters search problem involves calculations at the stage of device development in accordance with (17).

5   Use of Intergroup Preferences During Operation

The described tricriteria approach and the principles of using qualitative information about preferences allows one to automatically calculate the importance weight factors during the evice operation. Such analysis and calculations are possible when intergroup preferences are formulated automatically based on an analysis of the environment and conditions.

Based on the preliminary analysis, the following intergroup pairs of preferences can be formulated:

$$\{P \succcurlyeq Q\} \, or \, \{Q \succcurlyeq P\}\,;\,\{P \succcurlyeq R\} \, or \, \{R \succcurlyeq P\}\,;\,\{R \succcurlyeq Q\} \, or \, \{Q \succcurlyeq R\}. \tag{19}$$

The pairs of preferences must have the transitivity property, that is, the corresponding preference graph must not have a cycle.

From each pair high-quality information in transformed to pairwise preferences and the corresponding ratios, for example

$$\{P \succ Q\} \Rightarrow y_i \geqslant w_j, i = 1, ..., M; j = 1, ..., K. \tag{20}$$

This leads to a further narrowing of the weight coefficient admissible values range (15) and taking into account the intergroup preference relations when solving the problem (17).

6 Conclusion

This paper describes the tricriteria approach to the search optimal design of technical systems, that is dividing criteria into three subsets of "costs", "efficiency" and "reliability". The principles of constructing particular and generalized criteria of "reliability" are given. The tricriteria model for the multicriteria design of integrated radar and communication system with specific criteria of "reliability" has been developed. The following approach to wireless system design is suggested:

Step 1. The criteria are divided into three groups "efficiency", "cost" and "reliability". Within each group, preferences are established and importance weights are determined in accordance with the principle of a guaranteed result.

Step 2. The decision maker solve the problem of rational choice at the design stage by specifying the exact values of all particular criteria weight coefficients or specifying quality preferences.

Step 3. The decision maker formulates preferences within the criteria groups.

The article proposes an approach in which the selection problem is solved automatically during operation, taking into account environmental conditions.

References

1. Gergel, V.P., Kozinov, E.A.: Efficient multicriterial optimization based on intensive reuse of search information. J. Glob. Optim. **71**(1), 73–90 (2018)
2. Liu, Y., Liao, G., Yang, Z., Xu, J.: Multiobjective optimal waveform design for OFDM integrated radar and communication systems. Signal Process. **141**, 331–342 (2017). https://doi.org/10.1016/j.sigpro.2017.06.026
3. Batischev, D.: Optimal design methods. Radio and Communication, Moscow (1984)
4. Lea, P.: Internet of Things for Architects. Packt Publishing, Birmingham (2018)
5. Sclar, B.: Digital Communications: Fundamentals and Applications. Pearson Education, London (2016)
6. Cook, C., Bernfeld, M.: Radar Signals. An Introduction to Theory and Application. Academic Press, London (1967)
7. Scharf, L., Demeure, C.: Statistical Signal Processing: Detection, Estimation, and Time Series Analysis. Addison-Wesley Publications Co., Boston (1991)
8. Van Trees, H.L.: Detection, Estimation, and Modulation Theory. John Wiley and Sons, New York (2004)
9. Kryanev, A., Lukin, G., Udumyan, D.: Metric Analysis and Data Processing. Fizmatlit, Moscow (2018)

10. Batischev, D., Shaposhnikov, D.: Multi-criteria choice taking into account individual preferences. IAPRAS, Nizhni Novgorod (1994)
11. Makarova, J., Shaposhnikov, D.: Bicriterial problem of finding filter parameters using qualitative information on frequency interval preferences. In: Proceedings of the XXVI International Scientific and Technical Conference "Information Systems and Technology - 2020", pp. 836–847. NSTU, Nizhni Novgorod (2020)
12. Shaposhnikov, D.E., Makarova, J.M.: Multicriteria problem of calculation of wireless system parameters using qualitative information on the desicion maker's preference. In: Proceedings of the XX International Conference "Mathematical Modeling and Supercomputer Technologies", pp. 16–21. NNSU Publishing House, Nizhni Novgorod (2020)
13. Şahin, M.: A comprehensive analysis of weighting and multicriteria methods in the context of sustainable energy. Int. J. Environ. Sci. Technol. **18**(6), 1591–1616 (2020). https://doi.org/10.1007/s13762-020-02922-7

Application of Optimal Evaluation of Linear Time-Varying Systems Using Reachable Sets

Mariya Sorokina$^{(\boxtimes)}$ (ID)

Lobachevsky University, Nizhny Novgorod 603950, Russia
sorokina@itmm.unn.ru

Abstract. The paper is devoted to reachable sets of linear time-varying systems under uncertain initial states and disturbances with a bounded uncertainty measure. The uncertainty measure is the sum of a quadratic form of the initial state and the integral over the finite-time interval from a quadratic form of the disturbance. Method of evaluation of ellipsoidal reachable sets has been cosidered for such systems using matrix differential Riccati equation. Applying this method allows to find minimal ellipsoidal set that is defined by optimal observer. Besides, linear time-varying system with parametric time-varying uncertainty is being examined. Evaluation of ellipsoidal reachable sets is also given in the article. Applying the both methods is demonstrated with numerical modeling with the Mathieu-Hill equation for parametric vibrations and resonance illustrates this method and pendulum equation. Euler iterative method is applied to compute required evaluations.

Keywords: Reachable sets · Ellipsoidal sets · Optimal observer · Parametric uncertainty

1 Introduction

One of the main problems of dynamic system control theory is researching the opportunity of reaching this or that state under control. Reachable sets studying allows us to solve it.

Reachable sets play a large role in different parts of control theory. The main are optimal control problems, disturbance evaluation, etc. It makes them applicable to practically all spheres of activity: technical field (preservation of given trajectory by pilotless aircraft, taking into account the speed and direction of the wind [1] and building manipulator path [2]), economics [3], medicine [4], chemistry [5], etc.

In this article reachable sets application to solve unknown initial state problem and parametric uncertainty problem are researched.

This work was supported by Ministry of Science and Higher Education of the Russian Federation (project 0729-2020-0055) and the Scientific and Education Mathematical Center "Mathematics for Future Technologies" (project No. 075-02-2020-1483/1).

© Springer Nature Switzerland AG 2021
D. Balandin et al. (Eds.): MMST 2020, CCIS 1413, pp. 295–303, 2021.
https://doi.org/10.1007/978-3-030-78759-2_25

2   Reachable Sets

Let's consider a system:

$$\dot{x} = A(t)x(t) + B(t)v(t), \quad x(t_0) = x_0 \in M, t \in [t_0, T], \tag{1}$$

where $x \in R^{n_x}$ is system state; $v \in R^{n_v}$ is disturbance acting on system: $v = v(\sigma), \sigma \in [t_0, t]$; M is closed set in state space defining class of possible initial system states:

$$M(t, R) = \{(x, v(\sigma)) : x^T(t_0)R^{-1}x(t_0) + \int_{t_0}^t v^T(\sigma)G^{-1}v(\sigma) \le 1\}. \tag{2}$$

A class of trajectories emerges from each point of the M. These trajectories correspond to different values of disturbance. Reachable set for system at $t \ge t_0$ is class of all trajectory ends $x(t)$ emerging from M at $t \ge t_0$.

For the system under disturbance reachable set describes area in which the system comes under disturbance and allows us to evaluate accuracy of system hitting a finite state [6–10]. Reachable sets allow us to realise whether it is possible to put the system into the given state provided we add a control to the system.

3   Finding of Reachable Sets

3.1   Unknown Initial State Problem

Let's consider system (1). We suppose that $x(t_0)$ and $v = v(\sigma)$ are in the set:

$$\begin{aligned}
\mathbf{S}(t, t_0, R, G) = \{(x, v(\sigma)) : x = R^{1/2}\omega_1, v(\sigma) = G^{1/2}(\sigma)\omega_2(\sigma), \\
|\omega_1|^2 + \int_\tau^t ||\omega_2(\sigma)|^2|d\sigma \le 1\}
\end{aligned} \tag{3}$$

for given $R = R^T \ge 0$ and $G(\sigma) = G^T(\sigma) \ge 0$.

Theorem 1. *Reachable set for system (1) at $t \ge t_0$ in any initial states and disturbances satisfying (3) with $R \ge 0$ and $G(\sigma) \ge 0$, $\sigma \in [t_0, t], t \in [t_0, T]$ is ellipsoid $\mathcal{E}(Y(t))$ with its matrix satisfying equation:*

$$\dot{Y} = A(t)Y + YA^T(t) + B(t)G(t)B^T(t) \tag{4}$$

(proof given in [11]).

Example 1. Let's demonstrate the described method using the example of the Mathieu equation:

$$\ddot{x} + \omega_0^2(1 + \varepsilon \sin(\omega t))x = v, \quad x(0) = 0, \quad \dot{x}(0) = 0. \tag{5}$$

System matrixes:

$$A = \begin{pmatrix} 0 & 1 \\ -\omega_0^2(1 + \varepsilon \sin(\omega t)) & 0 \end{pmatrix}, \quad B = \begin{pmatrix} 0 \\ 1 \end{pmatrix} \tag{6}$$

Initial state matrix and constraint matrix:

$$R = \begin{pmatrix} 0 & 1 \\ 1 & 0 \end{pmatrix}, \quad B = \begin{pmatrix} 0 \\ 1 \end{pmatrix} \tag{7}$$

We solve the Eq. (4) for finding reachable sets. Let's choose parametric values: $\omega_0 = \pi, \quad \omega = 2\pi, \quad \varepsilon = 0.1$.

We solve the problem at time $[0, 1]$. Euler difference scheme is applied to solve the problem. The method of finding ellipsoids for special cases of uncertainty is described in [11].

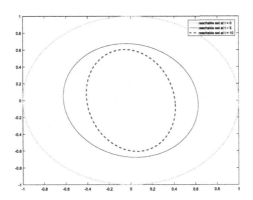

Fig. 1. Reachable sets for system with unknown initial state.

All reachable sets at $t > t_0$ contain inside reachable set at $t = t_0$ (Fig. 1).

3.2 Parametric Uncertainty Problem

Let's consider problem from [15]. It's of the form of

$$\dot{x} = \hat{A}(t)x(t), \hat{A} = A + F\Omega(t)E, x(t_0) = x_0, t \in [t_0, T] \tag{8}$$

where A is matrix of initial system; F, E are given constant matrixes; $\Omega(t)$ is unknown matrix function: $\Omega^T(t)\Omega(t) \leq I$.

We add disturbance and denote: $\omega(t) = \Omega(t)E(t)x(t)$.

$$\dot{x} = A(t)x(t) + B(t)v(t) + F(t)\omega(t), x(t_0) = x_0, t \in [t_0, T]$$
$$\omega^T(t)\omega(t) \leq z^T z, z = E(t)x(t); \tag{9}$$
$$x_0^T R^{-1} x_0 + \int_{t_0}^{T} v^T(t)v(t)dt \leq 1$$

Theorem 2. *Reachable set of (9) is contained in reachable set that is representable as ellipsoid* $\mathcal{E}(Y(t))$. *Its matrix satisfies inequality*[1]

$$\dot{Y} \geq YA^T + YA + BB^T + \mu^{-2}FF^T + \mu^2 YE^T EY \tag{10}$$

with initial state $Y(t_0) = R$.

---

[1] A differential matrix inequality is understood as the positive definiteness of the matrix $\dot{Y} - (YA^T + YA + BB^T + \mu^{-2}FF^T + \mu^2 YE^T EY) \geq 0$.

Proof. Let's consider quadratic form $V = x^T Y^{-1} x$ for system (9). Its matrix Y is positively defined symmetrical matrix and satisfies (10).

Let's find derivative in virtue of system of quadratic form:

$$\dot{V} = x^T[-Y^{-1}\dot{Y}Y^{-1} + A^T Y^{-1} + Y^{-1}A]x + v^T B^T Y^{-1}x + x^T Y^{-1}Bv \\ + \omega^T F^T Y^{-1}x + x^T Y^{-1}F\omega \quad (11)$$

Using (10) we obtain:

$$\dot{V} \leq v^T v + \mu^2(|\omega|^2 - |z|^2) - (v - v_*)^T(v - v_*) - (\mu\omega - \mu^{-1}\omega_*)^T(\mu\omega - \mu^{-1}\omega_*), \\ v_* = B^T Y^{-1}x, \omega_* = F^T Y^{-1}x$$

Having integrated at $[t_0, T]$ and using constraints (9) we obtain:

$$x^T(t)Y^{-1}(t)x(t) \leq 1 \quad (12)$$

So, at any $t \in [t_0, T]$ reachable set of system with parametric uncertainty (9) is inside ellipsoid (12).

Finding Function $\eta(t)$. Let's write down (10) in the form of:

$$\dot{Y} \geq YA^T + YA + BB^T + \eta(t)FF^T + \eta^{-1}(t)YE^T EY, Y(t_0) = R, \eta(t) = \mu^{-2}(t) \quad (13)$$

To find the function, we turn to the numerical characteristics of the matrices. Let's take the trace of the matrix as such a characteristic.

We denote $\dot{Y} = \Gamma$ then

$$trace(\Gamma) = trace(YA^T + AY + BB^T) + \eta\, trace(FF^T) + \eta^{-1}\, trace(YE^T EY) \quad (14)$$

Let's make the right part minimization by parameter η:

$$\eta_* = \sqrt{\frac{trace(YE^T EY)}{trace(FF^T)}}$$

It allows us to find the least ellipsoid defined by matrix $Y(t)$.

Example 2. Let's consider linear oscillator with floating stiffness coefficient:

$$\dot{x}_1 = x_2 \\ \dot{x}_2 = -\omega_0^2(1 + \varepsilon\Omega(t))x_1 \quad (15)$$

System matrixes:

$$A = \begin{pmatrix} 0 & 1 \\ -\omega_0^2 & 0 \end{pmatrix}, \quad B = \begin{pmatrix} 0 \\ 1 \end{pmatrix}, \quad F = \begin{pmatrix} 0 \\ \omega_0^2 \end{pmatrix}, \quad E = \begin{pmatrix} -\varepsilon & 0 \end{pmatrix} \quad (16)$$

Initial state matrix:

$$R = \begin{pmatrix} 1 & 0 \\ 0 & 1 \end{pmatrix}$$

Let's choose parametric values: $\omega_0 = \pi$, $\varepsilon = 0.1$.

We solve the problem with zero initial state at time $[0, 1]$. Euler difference scheme is applied to solve the problem. Ellipsoid including all possible reachable sets for system being satisfied (9) is find with the help of solution of inequality (10). The method of finding ellipsoids for special cases of uncertainty is described in [11].

The special case being considered in this article is $\Omega(t) = \sin(t + \varphi)$, $\varphi = [\pi/6; \pi/4; \pi/2; \pi; 2\pi]$.

Ellipsoid obtained from solution of inequality and ellipsoids for different values of φ are displayed in Fig. 2.

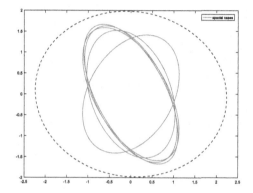

Fig. 2. Reachable sets for system with parametric uncertainty.

All reachable sets in special cases contain inside reachable set describing all possible states of system (8).

4   Reachable Sets Evaluation in the Unknown Initial State Problem

Let's consider the way of reachable sets optimal evaluation given in the article [11].

Let's examine a linear time-varying system

$$\dot{x} = A(t)x + B(t)v,$$
$$y = C(t)x + D(t)v \tag{17}$$

with unknown initial state $x(t_0)$.

Let initial state and disturbance are representable as

$$x(t_0) - x_* = R^{1/2}w_1, \; v(t) = G^{1/2}(t)w_2(t),$$
$$|w_1|^2 + \int_{t_0}^{t} |w_2(\sigma)|^2 d\sigma \leq 1, t \in [t_0, T]. \tag{18}$$

Let's consider evaluation problem of state $x(t)$ of system (17) by means of measurement of the output $y(\sigma)$, $\sigma \in [t_0, t]$ for given matrix $R = R^T > 0$ and matrix function $G^T(\sigma) = G(\sigma) > 0$.

Then:

$$|x(t_0) - x_*|_R^2 + ||v||_{G_{[t_0;t]}}^2 \le 1. \tag{19}$$

Let's derive a full-order observer [12]

$$\dot{\hat{x}} = A(t)\hat{x} + L(t)[y - C(t)\hat{x}], \quad \hat{x}(t_0) = x_*, \tag{20}$$

where $\hat{x}(t)$ is evaluation of state $x(t)$, $L(t)$ is observer parameters matrix to be defined.

Let's denote evaluation error as $\varepsilon(t) = x(t) - \hat{x}(t)$ that satisfies

$$\dot{\varepsilon} = A_c(t)\varepsilon + B_c(t)v, \quad \varepsilon(t_0) = x(t_0) - x_*, \tag{21}$$

where $A_c(t) = A(t) - L(t)C(t)$, $B_c(t) = B(t) - L(t)D(t)$.

We use the following theorem to find optimal ellipsoidal evaluation of state of (17) [11].

Theorem 3. *If $\det[D(\sigma)G(\sigma)D^T(\sigma)] \ne 0$, $\sigma \in [t_0, t]$, then optimal observer (20) guaranteeing optimal ellipsoidal evaluation $\mathcal{E}(Y_*(t), \hat{x}(t))$ of state of (17) at $t \ge t_0$ in any initial states and disturbances satisfying constraint (18) with $R \ge 0$ and $G(\sigma) \ge 0$, $\sigma \in [t_0, t]$ given by*

$$L_*(t) = [D(t)G(t)B^T(t) + C(t)Y_*(t)]^T[D(t)G(t)D^T(t)]^{-1}, \tag{22}$$

where matrix $Y_*(t) \ge 0$ is the solution of matrix differential Riccati equation

$$\dot{Y} = A(t)Y + YA^T(t) + B(t)G(t)B^T(t)$$
$$- [D(t)G(t)B^T(t) + C(t)Y]^T[D(t)G(t)D^T(t)]^{-1}[D(t)G(t)B^T(t) + C(t)Y] \tag{23}$$

with initial state $Y(t_0) = R$. Besides if $R > 0$ then $Y_*(t) > 0$, $t \in [t_0, T]$.

So, for system (20) reachable set is the ellipsoid $\mathcal{E}(Y(t))$ with its matrix satisfying equation

$$\dot{Y} = A_c(t)Y + YA_c^T(t) + B_c(t)G(t)B_c^T(t) \tag{24}$$

with initial state $Y(t_0) = R$.

In other words for system (17) state $x(t)$ is inside the ellipsoid $\mathcal{E}(Y(t))$ with its center in $\hat{x}(t)$ given by equation of observer (20).

The derived set allows us to find state $x(t)$ at any t.

Example 3. Let's demonstrate the described method using the example of the Mathieu-Hill dying-away equation [13, 14]:

$$\ddot{x} + \varepsilon\dot{x} + \omega_0^2(t)(1 + F(t))x = u + v, \quad F(t) = \frac{2\mu}{a + b\cos\omega t} \tag{25}$$

where $a > 0$, $b > 0$, $a > b$. Let's add output $y = x_1 + x_2 + v$ and solve this problem over time $[0, 4]$.

The system becomes:

$$A = \begin{pmatrix} 0 & 1 \\ -\omega_0^2(1 + F(t)) & \varepsilon \end{pmatrix}, \quad B = \begin{pmatrix} 0 \\ 1 \end{pmatrix}, \quad C = \begin{pmatrix} 1 & 1 \end{pmatrix}, \quad D = 1 \quad (26)$$

Matrixes for (18):

$$G = 1, \quad R = \begin{pmatrix} 0.2 & 0.1 \\ 0.1 & 0.3 \end{pmatrix} \quad (27)$$

Let's choose parameter values: $\omega_0 = \pi$, $\omega = 2\pi$, $\varepsilon = 0.1$, $a = 1$, $b = 0.5$, $\mu = 0.1$.

We solve Eq. (23) to find reachable set of the system under disturbance. Using its solution we calculate $L_*(t)$ with the help of (22) and matrixes of close-loop system A_c and B_c. After that, we find reachable set solving (24).

We plot trajectories of system and observer for clarity. To do that we add tyme-varying disturbance $v = 0.5 \sin \pi t$ to system (17) and establish its initial state in

$$x_0 = \begin{pmatrix} 0.01 \\ 0.01 \end{pmatrix}. \quad (28)$$

Then using (18) we calculate initial state for optimal observer (20):

$$x_0 - R^{1/2} \begin{pmatrix} 0 \\ 0.2 \end{pmatrix}. \quad (29)$$

Let's plot trajectory of system and location of observer. For this purpose we separately plot (x_1, x_2) and (x_3, x_4) (see Fig. 3). Reachable sets evolution at $t = 0$, $t = 2$ and $t = 4$ and their centers are displayed in Fig. 3 too.

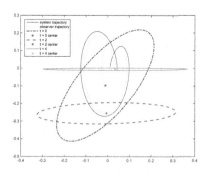

Fig. 3. System trajectories and reachable sets evolution.

Size of reachable sets reduce in time. They simultaneously contain system trajectory and observer location that is its center.

5 Conclusion

Optimal evaluation of reachable set for system with unknown initial state and parametric uncertainty problem have been considered in this article. Numerical experiments have been carried out for the assigned problems. The construction of reachable sets for systems with an unknown initial state has been demonstrated using the Mathieu equation as an example. Reachable set has been constructed, which includes the reachable sets for all admissible values of the parameters for the linear oscillator with floating stiffness coefficient equation. Finally, an evaluation of the reachable sets at different times has been constructed for the Mathieu-Hill dying-away equation. It is assumed that the problem will develop towards the evaluation of reachable sets for problems with parametric uncertainty.

References

1. Rogalev, A.N., Rogalev, A.A.: Controlling the path and reachable set estimations of unmanned air vehicle. Mathematical methods of modelling, control and data analysis (2017)
2. Holmes, P., Kousik, S., Zhang, B.: Reachable Sets for Safe, Real-Time Manipulator Trajectory Design (2020)
3. Lagosha, B.A., Apal'kova, T.G.: Optimal control in economics: theory and applications. Finance and statistics, Moscow (2008)
4. Bolodurina, I.P., Lygovskova, Y.P.: Optimal control of immunological reactions of the human body. Control Sci. **5**, 44–52 (2009)
5. Shatkhan, F.A.: Application of maximum principle to optimization problems of parallel chemical reactions. Avtomat. i Telemekh. **25**(3), 368–373 (1964)
6. Chernousko, F.L.: Estimation of the Phase State of Dynamic Systems. Nayka, Moscow (1988)
7. Kurzhanski, A., Valyi, I.: Ellipsoidal Calculus for Estimation and Control. IIASA, Laxenburg
8. Filippova, T.: Ellipsoidal Estimates of Reachable Sets for Control Systems with Nonlinear Terms. IFAC (2017)
9. Chernousko, F.L., Ovseevich, A.I.: Properties of the optimal ellipsoids approximating the reachable sets of uncertain systems. J. Optim. Theory Appl. **120**(2), 223–246 (2004)
10. Goncharova, E., Ovseevich, A.: Small-time reachable sets of linear systems with integral control constraints: birth of the shape of a reachable set. J. Optim. Theory Appl. **168**(2), 615–624 (2015). https://doi.org/10.1007/s10957-015-0754-4
11. Balandin, D.V., Biryukov, R.S., Kogan, M.M.: Ellipsoidal reachable sets of linear time-varying continuous and discrete systems in control and estimation problems. Automatica **116**, 1–8 (2020)
12. Kvakernaak, H., Sivan, R.: Linear Optimal Control Systems. Wiley, New York (1972)
13. Sorokina M.S.: Optimal evaluation of linear time-varying systems using reachable sets. In: 9th International Scientific Youth School-seminar <<Mathematical Modelling, Numerical Methods and Program Complexes>>, Saransk, pp. 252–258 (2020)

14. Sorokina, M.S.: Application of reachable sets in optimal estimationof linear tyme-varying systems. In: Mathematical Modelling and Supercomputer Technologies, Nizhny Novgorod, pp. 359–361 (2020)
15. Balandin, D.V., Kogan, M.M.: Synthesis of Control Laws Based on Linear Matrix Inequalities. Nauka, Moscow (2007)

Supercomputer Simulation

M2H3D Code: Moving Mesh Hydrodynamics by Means AVX-2 Technology

Igor Kulikov[1]([✉]), Igor Chernykh[1], Eduard Vorobyov[2], Vardan Elbakyan[3], and Lyudmila Vshivkova[1]

[1] Institute of Computational Mathematics and Mathematical Geophysics SB RAS, Novosibirsk, Russia
kulikov@ssd.sscc.ru, chernykh@parbz.sscc.ru, lyudmila.vshivkova@sscc.ru
[2] Department of Astrophysics, University of Vienna, Vienna, Austria
eduard.vorobiev@univie.ac.at
[3] Research Institute of Physics, Southern Federal University, Rostov-on-Don, Russia
vgelbakyan@sfedu.ru

Abstract. A new M2H3D code (Moving Mesh Hydrodynamics in 3D) was described. A new approach to vectorization of computational fluid dynamics algorithms adapted for astrophysical applications is proposed. A computational model is briefly described as an example of the approach. A review of papers on vectorization of calculations to simulate hydrodynamic processes and related problems is presented. The computational technology and vector instructions used to speed up the critical parts of the code are described. A performance of 90 gigaflops with a single Intel Cascade Lake processor using an AVX2 technology is achieved. Some numerical examples are given.

Keywords: HPC · Computational astrophysics · SIMD intrinsics

1 Introduction

Vector extensions can significantly speed up the calculations [1], of astrophysical problems in particular [2]. Vectorization is one of the basic components of Intel Xeon Phi processors that calls for a special design of the corresponding numerical methods and codes [3–5]. In recent years, vector calculations have been widely used to organize computations in numerous problems of computational astrophysics. Vectorization has been effectively employed to simulate hydrodynamic flows [6,7], N-body gravitational interaction [8–11], and plasma physics [12]. Some original methods for solving the corresponding Poisson equations [13] and hydrodynamic equations [14] have been developed. In the paper, we are described the vectorization of developed code from papers [15,16].

In Sect. 2, a numerical model of gravitational hydrodynamics is briefly described. Section 3 is devoted to a description of the code. Section 4 presents some numerical examples of using the code. Conclusions to the paper are given in Sect. 5.

© Springer Nature Switzerland AG 2021
D. Balandin et al. (Eds.): MMST 2020, CCIS 1413, pp. 307–319, 2021.
https://doi.org/10.1007/978-3-030-78759-2_26

2 Numerical Model

Consider the conservation laws of hydrodynamics in flux-conservative form

$$\frac{\partial}{\partial t}\begin{pmatrix} \rho \\ \rho u_x \\ \rho u_y \\ \rho u_z \\ \rho E \end{pmatrix} + \frac{\partial}{\partial x}\begin{pmatrix} \rho u_x \\ \rho u_x u_x + p \\ \rho u_y u_x \\ \rho u_z u_x \\ [\rho E + p]\, u_x \end{pmatrix} + \frac{\partial}{\partial y}\begin{pmatrix} \rho u_y \\ \rho u_x u_y \\ \rho u_y u_y + p \\ \rho u_z u_y \\ [\rho E + p]\, u_y \end{pmatrix}$$

$$+ \frac{\partial}{\partial z}\begin{pmatrix} \rho u_z \\ \rho u_x u_z \\ \rho u_y u_z \\ \rho u_z u_z + p \\ [\rho E + p]\, u_z \end{pmatrix} = \begin{pmatrix} 0 \\ -\rho\,\nabla_x\,\Phi \\ -\rho\,\nabla_y\,\Phi \\ -\rho\,\nabla_z\,\Phi \\ -\rho(\mathbf{u},\nabla\Phi) \end{pmatrix},$$

with the Poisson equation for the gravitational potential

$$\triangle\Phi = 4\pi G\rho,$$

where ρ is the gas density, $\mathbf{u} = (u_x, u_y, u_z)$ is the velocity vector, p is the gas pressure, $E = p/(\gamma - 1) + \rho\mathbf{u}^2/2$ is the total mechanical energy of the gas, γ is the adiabatic index, Φ is the gravitational potential, and G is the gravitational constant. To solve the equations of hydrodynamics, we use a Godunov-type scheme adapted for tetrahedral grids [15,16], and to solve a Riemann problem, a Rusanov-type scheme [4,5] with a piecewise linear reconstruction of the physical variables [17]. To solve the Poisson equation, a potential defined at the calculation grid nodes and a finite element method in a weak formulation for a Laplace operator discretization are employed. When constructing a high-order accurate scheme for solving the hydrodynamic equations, the density function, as well as the potential, are considered at the tetrahedron nodes. On the domain boundary, a fundamental solution of the Laplace equation is used in the form of boundary conditions of the first kind. The conjugate gradient method is used to solve the thus obtained sparse system of equations.

3 Code

In this section, the code design, main data structures, and vectorization of the calculations will be described. Some basic parts of the code in the C/C++ language or, for simplicity, in a pseudocode (when it is more convenient) will be presented. We hope that these examples would help the readers to implement the code.

3.1 Data Structures

The main feature of polygonal grids is the use of analytical geometry procedures. A list of appropriate procedures will be presented below. All procedures are based on a three-dimensional vector (see Listing 1.1):

Listing 1.1. 3D vector

```
struct vector3d
{
 double x, y, z;
};
```

Let us define a set of physical variables, such as density, velocity vector, pressure, and speed of sound $c_s = \sqrt{\frac{\gamma p}{\rho}}$. Although the speed of sound is easy to reconstruct, it is also more convenient to use it. A set of conservative variables is chosen depending on the equations to be solved (see Listing 1.2):

Listing 1.2. Physical and conservative variables

```
struct physics
{
 double density, pressure, sound,
        velocity_x, velocity_y, velocity_z;
};

struct conservative
{
 double density, energy,
        momentum_x, momentum_y, momentum_z;
};
```

A more complex construction for a new type called **conservative** was used in vector computations. A simpler hydrodynamic model without solving the equation for total energy will be used for vectorization. Such models (for instance, adiabatic or isothermal gas models) are widespread in astrophysics. For vectorization, we employed a C/C++ construction called **union**, in which the structure and the vector type of the data are used (see Listing 1.3):

Listing 1.3. Conservative variables with vectorization

```
struct scalar_conservative
{
 double density, momentum_x, momentum_y, momentum_z;
};

union conservative
{
 scalar_conservative scalar_vector;
 __m256d simd_vector;
};
```

To describe the nodes, we will store their coordinates, the number of neighboring nodes, an array of numbers of the neighboring nodes, the status (whether a node is a boundary one or not), the number of cells in which it is included, the set of the physical variables considered at the node, and the gravitational potential value (see Listing 1.4):

Listing 1.4. "Node" data structure

```
struct node
{
  vector3d point;          //node coordinates
  int number_of_neigh;    // number and numbers
  int *neighbours;        // of neighboring nodes
  int in_border;          // status ''on the boundary''
  int number_of_cells;    // number of adjacent cells
  physics hydro;          // physical variables
  double gravity;         // potential function
};
```

To reconstruct the potential by a finite element method, we need linear basic functions and the potential gradient (see Listing 1.5). Data of this type will be used in a cell that will be described at the end of this subsection. Assume that each tetrahedral cell $ABCD$ is constructed to be non-degenerate. We will use four basic functions, $\phi_i = a_i x + b_i y + c_i z + d_i$ and Φ_i, where $i = A, B, C, D$ are the cell nodes.

Listing 1.5. "Gravity" data structure

```
struct gravity_term
{
  double a[A|B|C|D],      // coefficients of linear functions
         b[A|B|C|D],
         c[A|B|C|D],
         d[A|B|C|D];
  double phi[A|B|C|D];    //gravity values at nodes
  double gradfix,         // potential gradient
         gradfiy,
         gradfiz;
};
```

The structure notation is slightly shortened by listing the variants in the square brackets. The description of the data structures ends with a description of a cell (see Listing 1.6):

Listing 1.6. "Cell" data structure

```
struct cell
{
  // number of nodes
  int numA, numB, numC, numD;
  // numbers of neighboring cells
  int ngh[ABC|ABD|ACD|BCD];
  // cell volume
```

```
double vol;
// face area
double sq[ABC|ABD|ACD|BCD];
// center of inscribed sphere
vector3d center;
// radius of inscribed sphere
double radius;

// vectors of normals
vector3d nrm[ABC|ABD|ACD|BCD];
// physical variables
physics hydro;
// conservative variables
conservative vector;
// gravity
gravity_term gravity;
// Riemann problem solutions
conservative flux[ABC|ABD|ACD|BCD];

// piecewise linear reconstruction
// of variables and fluxes
physics rechydro[ABC|ABD|ACD|BCD];
conservative vector[ABC|ABD|ACD|BCD];
conservative flux[ABC|ABD|ACD|BCD]_x;
conservative flux[ABC|ABD|ACD|BCD]_y;
conservative flux[ABC|ABD|ACD|BCD]_z;
};
```

Let us consider calculation grid reading.

3.2   Grid Format

A calculation grid format is constructed so that no procedures for finding nodes or cells in any computational situation are needed in the hydrodynamic code. The grid is stored in the following four files:

1. nodes.dat – node coordinates,
2. cells.dat – tetrahedra splitting the calculation domain,
3. portrait.dat – a grid portrait with node links,
4. border.dat – numbers of nodes lying on the domain boundary.

We used an explicit separation of the node geometry and the logic of work with nodes and cells. This approach simplifies the development, debugging, and subsequent maintenance of an application. The algorithm of work with these files in a pseudo-language is given below (see Listing 1.7):

Listing 1.7. Calculation grid loading

```
read file nodes.dat:
read n            # read number of calculation grid nodes
do i=1,n          # cycle over all nodes
 read x,y,z       # read node coordinates
enddo             # end of cycle

read file cells.dat:
read m            # read number of cells
do i=1,m          # cycle over all cells
 read a,b,c,d     # read tetrahedron node numbers
 read abc         # read cell number - neighbor on face abc
 read abd         # read cell number - neighbor on face abd
 read acd         # read cell number - neighbor on face acd
 read bcd         # read cell number - neighbor on face bcd
enddo             # end of cycle
                  # if the neighboring cell number is 1,
                  # the face is external

read file portrait.dat:
read n            # read number of calculation grid nodes
do i=1,n          # cycle over all nodes
 read l           # read number of adjacent (i-1) nodes
 do j=1,l         # cycle of adjacent nodes
  read k          # read adjacent node number
 enddo            # end of cycle of adjacent nodes
enddo             # end of cycle of nodes

read file border.dat:
read n            # read number of grid boundary nodes
do i=1,n          # cycle over boundary cells
 read k           # read boundary node number
enddo             # end of cycle of boundary cells
```

File portrait.dat is very important for constructing a sparse matrix portrait in which the numbers of adjacent nodes and the node in question are arranged in ascending order. File border.dat is used to take into account the boundary conditions of the Poisson equation. We used a Fortran-like grid reading procedure to shorten the program code, which is written in C/C++.

3.3 Geometry Subroutine

To implement analytical geometry procedures, we used the following functions:

1. **length** – distance from point to origin of coordinates,
2. **distance** – distance between two points,
3. **square** – triangle area by its three vertices,

4. **volume** – tetrahedron volume by its four vertices,
5. **scalar_dot** – Euclidean inner product,
6. **normal** – construct a normal to a triangle face given by its three vertices at a given tetrahedron center,
7. **center_of_sphere** – find the center of the sphere inscribed in a tetrahedron with its given vertices and face areas,
8. **radius_of_sphere** – find the radius of the sphere inscribed in a tetrahedron with its given face areas and volume,
9. **determinant4by4** – determinant of a 4×4 matrix.

These procedures, which are simple from the point of view of analytical geometry, are sufficient for implementing the program code.

3.4 Physical Variables Reconstruction

To develop a high-order accuracy method, the numerical solution is reconstructed. For this, we perform the following algorithm:

1. **for all nodes:** average the physical variables over cells adjacent to the node in question,
2. **for all cells and for each cell face:** calculate three values for each physical variable f: f_- – the value of the physical variable at the opposite cell node, f_0 – the value of the physical variable at the center of the cell in question, f_+ – the value of the physical variable in the cell adjacent to the face in question,
3. **for all cells and for each cell face:** recalculate the values of the physical variable with a piecewise linear reconstruction by the equation

$$f^0 = f_0 + \frac{min(max(f_{0-}, 0), max(f_{+0}, 0)) + max(min(f_{0-}, 0), min(f_{+0}, 0))}{2},$$

where $f_{0-} = f_0 - f_-$, $f_{+0} = f_+ - f_0$.
4. **for all cells and for each cell face:** calculate the fluxes for each cell with the recalculated values f^0, (to be used for solving a Riemann problem).

Skip this step if a first-order accuracy scheme is constructed.

3.5 Hydrodynamic Equations Solver

Let us rewrite the equations of gravitational hydrodynamics in vector form:

$$\frac{\partial u}{\partial t} + \frac{\partial f(u)}{\partial x} + \frac{\partial g(u)}{\partial y} + \frac{\partial h(u)}{\partial z} = q(u),$$

where u is the vector of conservative variables, $f(u)$, $g(u)$, $h(u)$ are the fluxes in corresponding directions, and $q(u)$ is the right-hand side vector. Consider an arbitrary tetrahedron cell i, and use index j to describe the set of adjacent cells. Determine the outer normals n_{ij} of cell i in the direction of cells j. The volume V_i of the tetrahedron and the areas S_{ij} of the triangles are found using the above procedures. The Godunov-type scheme to be used is

$$\frac{u_i^{n+1} - u_i^n}{\tau} + \sum_j \frac{S_{ij}}{V_i} \left(F_{ij} n_{ij}^x + G_{ij} n_{ij}^y + H_{ij} n_{ij}^z \right) = q_i^n,$$

where $F_{ij} = \mathcal{R}\left(f, u_i^n, u_j^n\right)$, $G_{ij} = \mathcal{R}\left(g, u_i^n, u_j^n\right)$, $H_{ij} = \mathcal{R}\left(h, u_i^n, u_j^n\right)$ is the solution of a Riemann problem $\mathcal{R}\left(w, u^L, u^R\right)$ for the equations

$$\frac{\partial u}{\partial t} + \frac{\partial w(u)}{\partial x} = 0,$$

with initial conditions $u(x,t) = u^L$ at $x < 0$ and $u(x,t) = u^R$ at $x \geq 0$. To solve the Riemann problem, a Rusanov-type scheme is used:

$$\mathcal{R}\left(w, u^L, u^R\right) = \frac{w\left(u^L\right) + w\left(u^R\right)}{2} + \frac{\left|\frac{\partial w}{\partial u}\right|}{2} \left(u^L - u^R\right).$$

Note that this procedure is the most computer time-consuming, and we will speed it up using vector instructions.

3.6 Poisson Solver

Recall that to solve the Poisson equation four basic functions are defined on each tetrahedron. Each of these functions is equal to unity only at a single node and zero at the other nodes:

$$\phi_i(x, y, z) = a_i x + b_i y + c_i z + d_i,$$

where $i = A, B, C, D$. The coefficients a_i, b_i, c_i, d_i are reconstructed by Cramer's method. For this, we introduced a function for calculating the determinant of a 4×4 matrix. The gradient of each of the basic functions in a tetrahedron has a simple analytical form:

$$\nabla \phi_i = (a_i, b_i, c_i).$$

The sum of these functions makes it possible to construct the gradient of the potential in a tetrahedron cell:

$$\nabla \Phi = \begin{pmatrix} \Phi_A a_A + \Phi_B a_B + \Phi_C a_C + \Phi_D a_D \\ \Phi_A b_A + \Phi_B b_B + \Phi_C b_C + \Phi_D b_D \\ \Phi_A c_A + \Phi_B c_B + \Phi_C c_C + \Phi_D c_D \end{pmatrix},$$

where Φ_i are the values of the potential at the corresponding nodes. Equations for the stiffness matrix S and the mass matrix M in a finite element statement for the tetrahedron in question can be found in [16]. On the domain boundary, a fundamental solution to the Laplace equations is used in the form of boundary conditions of the first kind. To solve the thus obtained sparse system of equations, the conjugate gradient method (or the GMRES method) is used.

3.7 Vectorization

As noted above, the most computer time-consuming part is the calculation by the finite-volume scheme. The following code implements the Godunov scheme using AVX2 intrinsics (see Listing 1.8):

Listing 1.8. Vectorization of the Godunov scheme

```
// OpenMP pragma adjustment
#pragma omp parallel for default(none)        \
        // shared memory objects
        shared(cells , tau)                    \
        // private variables
        private(i,tauvec, number_of_cells \
                // area to volume ratio
                sqdivvolabc , sqdivvolabd , \
                sqdivvolacd , sqdivvolbcd) \
        // number of OpenMP threads
        num_threads(MIC_NUM_THREADS)           \
        // task   assignment method
        schedule(dynamic)


// the Godunov scheme
for( i = 0 ; i < number_of_cells ; i++ )
{
  // loading of time step into vector
  tauvec       = _mm256_set1_pd(-tau);

  // volume of cell
  cvol         =   cells[i].vol;

  // loading of area to volume ratios into vectors
  sqdivvolabc = _mm256_set1_pd(cells[i].sqabc / cvol);
  sqdivvolabd = _mm256_set1_pd(cells[i].sqabd / cvol);
  sqdivvolacd = _mm256_set1_pd(cells[i].sqacd / cvol);
  sqdivvolbcd = _mm256_set1_pd(cells[i].sqbcd / cvol);

  // vector implementation of the scheme
  cells[i].vector.simd_vector =
    _mm256_mul_pd(
        tauvec ,
        _mm256_add_pd(
          _mm256_add_pd(
            _mm256_mul_pd(sqdivvolabc ,
                  cells[i].fluxabc.simd_vector),
            _mm256_mul_pd(sqdivvolabd ,
                  cells[i].fluxabd.simd_vector)
                  ),
          _mm256_add_pd(
            _mm256_mul_pd(sqdivvolacd ,
                  cells[i].fluxacd.simd_vector),
            _mm256_mul_pd(sqdivvolbcd ,
```

```
                    cells [ i ]. fluxbcd . simd_vector )
                        )
                    )
                );
}
```

Let us test the implementation on Intel Cascade Lake processors in a cluster called NKS-1P of the Siberian Supercomputer Center. For the study, a 48-core Intel Xeon Platinum 8268 processor was used. For this, the number of threads will be varied, and the performance will be measured in gigaflops. The following line was used for compilation:

mpiicc -xCORE-AVX2 -qopenmp -O2 -o hllk3d main.cpp -lm

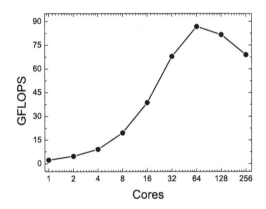

Fig. 1. The performance of vectorized code implementing the Godunov scheme with some numbers of OpenMP threads.

A performance of 90 gigaflops has been achieved (see Fig. 1), which corresponds to approximately 62% of that with AVX-512 technology [2], and is much greater than that in the first results of computational experiments on Intel Xeon Phi accelerators [3].

4   Numerical Example

4.1   Sedov Blast Wave

The Sedov point explosion problem is one of the main test problems for numerical methods and their software implementations in solving problems of supernova explosions based on core collapse. Consider a domain bounded by a radius $R = 0.5$. The adiabatic index $\gamma = 5/3$, which corresponds to neutral atomic hydrogen. The initial density in the domain $\rho_0 = 1$, and the initial pressure $p_0 = 10^{-5}$. An energy is injected at the initial time, $E_0 = 0.6$. The explosion area is limited by a radius $r = 0.01$. A density profile and an angular momentum profile are

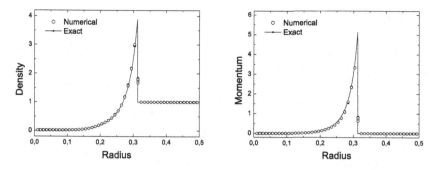

Fig. 2. Density and angular momentum profiles for the Sedov problem.

considered at time $t = 0.05$. The Sedov point explosion test is a standard test of the ability of a method and its implementation to reproduce strong shock waves with large Mach numbers, which takes place in explosions of spherical objects such as massive stars. The sound speed of the medium is rather small and, therefore, the Mach number can reach a value of $\mathcal{M} \approx 1500$. One can see that the above-developed numerical method reproduces the shock front quite well.

4.2 Evrard Collapse

The Evrard collapse problem is of interest in that first there is a short process of compression of the center, its rapid heating, and further expansion. To solve this problem, a non-rotating cloud is simulated with a dimensionless radius $R_0 = 1$, a density distribution within the radius $\rho(r) = 1/(2\pi r)$, adiabatic index $\gamma = 5/3$, and a total internal energy $u = 0.05$. The energy behavior is in quantitative and qualitative agreement (see Fig. 3) with the results obtained by other authors [18].

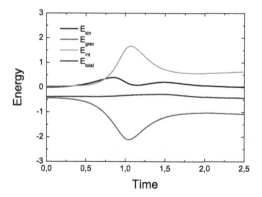

Fig. 3. Behavior of energy in simulation of Evrard collapse.

5   Conclusions

A new approach to the vectorization of computational fluid dynamics algorithms adapted for simulating astrophysical applications was presented. A new computational model of gravitational hydrodynamics was briefly described. With a vectorization procedure that implements a Godunov-type scheme, AVX2 tools were used to achieve a performance of 90 gigaflops with a single Intel Cascade Lake processor. Classical numerical examples (the Sedov problem and the Evrard collapse test) have been used to test the model. In future M2H3D code will should use to numerical simulation of star formation.

Acknowledgements. The reported study was funded by RFBR and FWF according to the research project 19-51-14002 (RFBR) and I4311 (FWF).

References

1. Amiri, H., Shahbahrami, A.: SIMD programming using Intel vector extensions. J. Parallel Distrib. Comput. **135**, 83–100 (2020). https://doi.org/10.1016/j.jpdc. 2019.09.012
2. Kulikov, I., Chernykh, I., Tutukov, A.: A new hydrodynamic code with explicit vectorization instructions optimizations that is dedicated to the numerical simulation of astrophysical gas flow. I. Numerical method, tests, and model problems. Astrophys. J. Suppl. Ser. **243**, Article Number 4 (2019). https://doi.org/10.3847/1538-4365/ab2237
3. Kulikov, I.M., Chernykh, I.G., Snytnikov, A.V., Glinskiy, B.M., Tutukov, A.V.: AstroPhi: a code for complex simulation of dynamics of astrophysical objects using hybrid supercomputers. Comput. Phys. Commun. **186**, 71–80 (2015). https://doi.org/10.1016/j.cpc.2014.09.004
4. Kulikov, I.M., Chernykh, I.G., Glinskiy, B.M., Protasov, V.A.: An efficient optimization of Hll method for the second generation of intel Xeon Phi processor. Lobachevskii J. Math. **39**(4), 543–551 (2018). https://doi.org/10.1134/S1995080218040091
5. Kulikov, I.M., Chernykh, I.G., Tutukov, A.V.: A new parallel intel xeon phi hydrodynamics code for massively parallel supercomputers. Lobachevskii J. Math. **39**(9), 1207–1216 (2018). https://doi.org/10.1134/S1995080218090135
6. Stone, J., Tomida, K., White, C., Felker, K.: The athena++ adaptive mesh refinement framework: design and magnetohydrodynamic solvers. Astrophys. J. Suppl. Ser. **249**, Article Number 4 (2020). https://doi.org/10.3847/1538-4365/ab929b
7. Mendygral, P., et al.: WOMBAT: a scalable and high-performance astrophysical magnetohydrodynamics code. Astrophys. J. Suppl. Ser. **228**(2), Article Number 23 (2017). https://doi.org/10.3847/1538-4365/aa5b9c
8. Yoshikawa, K., Tanikawa, A.: Phantom-GRAPE : a fast numerical library to perform n-body calculations. Res. Notes AAS **2**, Article Number 231 (2018). https://doi.org/10.3847/2515-5172/aaf7a2
9. Wang, L., et al.: NBODY6++GPU: ready for the gravitational million-body problem. Mon. Not. R. Astronom. Soc. **450**, 4070–4080 (2015). https://doi.org/10.1093/mnras/stv817

10. Rodriguez, C., Morscher, M., Wang, L., Chatterjee, S., Rasio, F., Spurzem, R.: Million-body star cluster simulations: comparisons between Monte Carlo and direct N-body. Mon. Not. R. Astronom. Soc. **463**, 2109–2118 (2016). https://doi.org/10. 1093/mnras/stw2121

11. Garrison, L., Eisenstein, D., Pinto, P.: A high-fidelity realization of the Euclid code comparison N-body simulation with ABACUS. Mon. Not. R. Astronom. Soc. **485**, 3370–3377 (2019). https://doi.org/10.1093/mnras/stz634

12. Surmin, I.A., Bastrakov, S.I., Efimenko, E.S., Gonoskov, A.A., Korzhimanov, A.V., Meyerov, I.B.: Particle-in-Cell laser-plasma simulation on Xeon Phi coprocessors. Comput. Phys. Commun. **202**, 204–210 (2016). https://doi.org/10.1016/j. cpc.2016.02.004

13. Khoperskov, S., Mastrobuono-Battisti, A., Di Matteo, P., Haywood, M.: Mergers, tidal interactions, and mass exchange in a population of disc globular clusters. Astron. Astrophys. **620**, Article Number A154 (2018). https://doi.org/10.1051/ 0004-6361/201833534

14. Hadade, I., di Mare, L.: Modern multicore and manycore architectures: modelling, optimisation and benchmarking a multiblock CFD code. Comput. Phys. Commun. **205**, 32–47 (2016). https://doi.org/10.1016/j.cpc.2016.04.006

15. Kulikov, I.M., Vorobyov, E.I., Chernykh, I.G., Elbakyan, V.G.: Application of geodesic grids for modeling the hydrodynamic processes in spherical objects. J. Appl. Ind. Math. **14**, 672–680 (2020). https://doi.org/10.1134/s1990478920040067

16. Kulikov, I.M., Vorobyov, E.I., Chernykh, I.G., Elbakyan, V.G.: Hydrodynamic modeling of self-gravitating astrophysical objects with the help of tetrahedron meshes. J. Phys. Conf. Ser. **1640**, Article Number 012003 (2020). https://doi. org/10.1088/1742-6596/1640/1/012003

17. Chen, G., Tang, H., Zhang, P.: Second-order accurate Godunov scheme for multi-component flows on moving triangular meshes. J. Sci. Comput. **34**, 64–86 (2008). https://doi.org/10.1007/s10915-007-9162-8

18. Springel, V.: E pur si muove: galilean-invariant cosmological hydrodynamical simulations on a moving mesh. Mon. Not. R. Astronom. Soc. **401**, 791–851 (2010). https://doi.org/10.1111/j.1365-2966.2009.15715.x

Comparison of AMD Zen 2 and Intel Cascade Lake on the Task of Modeling the Mammalian Cell Division

Maxim A. Krivov$^{(\boxtimes)}$ ⬥, Nikita G. Iroshnikov, Andrey A. Butylin, Anna E. Filippova, and Pavel S. Ivanov ⬥

Lomonosov Moscow State University, Moscow, Russia
m_krivov@cs.msu.su

Abstract. Modern architectures of central processors, in particular, AMD Zen 2 and Intel Cascade Lake, allow one to build shared memory systems with more than 100 computational cores. This paper presents the results of comparing the performance of these architectures shown on numerical modeling of mitosis in eukaryotes. The MiCoSi software that was developed by the authors and previously demonstrated a linear scalability when executed on cluster systems was used as a benchmark. The testing was performed on Amazon EC2 cloud nodes with 96 logical cores each. It is shown that, in relation to the problem of mitosis modeling, the two architectures under study have similar performance, while in some cases, Intel Cascade Lake bypasses its competitor AMD Zen 2 by 5–23%.

Keywords: Manycore processors · Shared memory · Parallel programming · Cell division · Mitosis · Prometaphase

1 Introduction

Modern processor architectures such as AMD Zen 2 and Intel Cascade Lake allow one to build computing systems with dozens or even hundreds of shared-memory computing cores. As a result, many theoretical and applied scientific problems that previously could be numerically solved on small clusters now can run on individual workstations.

This brings up the question of whether modern software packages can run on such systems in their natural SMP mode without any cluster emulation or separation of calculations by processes. This problem can relate both to architecture of the package itself, which does not allow to effectively use an increased number of cores, and to the specifics of memory organization, access to which is complicated due to the chiplet layout and multiprocessor configurations. As a result, the authors of numerous recently published papers assessed separately the

This work was supported by Russian Foundation for Basic Research (RFBR), grant 19-07-01164a (to M.K., N.I., A.F. and P.I.).

© Springer Nature Switzerland AG 2021
D. Balandin et al. (Eds.): MMST 2020, CCIS 1413, pp. 320–333, 2021.
https://doi.org/10.1007/978-3-030-78759-2_27

scalability of various algorithms, libraries, and packages when they are executed on systems with a really large number of cores. We pursue similar goals considering the problem of modeling cell division which is typical of such disciplines as biophysics and computational biology.

This paper has the following structure. Section 2 contains a comparison of the 48-core Amazon EC2 cloud nodes with processors based on AMD Zen 2 and Intel Cascade Lake architectures. Section 3 provides a brief description of the studied process of a eukaryotic cell division (mitosis) which results in the formation of two daughter cells. Section 4 is dedicated to the structure of the numerical algorithm and the chosen parallelization scheme. The specific settings of the virtual experiments and the characteristics of the MiCoSi package we are developing are given in Sect. 5. Finally, Sect. 6 contains conclusions about the performance of both architectures in relation to the considered subject area.

2   Computer Systems

2.1   Technical Specifications

The calculations were performed on the c5ad (AMD) and c5d (Intel) nodes of the Amazon EC2 cloud, which allow to create a computing system with 48 physical cores and a total memory of 192 GB. It is worth explaining that in Amazon EC2, one can only select the number of cores for which a certain level of performance is guaranteed. The specific processor models and even their architecture may change depending on the configuration of the virtual node and probably the selected data center.

Our measurements have shown that AMD EPYC 7R32 processors based on Zen 2 are always allocated for c5ad nodes. As for the c5d family, the architecture does vary. The maximum configuration with 48 cores (c5d.24xlarge) uses Cascade Lake processors, but when the number of cores decreases to eight (c5d.4xlarge), the former are replaced by a processor model of previous generation, based on a very similar, but still different Skylake-SP architecture.

Information about the technical characteristics of the nodes presented in Table 1 was obtained by summarizing the official documentation and the results of third-party tests. Since the processors in question were created specifically for Amazon, the values of similar models are given for unknown parameters.

It should be also noted that Zen 2 and Cascade Lake architectures support multiprocessor configurations, and the number of cores provided by Amazon EC2 cloud does not correspond to their uppermost capabilities. According to the documentation, for AMD EPYC processors, the maximum possible number of cores per node equals 128 (2 processors with 64 cores) while for Intel Xeon this figure reaches 224 (8 processors with 28 cores).

2.2   Comparison of Zen 2 with Cascade Lake

Due to their relative novelty, the Zen 2 and Cascade Lake architectures have not yet received significant coverage in the scientific literature. As a result, the

Table 1. Technical characteristics of the computer systems used.

Amazon EC2 node	c5ad.24xlarge	c5d.4xlarge	c5d.24xlarge
Processor	AMD EPYC 7R32	Intel Xeon Platinum 8124M	Intel Xeon Platinum 8275CL
Architecture	Zen 2	Skylake-SP	Cascade Lake
Year	2019*	2017	2019**
Technology, nm	7+14	14	14
TDP, W	280	240	240 × 2
Number of cores/threads	48/96	8/16 from 18/36	24/48 × 2
Base frequency, GHz	2.2*	3.0	3.0
Dynamic frequency, GHz	3.3	3.5	3.9
L1/L2 caches, KB/core	32+32/512	32+32/1024	32+32/1024
L3 cache, MB	192	24.75	35.75 × 2
Type of memory	DDR4-3200*	DDR4-2666	DDR4-2933**
Number of memory channels	8	6	6 × 2
Memory size, GB	192	32	192

*AMD EPYC 7552
**Intel Xeon Platinum 8270

comparison that follows is based on the information from technical presentations and documentation provided by their manufacturers, AMD and Intel.

AMD EPYC processors [1,2] are based on individual chiplets (Fig. 1), referred to as Core Complex Die (CCD). They are completely independent chips produced by 7 nm technology and subsequently combined into a single processor. Each such CCD chiplet, in turn, is divided into two groups of four cores (Core Complex, CCX), equipped with 16 MB of L3 cache memory. Another independent IO-chiplet is responsible for working with RAM and I/O interfaces. It is manufactured using 12 or 14 nm technology and adapted to a specific processor model. For example, the EPYC 7R32 processor uses an IO-chiplet manufactured by 14 nm technology and equipped with eight memory channels, which results in a total bandwidth of about 200 GB/s.

All CCD chiplets (and even CCX groups) cannot communicate with each other directly or address RAM explicitly. Corresponding requests must go through IO-chiplet to which they are connected by a high-speed bus. Depending on the model, the Zen 2-based processor can contain up to 8+1 chips, which in total provides up to 64 physical cores visible to the operating system. In addition to them, the server models contain another specialized ARM core responsible for implicit encryption of the contents of RAM.

Intel solutions [3] use a more classic monolithic architecture (Fig. 1), and the increase in the number of cores is provided by multiprocessor configurations. They may contain 10, 18, or 28 computing cores per chip but some of them may

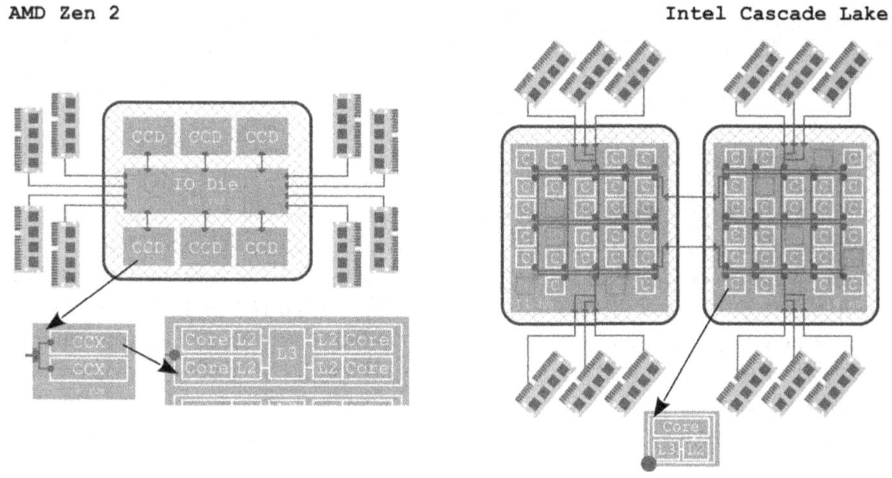

AMD Zen 2   **Intel Cascade Lake**

Fig. 1. The architecture of processors used in the c5ad.24xlarge (AMD Zen 2) and c5d.24xlarge (Intel Cascade Lake) nodes of Amazon EC2 cloud.

be disabled. For example, c5d.24xlarge nodes are based on processors with 28 cores, while only 24 of them are in operation. Each core has its own personal L2 cache of 1 MB, and also contains a portion of the total L3 cache (1.375 of 35.75 MB) available to all cores. Also, unlike Zen 2, the Cascade Lake architecture supports AVX-512 vector extensions, allowing, for example, the core to perform eight FP64 operations per CPU cycle.

All cores of a single processor are connected to each other by a high-speed on-chip interconnect with a topology in the form of rows and columns. Two memory controllers are implemented on each chip, providing a total of six channels and a bandwidth of about 128 GB/s per socket. Inter-chip interconnect referred to as Ultra Path Interconnect (UPI) and characterized by low latency and a bandwidth of about 60 GB/s, is used to communicate with cores and memory controllers from another socket.

It is almost impossible to determine which of these architectures is more preferable comparing only their technical characteristics. The main advantage of the chiplet approach is much larger number of transistors. For example, it allowed to significantly increase the total size of L3 cache (192 MB for Zen 2 versus 71.5 MB for Cascade Lake, see Table 1). However, due to the chiplet layout, L3 cache is no longer shared and a single core can use no more than 16 of 192 MB cache.

The situation with memory access looks quite similar. In Zen 2, all memory controllers are placed on a separate IO-chiplet, so the memory is unified, but at the cost of additional latency. In Cascade Lake, the access time depends on the exact location of memory controller that is processing the request, namely on the same chip as the core or in a neighboring socket. In the latter case, an additional data transfer is performed over the UPI links, which increases the latency and potentially can even reduce the throughput. On the other hand, this architecture allowed for higher peak throughput (256 GB/s for Cascade Lake versus 200 GB/s for Zen 2).

2.3   Related Works

Usually, manycore processors are considered only as a tool for solving scientific problems. This explains the comparative scarcity papers dedicated to a full–fledged comparison of the performance of SMP systems from Intel and AMD. For example, the authors of [4] evaluate the scalability with the number of cores for the modified NAS Parallel Benchmark Suite, but the measurements were carried out on the already outdated architectures Intel Haswell (18 cores × 4 sockets) and AMD Piledriver (16 cores × 4 sockets). It is also worth noting similar estimates for single-socket systems based on Intel Haswell/Broadwell and AMD Zen, but for calculations using the Vienna Ab initio Simulation Package [5].

A larger-scale comparison of the Intel Skylake and AMD Zen architectures was undertaken in [6] by performing hydrodynamic calculations on 100 nodes of two supercomputers using an in-house software Hydro3D as a benchmark. According to their results, the cluster based on Intel processors had better scalability with the number of nodes. However, the AMD Zen architecture, due to the larger number of cores (64 versus 40 per node), still provided the shortest calculation time.

Of particular interest is the work [7], dedicated to the analysis of data with CERN LHCb detector in real time, which used processors based on Intel Skylake (2 × 16 cores) and AMD Zen 2 (1 × 64 cores). The transition from AVX2 to AVX-512 vector extensions, which are supported only by Intel processors, resulted only in a slight (about 10%) increase in performance, while in general, the AMD Zen 2 architecture appeared to be more preferable. For the same number of cores, it beat Intel Skylake by about 16%, but when all 64 cores were used, the gap increased to 93%.

3   The Simulated Biological Process

The cell cycle of the vast majority of eukaryotic cells consists of four repeated stages, the key of which is mitotic division that results in the formation of two genetically identical daughter cells. Violation of this process can lead to cell death or, even worse, to the appearance of aneuploid daughter cells with an altered set of chromosomes, which in some cases can lead to the development of malignant neoplasms [8]. There are various protective mechanisms that suppress possible "errors" of cell division, and currently the understanding of their work is an important fundamental task [9].

The complexity of studying mitosis arises from the small physical sizes of the objects under study (fractions of a micrometer), as well as the variety of biochemical reactions involved in the process of cell division. As a result, biophysicists actively use the apparatus of mathematical modeling to quantitatively supplement and expand the experimental data. For example, authors [10] discussed the potentially exact formula for the polymerization rate of microtubules consisting of tubulin proteins when they start to interact with chromosomes during the first ten minutes of mitosis. An example of another area of research is the paper [11], in which a computer model of a yeast cell was used to evaluate such characteristics of living cells that have not yet been experimentally established.

The authors of this paper attempt to build their own complex mathematical model of mammalian cells division [12] which will holistically describe the course of three consecutive stages of mitosis, namely prometaphase, metaphase, and anaphase (Fig. 2). During the first stage, the nuclear membrane is destructed and chromosomes are released into the cytoplasm. Prometaphase is followed by metaphase with growth of tubulin microtubules and the appearance of their attachments to the chromosomes to pull the latter to the corresponding spindle poles. After some time, a balance of exerted forces aligns attached chromosomes in the equatorial plane of the cell. This stage is followed by anaphase, in which the centromere that binds the sister chromosomes breaks leading to the chromosomes distribution between two new daughter nuclei.

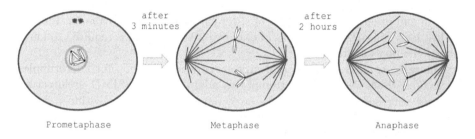

Fig. 2. Schematic representation of the stages of cell division accounted in the proposed model.

It is worth noting that there are numerous papers dedicated to the construction of mathematical models of the dividing cell and its individual parts, often based on mutually exclusive approaches to the description of mitosis. Many well-known models that were relevant as of 2012 are described and classified in the review [13].

4   Numerical Algorithm

The model of eukaryotic cell under consideration belongs to the class of the mechanical ones. Specifically, the chromosomes are represented by the closest geometric shapes (Fig. 3), which move both under the action of forces from the microtubules and according to some artificial laws that depend on the model settings. If the viscosity of cytoplasm is accounted for, the acceleration of chromosomes can be neglected. Knowing the forces acting on the chromosome at a particular time, one can calculate its linear and angular velocities. Further, by numerically integrating the velocities using the one-step Adams method, one can reproduce the trajectory of the chromosome motion during the modeled time interval $[t_{start}, t_{end}]$.

The structure of the numerical algorithm is shown in the listing in Table 2. From the point of view of biophysics, it is interesting to study not individual cells, but their ensembles consisting of 100 to 500 identical cells. Thus, the algorithm by its nature has a fairly good potential for parallelization (lines 2 and 5). To simulate a single cell, three loops at each time step should be performed. In the first (lines 7–14), the state of microtubules is updated and their collisions with chromosomes are searched. If the microtubule comes into contact with the chromosome, it can attach to its kinetochore (line 13), and also, depending on the model settings and probabilistic events, break off or begin to shorten.

The second loop (lines 16–20) is responsible for updating the position of the chromosomes. Each of them is affected by three types of forces (line 17) that originate from the dynamics of microtubules, Brownian motion of molecules, and the influence of the sister chromosome. Knowing the sum of these forces, one can construct a SLAE with a 6×6 matrix, the solution of which is the values of linear and angular velocities (line 18). After that, it is possible to update the position and orientation of the pair of sister chromosomes (lines 19–20). Finally, the last loop (lines 22–23) consists of only two iterations and is required to move the spindle poles artificially.

The main computational complexity arises when the loop for microtubules is executed (lines 7–14). The attempts to adapt it to the SIMD architecture were not very successful due to the large number of branches within a single iteration. For example, for the scenarios discussed in Sect. 5, the CUDA version of the algorithm provided an acceleration of about 2- to 6-fold relative to a single processor core. The application of techniques such as reordering iterations and using a hardware rasterizer allowed to increase the speed up to dozens of times, but significantly complicated the code and hindered the development of the mathematical model.

On the other hand, a more elegant solution turned out to be the parallelization of the loop by cells for the MIMD architecture (line 5), for which OpenMP+MPI technologies were used. Our measurements have shown [14] that when the ensembles of 1000 cells were simulated on ten nodes of the Lomonosov-2 cluster, an ideal scalability across cores was observed. Thus, this algorithm is a fairly illustrative example of the fact that manycore architectures such as Zen 2 or Cascade Lake can be in demand, including their usage in scientific modeling. The reason lies in the fact that some problems of computational biology are difficult to adapt to graphics accelerators while the computing power of classical clusters turns out to be redundant for them.

Table 2. Pseudocode of an algorithm that implements one of the versions of the cell division model proposed by the authors.

```
1    t ← t_start
2    cells[1..N] ← InitializeCells()
3
4    while t < t_end
5      parallel for each cell in cells
6
7        for each mt in cell.mtubules
8          ProbabilisticEvents(mt)
9          GrowOrShrink(mt, Δt)
10         if not mt.bound
11           for each chr in cell.chromosomes
12             if HasCollision(mt, chr) and MayAttach(mt, chr)
13               Bind(chr, mt)
14               break
15
16       for each chr in cell.chromosomes
17         force, torque ← ComputeForces(chr.bound_mtubules)
18         v, w ← ComputeVelocities(force, torque)
19         chr.pos ← chr.pos + v * Δt
20         chr.orient ← chr.orient + w * Δt
21
22       for each pole in cell.poles
23         MovePole(pole)
24     t ← t + Δt
```

5   Testing Methodology

To evaluate the performance of computing systems, we used the open source package MiCoSi [12]. It contains the implementation of the three-dimensional mathematical model of a dividing cell described in Sects. 3 and 4, and also has interfaces to flexibly configure numerical experiments and to measure physical quantities of interest. The solver version corresponded to the git revision eb11432, the package was built using the Visual C++ 2017 compiler and successfully passed all functional and unit tests.

The choice of the Amazon EC2 cloud and the Windows platform for testing was caused by the fact that MiCoSi software package is primarily addressed to users with modest programming skills. It is assumed that, using the user-friendly C# language, they will be able to create and debug a simple program on a local machine and then export the data they are interested in, for example, in the form of CSV tables for subsequent analysis using Excel or Python. In this case, to conduct large-scale computational experiments, it will be enough to transfer several files to the Amazon EC2 node, run calculations there, and then download tables with the results.

As a specific numerical experiment used as a benchmark, a program was developed to simulate prometaphase, a stage of cell division that lasts about 180 s and is characterized by the divergence of spindle poles in diametrically opposite parts of the cell. Such divergence generates a large number of interactions between tubulin microtubules and pairs of chromosomes that are freely floating in cytoplasm.

Two basic scenarios were considered. In the first, hereinafter referred to as 'Compute-bound' (Fig. 3a), the virtual cell consists of three pairs of chromosomes and 3000 microtubules. The simulation is performed in 0.1-s increments, and only the parameters of the final state of the cell are saved to the disk. This scenario generates a substantial computational load for performing geometric checks, as well as drawing up and solving SLAE for the subsequent determination of velocities. The second scenario, designated as 'IO-bound' (Fig. 3B), describes a simpler cell with one pair of chromosomes and 1000 microtubules. In addition, the states are unloaded every 0.01 s, resulting in about 628 MB of output data. As a result, the operations of serialization and packaging of information in the binary stream format *.cell begin to prevail.

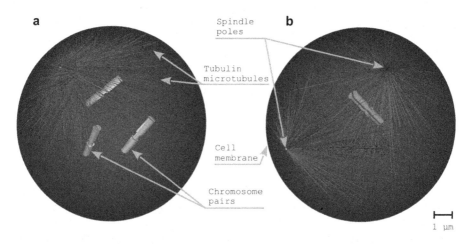

Fig. 3. Visualization of cells used for numerical simulation of prometaphase. (a) 'Compute-bound' scenario, 3 pairs of chromosomes and 3000 microtubules; (b) 'IO-bound' scenario, 1 pair of chromosomes and 1000 microtubules.

The size of the ensemble, depending on the type of test, was 48 and 384 virtual cells. The number of processor cores involved in the simulations was constrained by limiting the tasks, i.e. by subdividing the cells into groups of cells with exactly predetermined size. The total time spent executing the Launcher.StartAndWait() method was used as a metric. Testing was accomplished in the Amazon EC2 cloud on c5d (Intel Skylake-SP and Cascade Lake) and c5ad (AMD Zen 2) nodes with Windows Server 2019/20H2 installed. The software was launched from local NVMe disks that have a physical connection to the nodes.

6   Results and Discussion

6.1   Support for a Large Number of Cores Is Required Not Only from Applications, But Also from Operating Systems and Compilers

In a list of changes made over the past three years to the Visual C++ and gcc compilers, one can notice references to a lot of improvements in the architectures under study. In addition to the expected progressive enhancements, a support for chiplet processors is also declared, which should result in a noticeable performance gain for some patterns of memory operations. To evaluate this effect, the testing was performed using two versions of Visual C++ 2017 compiler, namely VC++ 14.1 (May 2017) and VC++ 14.16 (January 2021).

The situation with Windows Server operating system looks similar. According to the documentation from AMD, it is required to use versions that are based on at least the update 1903 (May 2019), as the new scheduler is better aware of the structure of caches, which increases the final speed by dozens of percent. As our tests have shown, these limitations are much more severe. The version based on the previous update 1809 (October 2018), installed in the Amazon EC2 cloud by default, forcibly sets the affinity mask and does not allow a single process to run on all the cores. Only 24 physical cores were used on the c5d.24xlarge (Cascade Lake) nodes, while only 32 physical cores were available on the c5ad.24xlarge (Zen 2) nodes. When the system was upgraded to version 20H2 (October 2020), these problems disappeared.

The total effect of the simultaneous update of the operating system and the compiler was indeed significant. When using 8–24 cores in the 'Compute-bound' scenario, the computation time was reduced by about 7–12% (Cascade Lake) and 9–18% (Zen 2). The greatest acceleration was observed for the 'IO-bound' scenario on the Zen 2 processor with the gain in the range 27–50%, which significantly improved the scalability with the number of cores.

6.2   For Intensive Computations on a Small Number of Cores, Cascade Lake and Zen 2 Demonstrate Similar Performance

Figure 4a shows the measurements at the c5d.4xlarge (Cascade Lake or Skylake-SP, depending on the launch) and c5ad.4xlarge (Zen 2) nodes, with eight physical cores each. In such scenarios, Intel Xeon processors can run at slightly higher clock speeds (3.9 GHz vs. 3.3 GHz), which may explain their better performance, although the gain was insignificant. In both cases, the scalability was near-linear (6-fold and 6.9-fold on eight cores). After switching to virtual cores, the Intel technology, which allows for the division of one physical core into two logical

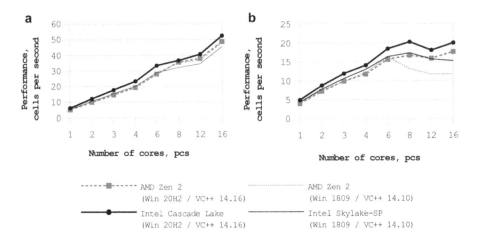

Fig. 4. Performance comparison when using c5ad.4xlarge (AMD Zen 2) and c5d.4xlarge (Intel Cascade Lake) nodes on an ensemble of 48 cells, more is better. (a) 'Compute-bound' scenario, without data unloading; (b) 'IO-bound' scenario, uploading data in 0.01 s increments.

ones, demonstrated slightly better results. For example, 16 Cascade Lake logical cores were 43% faster than 8 physical cores. For Zen 2, the speed up was 36%.

Unfortunately, the Cascade Lake and Skylake-SP architectures can hardly be compared objectively, since the nodes with Windows Server 20H2 have always been created only on the basis of Cascade Lake. Neglecting these differences, although it is not quite correct, the performance gain from switching to a new platform can be estimated as 13–17%.

6.3   On Active Memory Operations, Cascade Lake Is Slightly Ahead of Zen 2

The 'IO-bound' scenario (Fig. 4b) predictably demonstrated that from a certain point on, adding computational threads lead only to a decrease in performance. In both cases, this threshold turned out to be 6–8 cores. It is important to note that when operations with memory dominate over computations, the Cascade Lake architecture shows better results, consistently beating Zen 2 by 13–23%. These observations correlate well with technical specifications of these two architectures. At full load, Cascade Lake provides a larger number of memory channels per core (0.25 vs. 0.167 for Zen 2) and higher bandwidth per core (5.34 vs. 4.16 GB/s for Zen 2). It is also worth to account for the fact that the chiplet architecture of Zen 2 processors always causes additional latency, while Cascade Lake has cores and memory controllers both located on the same chip.

6.4   When the Second Socket Cores Are Used, Cascade Lake Sometimes Loses to Zen 2

The most interesting results have been obtained when the MiCoSi package was executed on all 96 logical cores of the c5d.24xlarge and c5ad.24xlarge nodes (Fig. 5). As noted in Sect. 6.1, these tests could only be performed on Windows Server 20H2, since older systems artificially limited the number of cores available to the process.

When the absolute performance is compared (Fig. 5a), the graph is clearly divided into two parts. In the range of 8–24 cores, Cascade Lake is the winner, consistently outperforming Zen 2 by 5–15%. However, for 32–96 cores, episodic failures begin to occur, thus allowing Zen 2 to significantly bypass its competitor. Such a picture is very characteristic of multi-socket systems which require a separate optimization of the executed program with special account for the physical heterogeneity of memory.

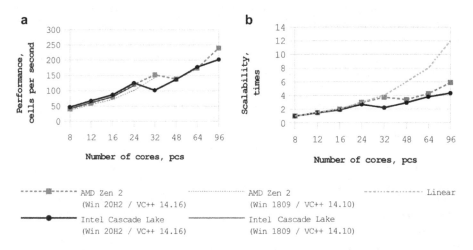

Fig. 5. Comparison of the performance of c5ad.24xlarge (AMD Zen 2) and c5d.24xlarge (Intel Cascade Lake) nodes with a predominance of computing load on an ensemble of 384 virtual cells, more is better.

Modern processors dynamically adjust the frequency even for a group of cores, remaining within the acceptable TDP level, which can somewhat "worsen" the scalability curve at their maximum load. Accordingly, the slight performance degradation observed for Zen 2 when switching from 32 to 48 cores may be of a similar nature. The base frequency of this processor is only 2.2 GHz, while for a group of cores it can increase up to 3.3 GHz.

It is also worth noting that in the relation to the effect of using logical cores, in the analogy with Sect. 6.2, a system based on AMD processors gets a greater benefit. For Zen 2, the acceleration brought about by the transition from 48 physical cores to 96 logical ones was 73%, while for Cascade Lake it appeared to be just 47%.

7 Conclusion

The use of synthetic benchmarks such as Linpack and HPCG is the most objective method of hardware platforms comparison, since they make it possible to measure the performance in absolute terms and approach the theoretically possible limits. Nevertheless, applied benchmarks that estimate the relative performance when solving a problem from an applied discipline are of particular interest. They are the ones that allow for making preliminary conclusions about the speedup one could expect when migrating to a new hardware platform.

The results of the comparison of Zen 2 and Cascade Lake in solving the considered problem of computational biology are an illustrative example of such an applied benchmark. The architectures considered are based on two alternative approaches: a chiplet layout and a large monolithic die. They can be easily compared based exclusively on technical characteristics or by extrapolating synthetic tests, while the latter is not quite correct. For example, Cascade Lake processors are produced using the formally outdated 14 nm technology and contain several times less L3 cache memory. Meanwhile, with the same number of cores, they have higher frequencies and support the new AVX-512 instruction set. Thus, it is quite difficult to make a full and objective comparison without solving applied problems from significantly different subject areas.

If one focuses only on the cost of renting the appropriate hardware, for the MiCoSi package, the Cascade Lake architecture is a preferred choice. The c5d and c5ad nodes presented in the Amazon EC2 cloud are positioned as interchangeable ones with the same rental price per core. At the same time, Intel solutions provide better performance in the scenario of intensive work with memory (13–23%) and slightly higher speed when accomplishing computations on the cores of a single die (5–15%). The only advantage of Zen 2-based processors is a much higher speed when running on 32 and 96 cores (19–50%). However, due to the lack of linear scalability with the number of cores, there are all reasons to believe that the observed 'performance spike' is caused by insufficient optimization of the MiCoSi package or by the chosen testing method.

References

1. Suggs, D., et al.: AMD "ZEN 2". In: 2019 IEEE Hot Chips 31 Symposium (HCS), pp. 1–24. IEEE Computer Society (2019)
2. Suggs, D., Subramony, M., Bouvier, D.: The AMD "Zen 2" processor. IEEE Micro **40**(2), 45–52 (2020)
3. Arafa, M., et al.: Cascade Lake: next generation Intel Xeon scalable processor. IEEE Micro **39**(2), 29–36 (2019)
4. Cho, Y., Oh, S., Egger, B.: Performance modeling of parallel loops on multi-socket platforms using queueing systems. IEEE Trans. Parallel Distrib. Syst. **31**(2), 318–331 (2019)
5. Stegailov, V., Vecher, V.: Efficiency analysis of Intel and AMD x86_64 architectures for Ab initio calculations: a case study of VASP. In: Voevodin, V., Sobolev, S. (eds.) RuSCDays 2017. CCIS, vol. 793, pp. 430–441. Springer, Cham (2017). https://doi.org/10.1007/978-3-319-71255-0_35

6. Ouro, P., Lopez-Novoa, U., Guest, M.: On the performance of a highly-scalable Computational Fluid Dynamics code on AMD, ARM and Intel processors. arXiv preprint arXiv:2010.07111 (2020)
7. Hennequin, A., et al.: A fast and efficient SIMD track reconstruction algorithm for the LHCb Upgrade 1 VELO-PIX detector. J. Instrum. **15**(06), P06018 (2020)
8. Ben-David, U., Amon, A.: Context is everything: aneuploidy in cancer. Nat. Rev. Genet. **21**, 44–62 (2020)
9. Cimini, D.: Detection and correction of merotelic kinetochore orientation by Aurora B and its partners. Cell Cycle **6**, 1558–1564 (2007)
10. Wollman, R., et al.: Efficient chromosome capture requires a bias in the 'search-and-capture' process during mitotic-spindle assembly. Curr. Biol. **15**, 828–832 (2005)
11. Edelmaier, C., et al.: Mechanisms of chromosome biorientation and bipolar spindle assembly analyzed by computational modeling. Elife **9**, e48787 (2020)
12. Krivov, M.A., Ataullakhanov, F.I., Ivanov, P.S.: Evaluation of the effect of cell parameters on the number of microtubule merotelic attachments in metaphase using a three-dimensional computer model. In: Panuccio, G., Rocha, M., Fdez-Riverola, F., Mohamad, M.S., Casado-Vara, R. (eds.) PACBB 2020. AISC, vol. 1240, pp. 144–154. Springer, Cham (2021). https://doi.org/10.1007/978-3-030-54568-0_15
13. McIntosh, R., et al.: Biophysics of mitosis. Q. Rev. Biophys. **45**, 147–207 (2012)
14. Krivov, M.A., et al.: Modeling the division of biological cells in the stage of metaphase on the "Lomonosov-2" supercomputer. Numer. Methods Program. **19**, 327–339 (2018). (in Russian)

Computing the Sparse Matrix-Vector Product in High-Precision Arithmetic for GPU Architectures

Konstantin Isupov[1]([✉]) [iD], Vladimir Knyazkov[2] [iD], Ivan Babeshko[1] [iD], and Alexander Krutikov[1] [iD]

[1] Vyatka State University, Kirov 610000, Russia
ks_isupov@vyatsu.ru
[2] Penza State University, Penza 440026, Russia
kniazkov@pnzgu.ru

Abstract. The multiplication of a sparse matrix by a vector (SpMV) is the main and most expensive component of iterative methods for sparse linear systems and eigenvalue problems. As rounding errors often lead to poor convergence of iterative methods, in this article we implement and evaluate the SpMV using high-precision arithmetic on graphics processing units (GPUs). We present two implementations that use the compressed sparse row (CSR) format. The first implementation is a scalar high-precision CSR kernel using one thread per matrix row. The second implementation consists of two steps. At the first step, the matrix and vector are multiplied element-by-element. The high efficiency of this step is achieved by using a residue number system, which allows all digits of a high-precision number to be computed in parallel using multiple threads. The second step is a segmented reduction of the intermediate results. Experimental evaluation demonstrates that with the same precision, our implementations are generally faster than CSR kernels built on top of existing high-precision general purpose libraries for GPUs.

Keywords: Sparse matrices · SpMV · GPU programming · High-precision arithmetic

1 Introduction

Single and double precision arithmetic is widely used in scientific computing and is natively supported by modern hardware and programming languages. However, floating-point operations introduce round-off errors that affect the results of calculations and in some cases cause problems. For example, in sparse linear algebra, iterative Krylov methods are widely used [17]. These methods converge in theory, but when using finite-precision floating-point arithmetic, they may converge slowly or even not at all [18]. High-precision arithmetic can be used to reduce the influence of round-off errors on numerical results.

This work was funded by the Russian Science Foundation grant number 20-71-00046.

ⓒ Springer Nature Switzerland AG 2021
D. Balandin et al. (Eds.): MMST 2020, CCIS 1413, pp. 334–345, 2021.
https://doi.org/10.1007/978-3-030-78759-2_28

Sparse matrix vector multiplication (SpMV), which calculates $y = Ax$ for a given sparse matrix A, occupies an important place in linear algebra algorithms and is the most important computational kernel for iterative linear solvers. A large body of research is devoted to the development of sparse matrix storage formats and native-precision SpMV implementations optimized for parallel computing platforms such as graphics processing units (GPUs) [1,3,12]. In turn, [5,7,14,18] offer high-precision versions of SpMV and other linear algebra kernels using double-double (DD) arithmetic that represents numbers by two double-precision floating-point numbers to emulate quadruple precision [6]. Well-known implementations of DD arithmetic are QD [6] and Lis [11]. In turn, the GQD library [13] is the GPU version of QD, while CAMPARY [10] supports not only the DD format, but also several other n-double formats for GPUs.

We discuss high-precision SpMV implementations from MPRES-BLAS[1], a library of high-precision linear algebra operations for CUDA-enabled GPUs [8]. MPRES-BLAS differs from existing high-precision libraries in that it uses a residue number system (RNS) rather than weighted number systems to provide high-precision capabilities. The RNS is interesting, in particular, because it provides efficient addition, subtraction and multiplication of large integers. These operations work on residues in parallel, and independently without carry propagation between them, instead of directly with the complete number [2].

MPRES-BLAS currently provides several different SpMV implementations in various sparse matrix formats. In particular, both scalar and vector high-precision kernels are available for the compressed sparse row (CSR) format. In addition, there is also a two-step SpMV, which is significantly different from the others. In this article, we discuss the scalar and two-step CSR implementations.

The rest of the article is structured as follows. In Sect. 2, we briefly describe the CSR matrix storage format. Section 3 contains the high-precision data types supported by the MPRES-BLAS library. The high-precision SpMV implementations are discussed in Sect. 4. An experimental evaluation is provided in Sect. 5, while conclusions are given in Sect. 6.

2 Compressed Sparse Row

The sparse matrix storage format has a significant impact on the performance of SpMV computation. There are several traditional formats such as Coordinate (COO), Diagonal (DIA), Compressed Sparse Row (CSR), ELLPACK/ITPACK, and Hybrid (HYB). Recently, many research efforts have been devoted to developing variations on traditional sparse storage formats optimized for parallel architectures. An extensive review of storage formats employed on modern GPUs that have appeared in the literature in recent years is given in [3].

In our high-precision SpMV implementations, we use the CSR format, which is the most common in iterative solvers. An example of the CSR format is shown in Fig. 1, where, similarly to [3], *AS* is an array of nonzero matrix coefficients, *JA* is an array of column indices and *IRP* is an array of row pointers.

---

[1] Available at https://github.com/kisupov/mpres-blas.

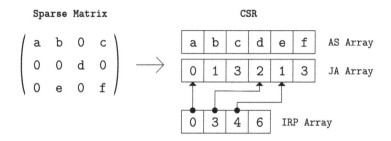

Fig. 1. Compressed Sparse Row (CSR) matrix storage format.

The main advantage of the CSR format is that it only stores nonzero matrix entries, without padding entries required in ELLPACK and DIA formats. This is especially important when the matrix entries are represented by high-precision numbers.

3   High-Precision Data Types of MPRES-BLAS

In MPRES-BLAS, a high-precision floating-point number x is represented as an object $x = \langle s, X, e, I(X/M) \rangle$, defined as follows:

- s is the sign of the number (either 0 or 1);
- $X = (x_1, x_2, \ldots, x_n)$ is the significand also called the mantissa;
- e is the integer exponent of the number;
- $I(X/M)$ is the extra part of the number needed to efficiently perform some complex operations in the RNS domain, such as sign determination, overflow detection, comparison, and rounding.

Being represented in the RNS [16] with a moduli set $\{m_1, m_2, \ldots, m_n\}$, the significand can take the values in the range from 0 to $M - 1$, where M is the product of all the m_i's. The size of the moduli set n specifies the number of digits in the significand, and each digit x_i is the least non-negative remainder when X is divided by m_i.

We can convert a high-precision number from the presented number format to the usual weighted form using the following formula:

$$x = (-1)^s \times \left| M_1 |x_1 w_1|_{m_1} + M_2 |x_2 w_2|_{m_2} + \cdots + M_n |x_n w_n|_{m_n} \right|_M \times 2^e,$$

where $M_i = M/m_i$, and w_i is the multiplicative inverse of M_i modulo m_i.

The additional attribute $I(X/M)$ included in the high-precision number format is called the interval evaluation of the significand. This evaluation represents an interval defined by its lower and upper bounds $\underline{X/M}$ and $\overline{X/M}$ that are finite precision floating-point numbers satisfying $\underline{X/M} \leq X/M \leq \overline{X/M}$. That is, $I(X/M)$ represents the range of possible values of $X = (x_1, x_2, \ldots, x_n)$ scaled by the moduli product M, and although the exact magnitude (i.e., weighted representation) of X remains unknown, the range provided is sufficient to efficiently perform many operations that are inherently difficult in the RNS.

The bounds $\underline{X/M}$ and $\overline{X/M}$ are native-precision floating-point numbers with an extended exponent range, which prevents numerical underflow in the case of large M (for $M < 2^{1000}$, the binary 64 format is enough for $\underline{X/M}$ and $\overline{X/M}$). The interval evaluation is calculated when converting a number to high-precision representation. When performing an arithmetic operation on high-precision numbers, the interval evaluation of the result is obtained in $O(1)$ time using interval arithmetic formulas. In addition, if required, $I(X/M)$ can be recalculated at any time from the residues $(x_1, x_2, \ldots, x_n)$ using only standard floating-point operations, i.e., without laborious RNS-to-binary conversion.

The high-precision floating-point data types in MPRES-BLAS is defined as C structures shown in Fig. 2.

```
typedef struct {              typedef struct {              typedef struct {
    int digits[n];                int *digits;                  int *digits;
    int sign;                     int *sign;                    int *sign;
    int exp;                      int *exp;                     int *exp;
    er_float_t eval[2];           er_float_t *eval;             er_float_t *eval;
} mp_float_t;                     int4 *buf;                } mp_collection_t;
                                  int  *len;
                              } mp_array_t;
```

Fig. 2. High-precision floating-point data types.

The mp_float_t type represents a single number, and the significand part is stored as a static array (digits[n]). For arrays, the mp_array_t type can be used, which stores a high-precision N-element array in a decomposed form, i.e., as a set of arrays representing separate parts of high-precision numbers. In particular, the significand parts are stored as an integer array of length $n \times N$. The mp_array_t type also includes the fields len (actual vector length) and buf (a buffer to transfer auxiliary variables between computational kernels). There is also the mp_collection_t type, which is a lightweight version of mp_array_t.

4   High-Precision SpMV for GPU

In this section, we discuss two implementations of high-precision matrix-vector product in CUDA. The input data for the implementations are a high-precision vector x of N elements and a high-precision sparse matrix A with NNZ nonzero elements, K rows and N columns. The result of the operation is a high-precision vector y of size K. We start with a scalar SpMV CSR kernel and then present a two-step implementation.

4.1   Scalar Kernel

The paper [1] describes a straightforward CUDA implementation of SpMV that uses one thread per matrix row to parallelize the computation; see Fig. 20 in [1]. If in this implementation, which is referred to as the scalar kernel, we replace the standard floating-point operations with the corresponding high-precision operations from the MPRES-BLAS library, we get our first high-precision SpMV. The pseudocode for this version is given in Algorithm 1.

Algorithm 1. High-precision scalar CSR kernel

1: row = threadIdx.x + blockIdx.x * blockDim.x
2: **while** row < K **do**
3:     dot = 0
4:     **for** i = IRP(row):IRP(row + 1) **do**
5:         mp_mul(prod, AS(i), x(JA(i)))   ▷ High-precision multiplication
6:         mp_add(dot, dot, prod)   ▷ High-precision addition
7:     **end for**
8:     y(row) = dot
9:     row += gridDim.x * blockDim.x
10: **end while**

In this kernel, each high-precision arithmetic routine is performed by one thread. The nonzero matrix entries are stored as an array of `mp_float_t` instances (AS). The input and output vectors are represented by similar arrays. For multiplying two high-precision numbers of the `mp_float_t` type, MPRES-BLAS provides the `mp_mul` function. In turn, the `mp_add` function performs high-precision addition of two numbers.

Advantages:
  - No coordination among threads within the same thread block/warp is required, so there is no need to use shared memory of the GPU. Note that in the case of high-precision arithmetic, shared memory can be a limiting factor for the occupancy of CUDA kernels (if the size of each high-precision number is too large).
  - Unlike the two-step implementation discussed in the next subsection, the scalar kernel does not require extra memory space to store intermediate results.

Drawbacks:
  - In RNS, all digits of a number can be computed in parallel by assigning one thread to compute one digit, however, in Algorithm 1, both high-precision addition and multiplication are computed by a single thread. If the precision (the number of RNS moduli) is quite high, sequential computation of all digits of the significand can take a long time, resulting in poor kernel performance, especially when the kernel is applied to a matrix with a many nonzero entries per row.

– In the CSR format, column indices (JA) and nonzero values (AS) are stored in row-major order, i.e., for a given row, these values are located contiguously in memory; however, they are not accessed simultaneously, which leads to an inefficient access pattern, since threads from the same warp access global memory with a stride. Moreover, the size of the stride depends on both the precision and the length of the matrix rows.

4.2 Two-Step Implementation

Our second SpMV CSR implementation uses the `mp_collection_t` type to store nonzero matrix entries, as shown in the example in Fig. 3, where $n = 4$, i.e., the significand of each nonzero value $\circ \in \{a, b, c, d, e, f\}$ consists of four digits, x_1, x_2, x_3, x_4. The symbol "." is used to access the parts of a high-precision number; l and u denote the lower and upper bounds of $I(X/M)$.

$$\begin{pmatrix} a & b & 0 & c \\ 0 & 0 & d & 0 \\ 0 & e & 0 & f \end{pmatrix} \longrightarrow$$

```
AS (mp_collection_t):
{
    digits = [ a.x₁ a.x₂ a.x₃ a.x₄ b.x₁ b.x₂ b.x₃ b.x₄
               c.x₁ c.x₂ c.x₃ c.x₄ d.x₁ d.x₂ d.x₃ d.x₄
               e.x₁ e.x₂ e.x₃ e.x₄ f.x₁ f.x₂ f.x₃ f.x₄ ]
    sign =   [ a.s b.s c.s d.s e.s f.s ]
    exp  =   [ a.e b.e c.e d.e e.e f.e ]
    eval =   [ a.l b.l c.l d.l e.l f.l
               a.u b.u c.u d.u e.u f.u ]
}
JA (int):   [ 0 1 3 2 1 3 ]
IRP (int):  [ 0 3 4 6 ]
```

Fig. 3. High-precision CSR format.

The input vector x and the output vector y are stored as `mp_array_t` instances. The implementation also requires a temporary global memory array BUF of size NNZ. The computation is a sequence of two steps:

1. The matrix A and vector x are multiplied element-by-element. The intermediate results are stored in the global memory array BUF.
2. A matrix-based segmented reduction of the BUF array is performed to produce the output vector y.

The first step is highly parallelizable thanks to the use of RNS. At this step, not only all elements of the intermediate array are computed in parallel, but also all the digits of each element. That is, n threads simultaneously compute one high-precision multiplication and the ith thread is assigned to calculate the ith digit modulo m_i. This approach is implemented using three kernel launches [9]:

1. Kernel #1—processing signs, exponents and additional information;
2. Kernel #2—processing digits;
3. Kernel #3—rounding.

The pseudocode of the kernels is presented in Algorithms 2, 3, and 4. In Algorithm 2, the functions rn_down and rnd_up indicate that the computation is performed in finite precision floating-point arithmetic with rounding down and rounding up, respectively. In Algorithm 4, the rnd_bits function calculates the number of rounding bits using the interval evaluation of the significand, and the pow2_scal function scales the significand by 2^{bits}. After scaling, the interval evaluation is recalculated using the calc_eval function.

Algorithm 2. Element-by-element multiplication of A and x — Kernel #1

```
1: i = blockDim.x * blockIdx.x + threadIdx.x
2: while i < NNZ do
3:     id = JA(i)
4:     BUF.sign(i) = AS.sign(i) xor x.sign(id)
5:     BUF.exp(i) = AS.exp(i) + x.exp(id)
6:     BUF.eval(i) = rnd_down(AS.eval(i) * x.eval(id) / ONE.upp)
7:     BUF.eval(NNZ + i) = rnd_up(AS.eval(NNZ + i) * x.eval(N + id) / ONE.low)
8:     i += gridDim.x * blockDim.x
9: end while
```

Algorithm 3. Element-by-element multiplication of A and x — Kernel #2

```
1: tid = threadIdx.x mod n                        ▷ n is the RNS moduli set size
2: m = MODULI(tid)
3: i = blockIdx.x * blockDim.x + threadIdx.x
4: j = (blockIdx.x * blockDim.x + threadIdx.x) / n
5: while i < NNZ * n do
6:     BUF.digits(i) = (AS.digits(i) * x.digits(JA(j) * n + tid)) mod m
7:     i += gridDim.x * blockDim.x
8:     j += gridDim.x * blockDim.x / n
9: end while
```

Algorithm 4. Element-by-element multiplication of A and x — Kernel #3

```
1: i = blockDim.x * blockIdx.x + threadIdx.x
2: while i < NNZ do
3:     bits = rnd_bits(BUF.eval(i))
4:     if bits > 0 then
5:         BUF.exp(i) += bits
6:         start = i * n
7:         end = n * (i + 1) - 1
8:         BUF.digits(start:end) = pow2_scal(BUF.digits(start:end), bits)
9:         BUF.eval(i) = calc_eval(BUF.digits(start:end))
10:    end if
11:    i += gridDim.x * blockDim.x
12: end while
```

The pseudocode of the segmented reduction kernel is given in Algorithm 5.

Algorithm 5. Segmented reduction of the intermediate array *BUF*

1: tid = threadIdx.x
2: row = threadIdx.x + blockIdx.x * blockDim.x
3: **if** row < K **then**
4:     sum(tid) = 0
5:     **for** i = IRP(row):IRP(row + 1) **do**
6:         *mp_add*(sum(tid), sum(tid), BUF(i))
7:     **end for**
8:     y(row) = sum(tid)   ▷ Output vector
9: **end if**

We summarize the advantages and drawbacks of the described two-step SpMV implementation as follows:

Advantages:
- The first step is highly parallelizable, since each digit of a high-precision number is computed by its own thread as shown in Algorithm 3.
- The element-by-element multiplication kernels (Algorithms 2 and 3) accesses the JA and AS arrays contiguously, producing the coalesced memory access pattern (this is not the case of the rounding kernel shown in Algorithm 4).
- The performance of the first step does not depend on the length of the matrix rows, since the matrix is treated as a one-dimensional array.

Drawbacks:
- Additional space must be allocated in the global GPU memory to store intermediate results (the *BUF* array).
- The approach leads to an increase in the number of global memory accesses.
- Unlike multiplication, one high-precision addition operation is computed by one thread, so the reduction kernel suffers from the same drawbacks as the scalar kernel from the previous subsection.

5   Performance Evaluation

We evaluated the performance of the presented SpMV implementations on matrices of different sparsity patterns from SuiteSparse Matrix Collection[2]. An overview of the matrices is presented in Table 1, where *NNZ* is total the number of nonzero values, while *MAXNZR* and *AVGNZR* is the maximum and average number of nonzero values per matrix row.

---

[2] https://sparse.tamu.edu.

Table 1. Matrices for experiments.

Name	Rows	*Nonzeros*	*MAXNZR*	*AVGNZR*
sme3Db	29 067	2 081 063	345	71
torso3	259 156	4 429 042	22	17
marine1	400 320	6 226 538	18	15
degme	185 501	8 127 528	624 079	43
atmosmodl	1 489 752	10 319 760	7	7
SiO2	155 331	11 283 503	2 749	72

The experiments were performed on a system with an NVIDIA RTX 2080 GPU (46 streaming multiprocessors, 8 GB of GDDR6 memory, compute capability version 7.5), an Intel Core i5 7500 processor and 16 GB of DDR4 RAM, running Ubuntu 20.04.1 LTS. We used CUDA Toolkit version 11.1.105 and NVIDIA driver version 455.32.00. The source code was compiled with the *-O3* option.

In the experiments, we evaluated the following high-precision SpMV implementations:

MPRES-BLAS (scalar)	— Scalar high-precision CSR kernel presented in Subsect. 4.1
MPRES-BLAS (two-step)	— Two-step high-precision CSR implementation presented in Subsect. 4.2
CAMPARY (scalar)	— Scalar CSR kernel using the CAMPARY high-precision library [10]
CUMP (scalar)	— Scalar CSR kernel using the CUMP high-precision library [15]
MPFR (OpenMP)	— Multicore CSR implementation using MPFR, a highly optimized library for high precision on CPUs [4]

The first four implementations were run on the GPU, while the last one was run in parallel on 4 CPU threads with 4 physical cores. The performance P is measured in high-precision operations per second: $P = 2 \times NNZ/T$, where T is the measured execution time. Note that each high-precision operation consists of several standard operations.

The evaluation results are shown in Table 2. We observe that the kernel using CAMPARY is the better choice for 106-bit precision. This is because the CAMPARY library provides optimized algorithms for DD arithmetic. In all other test cases, except for the "degme" matrix, the MPRES-BLAS implementations run faster.

Comparing the two MPRES-BLAS implementations, we see that the two-step one is less dependent on the precision than the scalar one. In particular, on the "marine1" matrix, when the precision increases by 8 times, the performance of the two-step SpMV decreases by 3.4 times, while the performance of the scalar SpMV decreases by 5.3 times, and the two-step version provides a speedup of up to 66% for 848-bit precision.

Unsurprisingly, the efficiency of high-precision SpMV depends on the matrix structure and sparsity pattern. The evaluated implementations exhibit fine-grained parallelism and hence perform better on the matrices "atmosmodl", "torso3", and "marine1", which have a large number of rows with few nonzero entries per row.

On the other hand, the "degme" matrix contains a number of very long rows, which increases the amount of computation performed by each GPU thread.

Table 2. Performance of various high-precision SpMV implementations in millions of high-precision op/s.

Matrix	Precision in bits	MPRES-BLAS (two-step)	MPRES-BLAS (scalar)	CAMPARY (scalar)	CUMP (scalar)	MPFR (OpenMP)
sme3Db	106	474.0	493.8	6813.3	410.4	83.1
	212	682.9	647.8	479.0	316.3	73.8
	424	446.7	372.5	92.5	225.0	69.2
	636	316.7	236.2	31.3	173.6	61.9
	848	259.6	172.3	16.0	129.4	54.4
torso3	106	1124.0	1032.8	11704.7	731.0	100.2
	212	998.6	1071.1	837.9	539.2	88.6
	424	604.7	579.8	155.0	375.5	38.1
	636	426.3	320.9	56.0	276.4	70.5
	848	333.3	239.3	26.4	208.1	70.1
marine1	106	1105.4	1023.9	10466.9	642.0	99.4
	212	939.0	922.8	784.6	439.7	82.7
	424	572.3	498.3	145.2	278.9	82.9
	636	417.5	282.8	53.8	200.2	71.2
	848	320.8	193.4	25.9	148.1	70.3
degme	106	14.4	8.1	79.4	9.5	77.1
	212	11.6	6.8	6.7	7.7	64.6
	424	8.2	4.5	1.0	7.3	64.2
	636	6.3	3.1	0.4	6.1	54.0
	848	4.1	2.3	0.2	5.3	50.5
atmosmodl	106	1209.4	1206.9	12377.4	961.9	87.7
	212	1052.7	1190.3	864.6	694.2	72.5
	424	575.3	579.8	163.3	467.5	77.1
	636	420.4	320.3	68.7	335.1	67.2
	848	319.0	218.5	31.0	249.5	66.8
SiO2	106	349.2	331.2	3194.1	385.7	64.2
	212	659.9	552.0	247.9	327.9	57.9
	424	412.0	325.6	49.6	221.2	51.9
	636	297.4	218.5	18.6	164.3	48.7
	848	223.9	180.3	7.0	119.2	47.1

Moreover, a very large stride between simultaneously addressed matrix entries leads to a drop in the effective memory bandwidth. The result is that the GPU implementations showed worse performance than the parallel CPU implementation using MPFR. For matrices with long rows, the GPU performance can be improved by assigning a group of threads to process each matrix row.

6 Conclusions

In this paper, we discussed two implementations of high-precision sparse matrix-vector multiplication in CUDA that are part of the MPRES-BLAS library. The high-precision capabilities of MPRES-BLAS are provided by using the residue number system, which has several advantages over weighted number systems. In particular, there is no carry propagation, which allows all digits of a high-precision number to be computed in parallel. This RNS benefit is exploited in our two-step SpMV routine, which is less dependent on the precision than the scalar CSR kernel. On the other hand, the two-step SpMV requires additional global memory space for intermediate results, while the scalar kernel does not.

It is worth noting that matrices that appear in real-world applications are usually represented in single or double precision and it is not necessary to convert them to higher precision. With this in mind, we will implement the multiplication of a sparse double-precision matrix by a high-precision vector and apply it to improve the convergence of iterative solvers.

References

1. Bell, N., Garland, M.: Efficient sparse matrix-vector multiplication on CUDA. NVIDIA Technical Report NVR-2008-004, NVIDIA Corporation (2008). https://www.nvidia.com/docs/IO/66889/nvr-2008-004.pdf
2. Bigou, K., Tisserand, A.: Improving modular inversion in RNS using the plus-minus method. In: Bertoni, G., Coron, J.-S. (eds.) CHES 2013. LNCS, vol. 8086, pp. 233–249. Springer, Heidelberg (2013). https://doi.org/10.1007/978-3-642-40349-1_14
3. Filippone, S., Cardellini, V., Barbieri, D., Fanfarillo, A.: Sparse matrix-vector multiplication on GPGPUs. ACM Trans. Math. Softw. 43(4), Article no. 30 (2017). https://doi.org/10.1145/3017994
4. Fousse, L., Hanrot, G., Lefèvre, V., Pélissier, P., Zimmermann, P.: MPFR: a multiple-precision binary floating-point library with correct rounding. ACM Trans. Math. Softw. 33(2), Article no. 13 (2007). https://doi.org/10.1145/1236463.1236468
5. Furuichi, M., May, D.A., Tackley, P.J.: Development of a stokes flow solver robust to large viscosity jumps using a Schur complement approach with mixed precision arithmetic. J. Comput. Phys. 230(24), 8835–8851 (2011). https://doi.org/10.1016/j.jcp.2011.09.007
6. Hida, Y., Li, X.S., Bailey, D.H.: Algorithms for quad-double precision floating point arithmetic. In: Proceedings of the 15th IEEE Symposium on Computer Arithmetic, Vail, CO, USA, pp. 155–162, June 2001. https://doi.org/10.1109/ARITH.2001.930115

7. Hishinuma, T., Hasegawa, H., Tanaka, T.: SIMD parallel sparse matrix-vector and transposed-matrix-vector multiplication in DD precision. In: Dutra, I., Camacho, R., Barbosa, J., Marques, O. (eds.) VECPAR 2016. LNCS, vol. 10150, pp. 21–34. Springer, Cham (2017). https://doi.org/10.1007/978-3-319-61982-8_4

8. Isupov, K., Knyazkov, V.: Multiple-precision BLAS library for graphics processing units. In: Voevodin, V., Sobolev, S. (eds.) RuSCDays 2020. CCIS, vol. 1331, pp. 37–49. Springer, Cham (2020). https://doi.org/10.1007/978-3-030-64616-5_4

9. Isupov, K., Knyazkov, V., Kuvaev, A.: Design and implementation of multiple-precision BLAS level 1 functions for graphics processing units. J. Parallel Distrib. Comput. **140**, 25–36 (2020). https://doi.org/10.1016/j.jpdc.2020.02.006

10. Joldes, M., Muller, J.-M., Popescu, V., Tucker, W.: CAMPARY: Cuda multiple precision arithmetic library and applications. In: Greuel, G.-M., Koch, T., Paule, P., Sommese, A. (eds.) ICMS 2016. LNCS, vol. 9725, pp. 232–240. Springer, Cham (2016). https://doi.org/10.1007/978-3-319-42432-3_29

11. LIS user guide. https://www.ssisc.org/lis/lis-ug-en.pdf. Accessed 24 Jan 2021

12. Liu, Y., Schmidt, B.: LightSpMV: faster CSR-based sparse matrix-vector multiplication on CUDA-enabled GPUs. In: 2015 IEEE 26th International Conference on Application-Specific Systems, Architectures and Processors (ASAP), pp. 82–89 (2015). https://doi.org/10.1109/ASAP.2015.7245713

13. Lu, M., He, B., Luo, Q.: Supporting extended precision on graphics processors. In: Sixth International Workshop on Data Management on New Hardware (DaMoN 2010), Indianapolis, Indiana, USA, pp. 19–26, June 2010. https://doi.org/10.1145/1869389.1869392

14. Masui, K., Ogino, M., Liu, L.: Multiple-precision iterative methods for solving complex symmetric electromagnetic systems. In: van Brummelen, H., Corsini, A., Perotto, S., Rozza, G. (eds.) Numerical Methods for Flows. LNCSE, vol. 132, pp. 321–329. Springer, Cham (2020). https://doi.org/10.1007/978-3-030-30705-9_28

15. Nakayama, T., Takahashi, D.: Implementation of multiple-precision floating-point arithmetic library for GPU computing. In: Proceedings of the 23rd IASTED International Conference on Parallel and Distributed Computing and Systems (PDCS 2011), Dallas, USA, pp. 343–349, December 2011. https://doi.org/10.2316/P.2011.757-041

16. Omondi, A., Premkumar, B.: Residue Number Systems: Theory and Implementation. Imperial College Press, London (2007)

17. Saad, Y.: Iterative Methods for Sparse Linear Systems, 2nd edn. SIAM, Philadelphia (2003)

18. Saito, T., Ishiwata, E., Hasegawa, H.: Analysis of the GCR method with mixed precision arithmetic using QuPAT. J. Comput. Sci. **3**(3), 87–91 (2012)

Performance Analysis of Deep Learning Inference in Convolutional Neural Networks on Intel Cascade Lake CPUs

Evgenii P. Vasiliev[1]([⊠]) [iD], Valentina D. Kustikova[1] [iD],
Valentin D. Volokitin[1,2] [iD], Evgeny A. Kozinov[1,2] [iD], and Iosif B. Meyerov[1,2] [iD]

[1] Department of Mathematical Software and Supercomputing Technologies,
Lobachevsky University, Nizhny Novgorod, Russia
[2] Mathematical Center, Lobachevsky University, Nizhny Novgorod, Russia

Abstract. The paper aims to compare the performance of deep convolutional network inference. Experiments are carried out on a high-end server with two Intel Xeon Platinum 8260L 2.4 GHz CPUs (48 cores in total). Performance analysis is done using the ResNet-50 and GoogleNet-v3 models. The inference is implemented employing the commonly used software libraries, namely Intel Distribution of Caffe, TensorFlow, PyTorch, MXNet, OpenCV, and the Intel Distribution of OpenVINO toolkit. We compare total run time and the number of processed frames per second and examine the strong scaling efficiency when using up to 48 CPU cores. Experiments have shown that OpenVINO provides the best performance and scales well up to 48 cores. We also observe that OpenVINO in the Throughput mode compared to latency mode accelerates inference from 4.9x for an image batch size of 1 to 1.4x for an image batch size of 32. We found that INT8 quantization in OpenVINO substantially improves the inference performance while maintaining almost the same classification quality.

Keywords: Deep learning inference · Convolutional neural networks · Performance analysis · Low-precision computations · Scaling efficiency

1 Introduction

Deep learning (DL) methods and models are commonly used in many research areas, namely pattern recognition [18,28,30], video analysis [13], natural language processing [33], bioinformatics [25], computational physics [15,16,24], and many others. The life cycle of DL models consists of three stages. At the first stage, the topology of the model is developed and trained. At the second stage, the constructed model is verified. At the last stage, the model is deployed into commercial products. The training procedure is often very computationally

The paper is recommended for publication by the Program Committee of the international conference Mathematical Modelling and Supercomputing Technologies-2020.

D. Balandin et al. (Eds.): MMST 2020, CCIS 1413, pp. 346–360, 2021.
https://doi.org/10.1007/978-3-030-78759-2_29

intensive and, as a rule, is performed only once on high-performance servers. On the contrary, employing the constructed model (deep learning inference) is reduced to a direct pass through the previously tuned neural network, which does not require high-performance hardware. However, there is a need to perform such inference multiple times for different input data, which explains the need to optimize inference implementations for modern and upcoming computing architectures. Intel Corporation made significant progress in this direction. A new toolkit for high performance deep learning inference, called OpenVINO, contains a wide range of inference tools optimized for Intel platforms. The purpose of this paper is to investigate the inference performance for two convolutional neural network models using state-of-the-art deep learning libraries. As in other studies [19, 26, 32], performance analysis is made on Intel CPU.

2   Related Work

Compared to classical machine learning methods, DL algorithms are computationally intensive both in training and in the use of deep models. During the last decade, pre-trained deep neural networks have become commonly used in various industries, including real-time systems, for which performance, latency, and throughput critically affect their applicability. Therefore, benchmarking the inference of deep models comes to the fore. There are specific software frameworks designed for benchmarking deep models. Such systems assess the DL training and inference performance of several widely used deep models on various hardware platforms.

MLPerf [26] is one of the commonly used benchmarks for measuring the performance of deep models on CPUs, GPUs, ASICs, and mobile devices. The results presented on the official project page are mainly obtained using NVidia GPUs, but there are also configurations with Intel Xeon CPUs and various neural accelerators. On the project web site, we can compare the performance in training and inference of deep models. DAWNBench [14] is another widely used benchmark. It contains performance data for three well-known DL models on different hardware. The results are collected by the user community. Deep Learning Workbench [17] is a GUI application for benchmarking and tuning the performance of deep models on various types of target devices.

Deep models performance analysis is a state-of-the-art topic. In [19], the inference performance results of deep models trained using TensorFlow on desktops (CPUs, GPUs, mobile SoC) are presented. The authors demonstrate an increase in inference performance as the next generation of hardware becomes available. In [32], the performance of manycore Intel Xeon Phi processors is assessed in training and inference of models using the Caffe and TensorFlow software libraries, including distributed training on two nodes with Intel Xeon Phi 7210. In [29], inference performance is compared for mobile and edge devices. The authors also provide performance data for the Intel Neural Compute Stick 2 neural coprocessor, which acts as an external USB device. Another research area is the selection of optimal parameter values to obtain the best performance on the available hardware configurations. When processing large amounts of data, fine-tuned parameters can save thousands of hours of CPU time. Paper [23] explores how DL algorithms load servers compared to other applications. The authors provide performance analysis data for DL models used in Facebook. The results of profiling individual operators of deep models are studied.

Research [22], supported by VMWare, analyzes the performance of multi-socket systems based on Gen 2 Intel Cascade Lake CPUs and the impact of virtualization with vSphere on performance. The paper presents a promising approach to improve performance, namely the creation of several virtual machines (VM) on one server when each VM processes its part of a dataset independently.

One of the commonly used methods for improving the performance of deep models is their quantization. This procedure is based on the employing of low-precision arithmetic. Quantization provides memory usage gains, while hardware support for INT8 data types leads to performance gains. The paper [20] reveals the essence of quantizing deep models, considers key problems of employing quantization, and compares the main features of several frameworks, supporting quantization. The authors also report on the performance gains achieved empirically by quantization. In paper [31], the theoretical basis of quantization algorithms is given in more detail, and the quality of quantization of deep models in several applications is empirically estimated.

We compare the performance and scaling efficiency of six DL frameworks, compute performance metrics and describe how to select the optimal run parameters. The tests are carried out on a high-end server with two Intel Xeon CPUs (48 cores and 96 threads in total), which support INT8 computations at the hardware level. We also study the impact of different modes and settings of the OpenVINO toolkit on the performance and scaling efficiency of DL inference.

3   Deep Learning Frameworks

We present a comparative analysis of the performance of the following DL inference frameworks: Intel Distribution of Caffe [5], TensorFlow [10], PyTorch [9], MXNet [1], OpenCV [8], and the Intel Distribution of OpenVINO toolkit [6]. Intel Distribution of Caffe is an implementation of the well-established Caffe library [2], optimized for Intel hardware. TensorFlow and PyTorch are among the commonly employed software libraries for solving research problems. The MXNet package is a cross-platform, high-performance library. It has a wide range of programming interfaces, employs symbolic computations, and implements a distributed training procedure. OpenCV is a commonly used library of CV algorithms, which includes a module that provides inference of deep models trained using a variety of frameworks. Typically, the library is used for the rapid development of image/video processing and analysis applications. OpenVINO is a relatively new framework that implements DL inference.

Note that many of the above toolkits have been already comprehensively described, therefore we take a closer look at OpenVINO's DL inference tools. OpenVINO implements optimizing DL models and allows building high-performance applications using deep models. The inference is optimized for various Intel hardware: CPUs, Intel Processor Graphics, Intel Movidius VPU, FPGA, Intel Gaussian & Neural Accelerator. OpenVINO supports the inference of models trained using a large number of commonly employed DL frameworks by converting pre-trained models to an intermediate representation and optimizing the model structure internally. DL inference involves a direct pass through the neural network for a set of input data, hereinafter referred to as a "batch". Such a single direct pass for a batch will be referred to as a "request".

OpenVINO supports two inference modes [21]. The first mode minimizes an execution time of a single request (*latency mode*). This mode assumes creating and executing one inference request on the selected device. The following request is executed only when the previous one completes. Reducing the execution time of a single request is achieved by parallelizing computations on separate network layers. This mode is synchronous and mainly aims to speed up *the execution of individual requests*. The second mode maximizes throughput (*throughput mode*). This mode assumes the creation of a set of inference requests. The order of execution of requests can be arbitrary. Run time improvement is provided due to the parallel processing of several requests during their execution. This mode is used in applications for which it is important to reduce the system's response time, for example, for CV or robotics applications such as analyzing the traffic situation from a stationary camera or video recorder, for multimedia applications, such as suppression of noise in an audio stream. The maximum throughput mode is used for problems that do not have strict restrictions on the processing time of one request. This mode allows faster completion of the processing of a large number of independent requests, for example, in an analysis of a text of search queries to a server. Inference in this mode can be asynchronous. The mode aims to increase *the throughput of the inference system*.

Applying low-precision computations is another resource for reducing the computation time of DL applications. Low-precision DL model training is widely used and provides excellent performance due to hardware support. However, employing low-precision computations can also improve the performance of DL inference. In this regard, some DL packages support the so-called *INT8-quantization* procedure. Model quantization in OpenVINO is implemented by adding specific quantization layers to a model. Firstly, the quantization of weights is performed. During this step, ranges of values of weights and activation functions are calculated, after which these ranges are rounded to the closest integer value [11]. Secondly, to improve the quality of the model, additional training of the quantized weights can be fine-tuned on a small subset of the training dataset.

4 Experimental Settings

4.1 Computational Infrastructure

The experiments are performed on a high-end two-socket system with the following characteristics: 2x Intel Xeon Platinum 8260L 2.4 GHz (2×24 cores and 2×48 threads overall) processors, TurboBoost is off, 196 GB RAM, CentOS 7.

The following DL libraries are used to compare performance: Intel Distribution of Caffe 1.1.0, Tensorflow 1.14.0, PyTorch 1.3.0, MXNet 1.5.0, OpenCV 4.1.1, Intel Distribution of OpenVINO toolkit 2020.2. We employ precompiled versions of frameworks, included in Anaconda 4.5.12, the codes are compiled with MKL-DNN support [7]. The performance testing infrastructure is developed in Python. The source codes are publicly available on GitHub [4].

4.2 Models and Data

We test performance and scaling efficiency using two state-of-the-art deep models: ResNet-50 [18] and Inception-v3 (GoogleNet-v3) [28]. These models allow solving the image classification problem with a large number of categories. The selected models contain 25 and 23 million parameters, respectively. We choose these two networks for the following reasons. Firstly, a large number of existing deep model topologies are developed based on the ResNet and Inception architectures, which allows using the performance results obtained in the article as an approximate estimate of the performance of many applications. Secondly, there exist pre-trained models for all selected DL frameworks. We also note that the ResNet-50 model is commonly used in research on the performance of deep models [12,14,26,29]. The quality of the models is assessed on the validation set (50 000 images) of the ImageNet dataset (ILSVRC2012) [27]. The inference performance of deep models is analyzed on a subset of the first 12 288 images of the ImageNet dataset. *In all intermediate figures, we present results for the ResNet-50 model, since the results for GoogleNet-v3 differ only in absolute values, but reflect the same trends.*

4.3 Performance Metrics

In all experiments, we assume that the set of test images is split into batches of equal size. The batch size is a parameter of the experiment. It will be shown below that choosing a relevant batch size critically affects performance. During the experiment, the total processing time of all batches is calculated. This metric reflects the total time for solving the problem; it is valid for all the DL frameworks under consideration. Along with this, the following metrics are calculated for Intel Distribution of Caffe, TensorFlow, MXNet, PyTorch, and for OpenVINO in the latency mode: 1) *Latency*. It is the median of the processing times of one batch. 2) The average *number of frames processed per second* (Frames per Second, FPS). It is the ratio of the batch size to the latency.

For the throughput mode in OpenVINO, we also calculate the FPS metric, which is defined as the ratio of the product of the image batch size and the number of iterations to the execution time of all requests.

5 Results and Discussion

5.1 Performance and Scalability Analysis of DL Inference

The DL inference is parallelized for shared memory systems employing the OpenMP and TBB technologies. Inference performance highly depends on the selection of the relevant values of parameters, in particular, the number of threads and the size of a batch of processed data. This section describes how we select the optimal values of such parameters when inferring the ResNet-50 and GoogleNet-v3 models. We also compute run time and scaling efficiency of the selected frameworks when the batch size changes. The frameworks under consideration use the MKL-DNN library to implement computations on neural network layers. While selecting the values of the parameters, the number of threads varies from 1 to 96, the size of the batch of images (batch size) is chosen from the set {1, 48, 96, 192, 384}. We also empirically find the best settings of affinity mask when a framework employs MKL-DNN parallelized with OpenMP.

Firstly, we tune the inference parameters for Intel Distribution of Caffe. The best performance results were obtained for the value of the parameter AFFINITY set to "compact, 1.0". We found that the optimal number of threads is equal to 48, which corresponds to the number of physical cores in the system (Fig. 1(a)). It is also shown that using 48 threads provides the best FPS for all batch size values, namely over 450 and 300 frames per second for ResNet-50 and GoogleNet-v3, respectively. Also, for batch sizes greater than 1, the speedup and FPS do not depend on the amount of input data being processed (the corresponding curves practically coincide). The framework demonstrates over 63% scaling efficiency in experiments with a large enough value of the batch size. In this case, the processing is performed in real-time for all enumerated parameters when the inference is run in parallel with the number of threads exceeding two.

Secondly, we run the same experiments for MXNet, also based on the MKL-DNN library, parallelized with OpenMP (Fig. 1(b)). Therefore, the best performance is achieved with the same affinity mask properties. The general conclusions drawn for Caffe are also true for MXNet. The framework demonstrates a scaling efficiency of 55% for the ResNet-50 model and over 63% for the GoogleNet-v3 model. The FPS value is approximately 300 frames per second for both models.

Next, we performed the same experiments using TensorFlow (Fig. 1(c)). TensorFlow also takes advantage of the MKL-DNN library parallelized on OpenMP. The best results are obtained with the "AFFINITY = scatter" setting. The inference works with FPS equal to 242 for ResNet-50 and 116 for GoogleNet-v3. Scaling efficiency when using the maximum number of cores is not as high as for Caffe and MXNet. It is equal to 37% for ResNet-50 and 23% for GoogleNet-v3.

Figure 2(a) shows the performance results for the PyTorch framework built using MKL-DNN. The best results in most runs for the ResNet-50 model are obtained with the setting "AFFINITY = scatter", they are shown in the diagrams. The inference of the GoogleNet-v3 model in the PyTorch 1.3.0 framework is run without MKL-DNN support due to incomplete support for all layers.

(a)

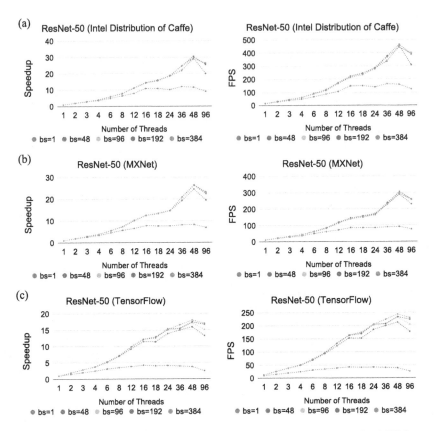

Fig. 1. FPS and speedup of DL inference in Intel Distribution of Caffe, MXNet and TensorFlow on the ResNet-50 model. The number of images in one batch is chosen from the set {1, 48, 96, 192, 384}. The number of threads varies from 1 to 96. The results on the GoogleNet-v3 model reflect the same trends.

Therefore, the maximum performance is not achieved. For the same reason, the demonstrated speedup and FPS values for sequential inference are much lower than those in other frameworks. In such limitations, PyTorch demonstrates scaling efficiency above 45% for the ResNet-50 model using MKL-DNN and up to 17% for the GoogleNet-v3 model without MKL-DNN. The best FPS values are equal to 230 and 44 fps for ResNet-50 and GoogleNet-v3, respectively.

DL inference in the OpenCV library can be run in two modes. The first mode uses an internal implementation of the layers in the library, the second one— an implementation from OpenVINO. Consider the results obtained in the first mode. Performance data are shown in Fig. 2(b). The OpenCV library scales well with an efficiency of 73% for ResNet-50 and 60% for GoogleNet-v3. Not that while scaling efficiency is good, the best FPS values are not very high (189 and 98 fps for ResNet-50 and GoogleNet-v3, respectively). Unlike other frameworks, the optimal number of threads for OpenCV is equal to the number of logical cores in a dual-processor system.

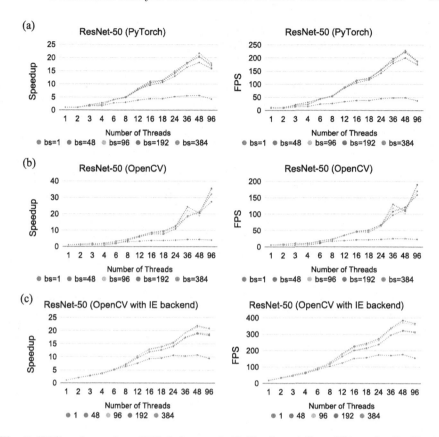

Fig. 2. FPS and speedup of DL inference in PyTorch and OpenCV on the ResNet-50 model. The number of images in one batch is chosen from the set {1, 48, 96, 192, 384}. The number of threads varies from 1 to 96. The results on the GoogleNet-v3 model reflect the same trends.

To run the inference using OpenCV in the second mode, we use a version of the library that supports the inference of OpenVINO models using the Open-VINO Inference Engine (Fig. 2(c)). The conclusions regarding the selection of the optimal parameters made for Caffe are also valid for this run mode of the OpenCV library. OpenCV demonstrates scaling efficiency above 45% in experiments with large values of batch size, the maximum values of FPS are 384 for ResNet-50 and 235 for GoogleNet-v3.

Finally, we perform the same experiments using OpenVINO. To run the inference, we employ the ResNet-50 and GoogleNet-v3 models trained using Caffe [2] and TensorFlow, respectively. Models are converted and optimized for Intel CPUs into an intermediate representation format using a specific tool from OpenVINO. Further, the converted models are used for inference. It is shown (Fig. 3) that OpenVINO outperforms other frameworks. The optimal value of the number of threads coincides with the number of physical cores, regardless of the batch size value. The framework demonstrates scaling efficiency over 59% and inference performance up to 500 fps.

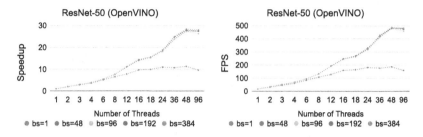

Fig. 3. FPS and speedup of DL inference in OpenVINO on ResNet-50. The number of images in one batch is chosen from the set {1, 48, 96, 192, 384}. The number of threads varies from 1 to 96. The results on GoogleNet-v3 reflect the same trends.

Note that OpenVINO can choose the relevant number of threads when inferring deep models. Figure 4 illustrates this fact. It is shown that when launched with the default parameters, OpenVINO chooses the number of threads so that we could not manually improve performance.

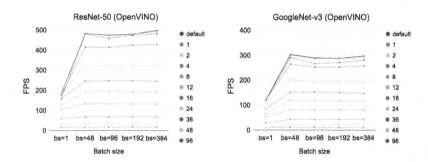

Fig. 4. Performance of the OpenVINO inference depending on the number of threads for two models: ResNet-50 and GoogleNet-v3. Each color corresponds to the number of threads. The red line shows the performance results when executing OpenVINO with default settings. It corresponds to the best results.

Experiments show that all considered frameworks demonstrate high performance and scalability when using 48 computational cores due to employing a highly optimized MKL-DNN library. The best performance is achieved by the OpenVINO and Intel Distribution for Caffe frameworks (about 500 fps for ResNet-50 and 300 fps for GoogleNet-v3). To achieve optimal performance, it is enough to choose the value of batch size greater or equal to the number of cores. Further increasing this setting does not change the results.

5.2   Performance and Accuracy of Quantized Models in OpenVINO

The quantization procedure in OpenVINO is implemented as follows. First, we need to download the original model trained in the FP32 or FP16 data types. Next, the model is converted to an intermediate representation of OpenVINO using the Model Optimizer tool. The model is then calibrated to INT8 representation using the Calibration Tool from OpenVINO. We used the DefaultQuantization algorithm [3] with default parameters. This algorithm sequentially applies Activation Channel Alignment, Min-max Quantization and Fast Bias Correction methods [3]. The model is trained for 2% of images from the validation dataset. The selection of the optimal parameters for employing the inference of INT8 models in the latency mode is performed as described in the previous section. The experimental results (Fig. 5) show that all our conclusions are also valid for calibrated models. The optimal number of threads corresponds to the number of physical cores, and the batch size should be taken greater or equal to 48. OpenVINO demonstrates scaling efficiency over 52% for the ResNet-50 model and up to 45% for the GoogleNet-v3 model. The inference performance of calibrated models grows by about 3 times and reaches 1604 fps for ResNet-50 and 894 fps for GoogleNet-v3. The substantial performance gain can be attributed to both the reduction in the size of the model weights by 4 times, which allows for more optimal data caching and the use of new vector neural network instructions (VNNI) from the AVX-512 instruction set.

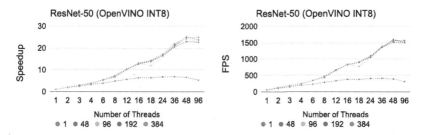

Fig. 5. Performance of the OpenVINO inference in the latency mode after INT8 calibration of ResNet-50. The results on GoogleNet-v3 reflect the same trends.

To assess the applicability of calibrated models, it is necessary to check their accuracy. This paper operates with pre-trained models from the Open Model Zoo or the official model repositories for each framework. Figure 6 reports the obtained values of the commonly employed quality indicators of image classification top-1 and top-5. These indicators have the following meaning. A DL inference procedure forms a vector whose elements contain probabilities of the image belonging to corresponding classes. The top-N metric is equal to the ratio of the number of images for which the true class is among the N maximum confidences of the predicted classes to the total number of processed images. We employ the Accuracy Checker tool, which is part of the Intel Distribution of OpenVINO toolkit. From the computed metrics, we conclude that the models classify images with nearly the same quality as before quantization and the achieved values correspond to the published ones. There are only slight deviations from the original results. Such deviations arise because of the following reasons. Firstly, each training experiment is unique due to a random initial state of the weights. Secondly, frameworks can have different implementations of the basic layers. It can also affect the results of performing elementary matrix operations that prevail in the inference process. Lastly, small deviations in the accuracy could also be a consequence of the weights quantization. However, we do not observe significant errors.

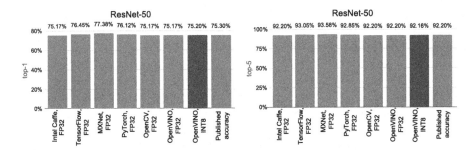

Fig. 6. The values of top-1 and top-5 accuracy metrics for original and INT8 quantized models.

5.3   Performance Analysis of the Throughput Mode in OpenVINO

The throughput mode allows for improving inference performance without increasing the batch size. OpenVINO splits computational resources into groups, called *streams*, in which computations can be performed simultaneously. Each stream processes one inference request taken from the queue of available requests. Consequently, OpenVINO can efficiently run multiple inference requests on the CPU (and other devices) in parallel, improving the throughput and better utilizing manycore CPUs.

Our experiment is as follows. Firstly, we choose the batch size and the number of requests that OpenVINO should process in parallel. Secondly, we set the number of streams corresponding to the size of the queue of available requests. The number of threads is taken by default. The dataset is then split into batches of a given size. The batches are added to the queue using the Round Robin algorithm. The number of batches equal to the number of requests is processed simultaneously. We vary the batch size and the number of requests to find the best values in terms of performance.

Figure 7(a) shows the experimental results. The maximum performance was obtained when using batches of 4 and 8 images when processing 24 or 48 requests simultaneously. As the number of images in each batch increases further, performance starts to degrade slightly. An important advantage of image processing in the throughput mode is sufficient performance even for requests with a batch of one image, provided there are enough requests. Anyway, the throughput mode greatly outperforms the latency mode in terms of total time.

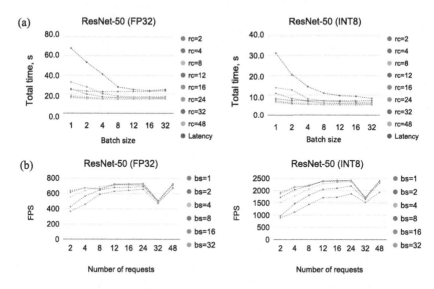

Fig. 7. DL inference performance in the throughput mode of OpenVINO depending on the batch size and the number of requests when using the FP32 and INT8 data types. The results on GoogleNet-v3 reflect the same trends.

By varying the number of simultaneous inference requests, we can draw the following conclusion (Fig. 7(b)). The number of requests should be a divisor of the number of CPU cores. For example, when using 32 requests, there is a loss of up to 20% of performance. Hence, for the considered models the best number of requests is equal to 24 (2 cores per 1 request) or 48 (1 core per 1 request).

5.4   Final Comparison

The best found parameters are presented in Table 1. Figure 8 shows the total processing time for a test set of 12288 images when using six DL frameworks. OpenVINO in the latency mode and Intel Distribution for Caffe, both working in the FP32 mode, demonstrate the best performance, while MXNet achieves similar results. Using the throughput mode and INT8 quantization in Open-VINO substantially reduces the run time. Note that the GoogleNet-v3 model in PyTorch 1.3.0 was implemented without MKL-DNN due to incomplete support for all layers, so maximum performance was not achieved.

Table 1. Empirically the best values of the parameters for DL inference of two models on two 24-cores CPUs (th – the number of threads; bs – the batch size; sn – the number of streams; rn – the number of requests).

	Precision	ResNet-50	GoogleNet-v3
Intel Caffe	FP32	th = 48, bs = 512	th = 48, bs = 96
OpenCV	FP32	th = 96, bs = 512	th = 48, bs = 512
TensorFlow	FP32	th = 96, bs = 512	th = 96, bs = 384
MXNet	FP32	th = 48, bs = 128	th = 48, bs = 192
PyTorch	FP32	th = 48, bs = 96	th = 48, bs = 192
OpenVINO latency mode	FP32	th = default(48), bs = 128	th = default(48), bs = 384
OpenVINO latency mode	INT8	th = default(48), bs = 64	th = default(48), bs = 96
OpenVINO throughput mode	FP32	th = default(48), bs = 16 sn = 24, rn = 24	th = 48, bs = 24, sn = 24, rn = 48
OpenVINO throughput mode	INT8	th = default(48), bs = 8 sn = 48, rn = 48	th = default(48), bs = 12 sn = 48, rn = 48

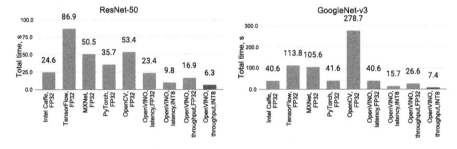

Fig. 8. Total run time for inference of two deep models: ResNet-50 and GoogleNet-v3. The dataset contains 12 288 images. Six DL frameworks are tested.

6 Conclusions

The paper analyzes the performance of DL libraries for a dual-processor server based on the Intel Cascade Lake generation CPUs. We found that the choice of parameters (binding threads to computational cores strategy, the number of threads, the number of images in the batch, the operating mode of the framework, the mode-specific settings) substantially affect performance. As a result of experiments with two DL models, we have formulated recommendations for choosing parameter values. Note that the OpenVINO toolkit partially solves this problem by successfully selecting some of the settings by default.

Experiments have shown that all six frameworks show sufficient performance and can utilize dozens of computational cores. The best results were shown by the OpenVINO toolkit and Intel Distribution for Caffe. INT8 quantization in OpenVINO allowed us to get the 3x performance gain while achieving almost the same accuracy. Employing the throughput mode of OpenVINO leads to further reducing the run time due to better utilization of a manycore system, from 4.9x with a batch size of 1 to 1.4x with a batch size of 32.

Acknowledgements. I.M. and V.V. acknowledge support of Russian Government Grant No. 0729-2020-0055. E.K., E.V., and V.K. acknowledge support of Intel Corporation. The authors are grateful to N. Ageeva, Yu. Gorbachev, K. Korniakov, and Z. Matveev for valuable comments. The experiments were performed on the Intel Endeavor supercomputer at Intel and the Lobachevsky supercomputer at UNN.

References

1. Apache MXNet. https://mxnet.apache.org
2. Caffe. http://caffe.berkeleyvision.org
3. Default Quantization algorithm in Intel Distribution of OpenVINO Toolkit. https://docs.openvinotoolkit.org/latest/pot_compression_algorithms_quantization_default_README.html
4. Inference Performance Analysis repository. https://github.com/itlab-vision/inference_performance_analysis
5. Intel Distribution of Caffe. https://github.com/intel/caffe
6. Intel Distribution of OpenVINO toolkit. https://software.intel.com/en-us/openvino-toolkit
7. OneAPI Deep Neural Network Library. https://github.com/oneapi-src/oneDNN
8. OpenCV. https://opencv.org
9. PyTorch. https://pytorch.org
10. TensorFlow. https://www.tensorflow.org
11. Uniform Quantization in the Intel Distribution of OpenVINO Toolkil. https://docs.openvinotoolkit.org/latest/po_compression_algorithms_quantization_README.html
12. Abts, D., et al.: Think fast: a tensor streaming processor (TSP) for accelerating deep learning workloads. In: Proceedings of the Symposium on Computer Architecture, pp. 145–158 (2020). https://doi.org/10.1109/ISCA45697.2020.00023
13. Ciaparrone, G., et al.: Deep learning in video multi-object tracking: a survey. Neurocomputing **381**, 61–88 (2020). https://doi.org/10.1016/j.neucom.2019.11.023
14. Coleman, C., et al.: DAWNBench: an end-to-end deep learning benchmark and competition. In: NIPS ML Systems Workshop, pp. 1–10 (2017). https://dawn.cs.stanford.edu/benchmark/papers/nips17-dawnbench.pdf
15. George, D., Huerta, E.A.: Deep Learning for real-time gravitational wave detection and parameter estimation: results with advanced LIGO data. Phys. Lett. B **778**, 64–70 (2018). https://doi.org/10.1016/j.physletb.2017.12.053
16. Gonoskov, A., et al.: Employing machine learning for theory validation and identification of experimental conditions in laser-plasma physics. Sci. Rep. **9**(1), 1–15 (2019). https://doi.org/10.1038/s41598-019-43465-3
17. Gorbachev, Y., et al.: OpenVINO deep learning workbench: comprehensive analysis and tuning of neural networks inference. In: Proceedings of the IEEE/ICCV Workshops (2019)
18. He, K., et al.: Deep residual learning for image recognition. In: Proceedings of the IEEE Computer Society Conference on Computer Vision and Pattern Recognition, pp. 770–778 (2016). https://doi.org/10.1109/CVPR.2016.90
19. Ignatov, A., et al.: AI benchmark: all about deep learning on smartphones in 2019, pp. 3617–3635, October 2019. https://doi.org/10.1109/ICCVW.2019.00447
20. Jain, A., et al.: Efficient execution of quantized deep learning models: a compiler approach. arxiv preprint arXiv:2006.10226 (2020)

21. Kustikova, V., Vasiliev, E., Khvatov, A., Kumbrasiev, P., Rybkin, R., Kogteva, N.: DLI: deep learning inference benchmark. In: Voevodin, V., Sobolev, S. (eds.) RuSCDays 2019. CCIS, vol. 1129, pp. 542–553. Springer, Cham (2019). https:// doi.org/10.1007/978-3-030-36592-9_44

22. March, P.S.: Optimize Virtualized Deep Learning Performance with New Intel Architectures (2020). https://www.vmware.com/techpapers/2020/virtualized-vnni-perf.html

23. Park, J., et al.: Deep learning inference in Facebook data centers: characterization, performance optimizations and hardware implications. arXiv preprint arXiv:1811.09886 (2018)

24. Raissi, M., et al.: Physics-informed neural networks: a deep learning framework for solving forward and inverse problems involving nonlinear partial differential equations. J. Comput. Phys. **378**, 686–707 (2019). https://doi.org/10.1016/j.jcp. 2018.10.045

25. Ravi, D., et al.: Deep learning for health informatics. IEEE J. Biomed. Health Inform. **21**(1), 4–21 (2017). https://doi.org/10.1109/JBHI.2016.2636665

26. Reddi, V.J.: MLPerf inference benchmark. In: Proceedings of the Symposium on Computer Architecture, pp. 446–459 (2020). https://doi.org/10.1109/ISCA45697. 2020.00045

27. Russakovsky, O., et al.: ImageNet large scale visual recognition challenge. Int. J. Comput. Vis. **115**(3), 211–252 (2015). https://doi.org/10.1007/s11263-015-0816-y

28. Szegedy, C., et al.: Rethinking the inception architecture for computer vision. In: Proceedings of the IEEE Computer Society Conference on CV and Pattern Recognition, pp. 2818–2826 (2016). https://doi.org/10.1109/CVPR.2016.308

29. Torelli, P., Bangale, M.: Measuring Inference Performance of Machine-Learning Frameworks on Edge-class Devices with the MLMark^{TM} Benchmark. https://www. eembc.org/techlit/articles/MLMARK-WHITEPAPER-FINAL-1.pdf

30. Voulodimos, A., et al.: Deep learning for computer vision: a brief review. Comput. Intell. Neurosci. (2018). https://doi.org/10.1155/2018/7068349

31. Wu, H., et al.: Integer quantization for deep learning inference: principles and empirical evaluation. arXiv preprint arXiv:2004.09602 (2020)

32. Yang, C.T., et al.: Performance benchmarking of deep learning framework on Intel Xeon Phi. J. Supercomput. (2020). https://doi.org/10.1007/s11227-020-03362-3

33. Young, T., et al.: Recent trends in deep learning based natural language processing [Review Article]. IEEE Comput. Intell. Mag. **13**(3), 55–75 (2018). https://doi.org/ 10.1109/MCI.2018.2840738

On the Problem of Choosing the Optimal Parameters for the Wind Farm in the Arctic Town of Tiksi

Sergei V. Strijhak[1,2] , Victor P. Gergel[3] , Aleksandr V. Ivanov[1(✉)] ,
and Sebastien Zh. Gadal[4,5]

[1] Ivannikov Institute for System Programming of the RAS, 109004 Moscow, Russia
{s.strijhak,av.ivanov}@ispras.ru
[2] Moscow Aviation Institute, 125993 Moscow, Russia
[3] Lobachevsky State University of Nizhny Novgorod,
603950 Nizhnij Novgorod, Russia
gergel@unn.ru
[4] Aix-Marseille Univ, CNRS, ESPACE UMR 7300, Univ Nice Sophia Antipolis,
Avignon Univ, 13545 Aix-en-Provence, France
sebastien.gadal@univ-amu.fr
[5] North-Eastern Federal University, 67000 Yakutsk,
Republic of Sakha Yakutia, Russia

Abstract. The paper considers the problem of choosing the optimal parameters for the operation of 3 horizontal wind turbines of the wind farm in the town Tiksi, in the Sakha Republic. The open-source WRF-ARW and FLORIS packages are used to calculate the physical parameters in the wind farm. During the calculation, the values of wind velocity, temperature, pressure fields, and the value of the generated power of wind power plants were obtained. The optimization problem is formulated, the objective function is defined, and 2 variable parameters are selected, namely, yaw angle and angle of attack. The calculations were carried out on the computing cluster of ISP RAS.

Keywords: Wind turbine · Wind farm · Calculation · Domain · Grid · Wind velocity · Objective function · Wind turbine parameters · Yaw angle · Power

1 Introduction

The design and selection of optimal parameters for horizontal wind turbines in the wind farm is an urgent task, especially in the case of the construction of new wind farms in the Russian Federation. One of the interesting areas for research may be the problem of choosing the best operating modes of the wind farm, taking into account the influence of local weather conditions, as well as the relative location of the wind turbine in a real area. The variable parameters include the angular rotation velocity of the wind turbine, the yaw angle, the angle of rotation of the blade with respect to the vector of the incoming flow, and others.

© Springer Nature Switzerland AG 2021
D. Balandin et al. (Eds.): MMST 2020, CCIS 1413, pp. 361–375, 2021.
https://doi.org/10.1007/978-3-030-78759-2_30

2   Wind Farm in the Town Tiksi

The wind farm built by RusHydro together with its Japanese partners in the Arctic town Tiksi, which is home to more than 4.6 thousand people, has demonstrated high efficiency and reliability in the harsh Arctic climate. In the winter of 2018–2020, the air temperature in Tiksi dropped to minus 42°, and the wind velocity reached 30 m/s.

In November 2018, in the Arctic urban-type settlement of Tiksi, which is an important transport hub for the Northern Sea Route, 3 wind turbines from Komaihaltec company from Japan with a capacity of 0.3 MW each were put into operation with the participation of Sakhaenergo and RusHydro. Now the wind farm works together with the existing diesel power plant of the town.

The wind turbines (WT) of the Tiksi wind farm are located at an altitude of about 120 m above sea level, 4 km from the town Tiksi and 2.7 km from the Laptev Sea, Fig. 1.

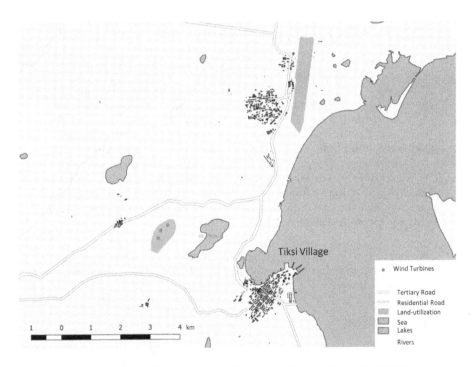

Fig. 1. The map of the town in Tiksi. Open Street Map, 11.09.2019.

Turbines coordinates are: 71°39′24.8″N 128°46′17.6″E (WT1), 71°39′20.″N 128°45′55.7″E (WT2), 71°39′12.02″N 128°45′43.57″E (WT3), see Fig. 2.

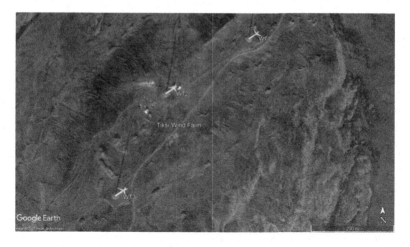

Fig. 2. Satellite image of the wind turbines location. Google Earth, 25.09.2020.

2.1 Mathematical Model in WRF-ARW Code

The calculation code WRF-ARW (WRF – Weather Research and Forecasting, ARW – Advanced Research WRF) is actively used to predict the capacity of the wind farm. The WRF-ARW model is a model of a regional weather forecast, for the construction of which it needs boundary and initial conditions [1,2]. Data sources in this case can be global models, observational data, or forecast reanalysis data. As a parent model for WRF-ARW, data from calculations based on the Global Forecast System (GFS) model is usually used. Figure 3 shows an algorithm for the interaction of different calculation models for different levels of calculation of physical quantities.

The WRF-ARW model allows one to obtain a large range of physical quantities that describe weather conditions. As a rule, the main parameters are the distribution of wind velocity, temperature, and pressure. Secondary parameters may include air humidity, precipitation distribution, precipitation type, and much more. In WRF-ARW, there are various parameterization models, including the model for parameterizing the operation of a wind turbine – the Actuator Disk Model [3].

The WRF-ARW model is based on non-hydrostatic equations for a compressible fluid written in Cartesian coordinates horizontally and using the orographic coordinate η vertically. In older versions of the package, η was determined using the hydrostatic pressure p_h:

$$\eta = \frac{p_d - p_t}{p_s - p_t}, \tag{1}$$

where p_d is the hydrostatic component of dry air pressure, p_s and p_h refer to values of p_d along the surface and top boundaries respectively. In WRF-ARW version 4, the vertical coordinate was generalized to reduce the influence of the surface on the coordinate grid with increasing height:

$$p_d = B(\eta)(p_s - p_h) + [\eta - B(\eta)](p_0 - p_h) + p_h,$$

where p_0 is a reference sea-level pressure. Here $B(\eta)$ defines the relative weighting between the terrain-following sigma coordinate and a pure pressure coordinate, such that η corresponds to the sigma coordinate (1) for $B(\eta) = \eta$ and reverts to a hydrostatic pressure coordinate for $B(\eta) = 0$. For a smooth transition from the sigma coordinate near the surface to the pressure coordinate at the upper levels, $B(\eta)$ is defined by a third order polynomial.

The governing equations system is represented as follows. The momentum equations are written as

$$\partial_t U + m_x[\partial_x(Uu) + \partial_y(Vu)] + \partial_\eta(\Omega u) \tag{2}$$
$$+ (m_x/m_y)(\alpha/\alpha_d)[\mu_d(\partial_x\phi' + \alpha_d\partial_x p' + \alpha_d'\partial_x\overline{p}) + \partial_x\phi(\partial_\eta p' - \mu_d')] = F_U,$$
$$\partial_t V + m_y[\partial_x(Uv) + \partial_y(Vv)] + (m_y/m_x)\partial_\eta(\Omega u) \tag{3}$$
$$+ (m_y/m_x)(\alpha/\alpha_d)[\mu_d(\partial_y\phi' + \alpha_d\partial_y p' + \alpha_d'\partial_y\overline{p}) + \partial_y\phi(\partial_\eta p' - \mu_d')] = F_V,$$
$$\partial_t W + m_x[\partial_x(Uw) + \partial_y(Vw)] + \partial_\eta(\Omega w) \tag{4}$$
$$- m_y^{-1}g(\alpha/\alpha_d)[\partial_\eta p' - \overline{\mu}_d(q_v + q_c + q_r)] + m_y^{-1}\mu_d'g = F_W,$$

and the mass conservation equation and geopotential equation is given by

$$\partial_t\mu_d' + m_x m_y[\partial_x U + \partial_y V] + m_y\partial_\eta\Omega = 0, \tag{5}$$
$$\partial_t\phi' + \mu_d^{-1}[m_x m_y(U\partial_x\phi + V\partial_y\phi) + m_y\Omega\partial_\eta\phi - m_y gW] = 0. \tag{6}$$

The conservation equations for the potential temperature Θ_m and the scalar moisture Q_m equations:

$$\partial_t\Theta_m + m_x m_y[\partial_x(U\theta_m) + \partial_y(V\theta_m)] + m_y\partial_\eta(\Omega\theta_m) = F_{\Theta_m}, \tag{7}$$
$$\partial_t Q_m + m_x m_y[\partial_x(Uq_m) + \partial_y(Vq_m)] + m_y\partial_\eta(\Omega q_m) = F_{Q_m}, \tag{8}$$

and the diagnostic equation for dry hydrostatic pressure:

$$\partial_\eta\phi' = -\overline{\mu}_d\alpha_d' - \alpha_d\mu_d', \tag{9}$$

with the diagnostic relation for the full pressure (dry air plus water vapor):

$$p = p_0\left(\frac{R_d\theta_m}{p_0\alpha_d}\right)^\gamma. \tag{10}$$

The following notations is used in the system (2)–(10):

$$U = \mu_d u/m_y, \quad V = \mu_d v/m_x, \quad W = \mu_d w/m_y, \quad \Omega = \mu_d\omega/m_y,$$

$$\Theta_m = \mu_d\theta_m, \quad Q_m = \mu_d q_m,$$

where $\mathbf{v} = (u, v, w)$ – velocity vector, $\omega = \dot{\eta}$ – vertical velocity in terms of orographic coordinates, μ_d defines mass of the dry air column, ϕ – geopotential, m_x, m_y – map scale factors. $\theta_m = \theta(1 + (R_v/R_d)q_v) \approx \theta(1 + 1.61q_v)$ is the moist potential temperature and $q_m = q_v, q_c, q_r, q_i...$ represents the mixing ratios of moisture variables (water vapor, cloud water, rain water, ...). α_d is the inverse density of the dry air $(1/\rho_d)$ and α is the inverse density taking into account the full parcel density $\alpha = \alpha_d(1 + q_v + q_c + q_r + q_i + ...)^{-1}$. The right-hand-side terms $F_U, F_V, F_W, F_{\Theta_m}$ and F_{Q_m} represent forcing terms arising from physical models, turbulent mixing, spherical projections, and the earth's rotation, g – acceleration due to the gravity. Hydrostatically-balanced reference state variables (denoted by overbars) are a function of height only and satisfy the governing equations for an atmosphere at rest.

Macro-scale $\rightarrow$ Meso-scale $\rightarrow$ Micro-scale

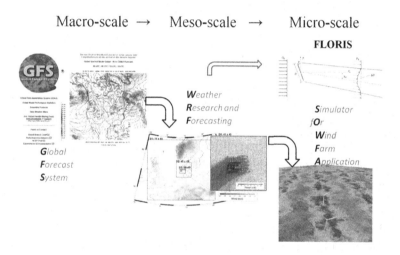

Fig. 3. How to model the operation of a wind farm.

WRF-ARW code supports the configuration of the shared and distributed memory of server. The code could be compiled with OpenMP and OpenMPI libraries for parallel calculations. The goal of parallel option is to speed-up the computing time on supercomputer.

2.2 Numerical Domain, Grid, and Results of Calculation

The layout of the numerical domain for the area of the town Tiksi with the image of nested domains is shown in Fig. 4. The calculations in the WRF-ARW package were performed for the following calendar dates: 29.09.2019, 30.09.2019, 01.10.2019.

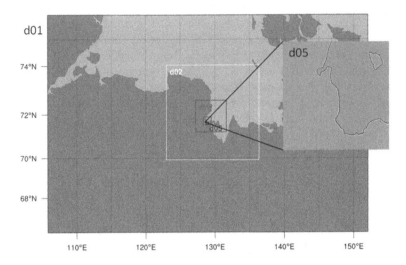

Fig. 4. Layout of the main area with the image of nested domains.

To compare the results, the Tiksi weather station was selected: 71°34′48″ N 128°54′E, altitude 7 m above sea level. Data on the temperature at the level of 2 m above the surface, the pressure, and wind velocity at the surface (at the level of the weather station) were displayed at the point closest to the location of the weather station in the d05 region. The graph of the velocity comparison is shown in Fig. 6. The model configuration and simulation results are given in [4].

In general, the results of the model correspond to real data, in any case, they repeat the profile of changes in indicators. Also, the wind farm energy generation was modeled using the built-in model, which was analyzed in [3]. The characteristics of the wind turbine, namely: the power curve and the thrust coefficient as a function of velocity, were taken from the parameters of similar wind turbines from open sources on the Internet (Fig. 5). The result of the simulation is the dynamics of the power of the Tiksi wind farm in a given period, Fig. 7. One can see that the results correspond to the average wind value in the specified area.

The results of the calculation showed that the efficiency of the wind farm does not exceed 20% for the selected days. In this regard, there is a need to choose the optimal parameters for the operation of wind turbines. This parameter can be the angle of rotation of the nacelle (the yaw angle).

The calculations were performed using 12-cores computer node. Several cases with different configuration of physical parameters were launched simultaneously. The average computation time for one case was about 24 h.

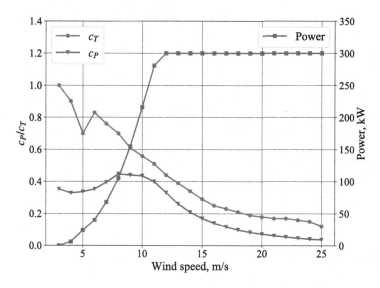

Fig. 5. The power curve and c_P, c_T coefficients for a single wind turbine.

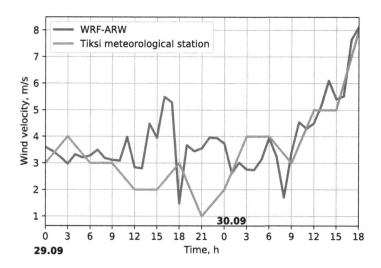

Fig. 6. Comparison of wind velocity at the location of the Tiksi weather station for model and real data.

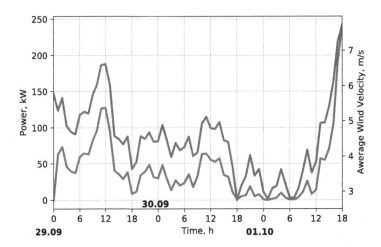

Fig. 7. Simulation change in the total power generation capacity of the Tiksi wind farm on 29.09–01.10.

The simulations were done using the resources of the UniHUB HPC cluster of ISP RAS with 32 computer nodes.

2.3 Setting Task of Optimizing the Generated Capacity of the Wind Farm

It is known that the front wind turbines have a significant impact on the operation of the rear wind turbines. There is a concept of velocity deficit. The vortex trail behind the wind turbine may deviate.

Data from a weather station located 10 km from the wind farm shows that the wind direction changes in the range of angles from $\theta_1 = 0°$ to $\theta_2 = 90°$. Thus, there is the problem of choosing the optimal angle of rotation of the engine nacelle of the wind turbine (yaw rate), the angle of rotation of the blade to obtain maximum power capacity of wind turbines.

Work in the field of optimization calculations for wind farms and the design of wind turbines was practically not carried out earlier in the Russian Federation. Among the international teams that deal with optimization issues in the design of new wind farms, there are research groups from JHU, USA (Prof. C. Meneveau); KU Leuven, Belgium (Dr. Johan Meyers); EPFL, Lausanne, Switzerland (Prof. Fernando Porte-Agel); NREL (Dr. P. A. Fleming); Stanford University (Dr. J. Park); Aalborg University, Aalborg, Denmark (Prof. Zheng Chen) [5–14]. These teams considered the following tasks:

1. optimization of the location of wind turbines in the wind farm [5–7];
2. selection of optimal parameters (yaw angle of the wind wheel axis, angular rotation velocity, control parameters, etc.) [10–14];

3. optimization of the geometric shape of the wind turbine blade in order to reduce the noise level on the ground and increase the durability of the structure made of composite material.

To solve such problems, an objective function was formed, which took into account various criteria and restrictions on the parameters to be changed. Among the optimization methods used, one can distinguish:

- genetic algorithms;
- sequential quadratic programming;
- cross-entropy method;
- an approach based on a combination of the conjugate gradient method, the Polak-Ribière method for determining weight coefficients, and the Brent search algorithm [8,9];
- machine learning methods based on "Bayesian ascent" [10,11];
- the particle swarm method [12,13].

The yaw angle's optimization problem for the wind wheel axis was studied in papers [15–17].

The authors of the work [9] solved the optimization problem using an approach with a cooperative game. All agents (participants) strive to maximize the overall goal. Interaction between agents is taken into account.

The vector maximization problem is solved:

$$x^* = \arg\max_x \sum_{i=1}^{N} \varphi_i f_i(x), \quad \varphi_i = 1, i = \overline{1, N}; \tag{11}$$

$$\text{maximize}_x\, f(x) \triangleq \sum_{i=1}^{N} P_i(\alpha, o, U, \theta^W), \quad \text{subject to } x^l \leqslant x \leqslant x^u, \tag{12}$$

where x^l, x^u – lower and upper bounds, φ_i represents the weighting coefficient on the objective function $f_i(x)$.

Because it is difficult to accurately determine the analytical function for the cardinality, the authors used Bayesian optimization (BO).

The problem is solved:

$$x^* = \arg\max_x f(x) \tag{13}$$

Observing for the behavior of the function:

$$y = f(x) + \epsilon. \tag{14}$$

The BO uses 2 phases: training and optimization.

In this paper, the objective function $f(x)$ associated with the total power generation was formulated.

It is known that the power for a single wind turbine is equal to:

$$P = \frac{1}{2}\rho A U^3 C_p(\alpha, 0). \tag{15}$$

An arbitrary set of parameters can be considered in an optimization problem:

$$\boldsymbol{x} = (x_1, \ldots, x_i, \ldots, x_N). \tag{16}$$

The target function for the total capacity of the wind farm is defined as:

$$f_i(\boldsymbol{x}) = P_i(\boldsymbol{\alpha}, \boldsymbol{o}, U, \theta^W) \tag{17}$$

The goal of any optimization is to achieve the maximum values of performance indicators while meeting the specified limits. At the stage of setting the problem, the researcher determines a set of parameters that need to be maximized, minimized, or limited. The indicators are optimized by varying the input parameters. The researcher sets the composition of the variable variables, the ranges of their changes, and, possibly, the relationship with other input parameters.

Usually, when conducting optimization studies, it is necessary to solve not one, but several optimization problems that differ:

- the composition of variable variables (from minimal changes in the project to a complete redesign);
- values of constraints (analysis of the possibility of relaxing individual requirements for the project);
- the number and composition of optimization criteria (from a complete set of alternative projects to a single option).

In the course of solving the problem, it is necessary to solve the problem of finding the maximum of the objective function $f(x)$:

$$\max_{\boldsymbol{x}} f(\boldsymbol{x}) \triangleq \sum_{i=1}^{N} P_i(\boldsymbol{\alpha}, \boldsymbol{o}, U, \theta^W), \quad \boldsymbol{x}^l \leq \boldsymbol{x} \leq \boldsymbol{x}^u, \tag{18}$$

where $\boldsymbol{x} = (\alpha_1, o_1, \ldots, \alpha_N, o_N)$, N – the number of wind turbines.

The generated power of wind turbines depends on 4 parameters: the angle of rotation of the blades (angle of attack), the values of the flow velocity, the angle of direction of the vector flow velocity, yaw rate. The range of changes in 2 selected parameters out of 4 set according to changes in weather data in the area of the location of the wind farm in the town Tiksi, as well as the technical characteristics of a wind turbine with a power of $P = 0.3\,\mathrm{MW}$. The range for changing all parameters is known, the yaw angle $\boldsymbol{o}$ varies from $0°$ to $25°$, the angle of rotation of the blade α from $-10°$ to $10°$.

3   Analytical Models for Calculation of Wake Parameters

Analytical models in comparison with models based on the solution of partial differential equations are simpler and require less computing resources of the computer. The open package FLORIS (FLOw Redirection and Induction in Steady-state) contains various analytical models for calculating the vortex traces of wind turbines in the wind farm, [18]. The FLORIS package was developed in the Python programming language at NREL (USA), TU Delft.

One of the first analytical models for calculating the parameters of the vortex wake was developed by N. Jensen in Denmark in 1983 [19]. In this paper, it was proposed to use the "top-hat shape" model for the cylinder to calculate the velocity deficit in the track (see Fig. 8) and written in the form:

$$\frac{\Delta U}{U_\infty} = \left(1 - \sqrt{1 - c_T}\right) / \left(1 + \frac{2k_{wake}x}{d_0}\right)^2, \tag{19}$$

where $\dfrac{\Delta U}{U_\infty}$ is a normalized (dimensionless) velocity deficit.

$$\frac{\Delta U}{U_\infty} = \frac{U_\infty - U_W}{U_\infty}. \tag{20}$$

The Eq. (19) has been widely used in the literature and has been implemented in commercial software (WasP, WindPRO, WindSim). However, there were two limitations to this model: 1) the assumption of the distribution of the velocity deficit was unrealistic; 2) only the law of conservation of mass was used to derive Eq. (19).

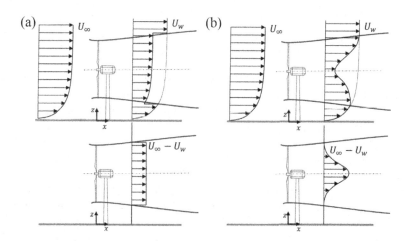

Fig. 8. Velocity profile for the wind turbine according to the a) "top-hat shape" model; b) Gaussian distribution model. Picture is taken from [20].

Subsequently, Frandsen [21] applied the equations of conservation of mass and amount of motion for the control volume (Fig. 9) around the wind turbine and proposed the following expression:

$$\frac{\Delta U}{U_\infty} = \frac{1}{2}\left(1 - \sqrt{1 - 2\frac{A_0}{A_W}c_T}\right). \tag{21}$$

(a)   (b)

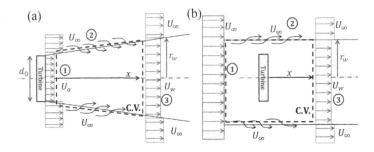

Fig. 9. Example of using the control volume: a) quadrilateral; b) rectangle. Picture is taken from [20].

Vortex traces behind bodies in free flow have been actively studied in the theory of shear flows. In these studies, a self-similar Gaussian profile was found to calculate the velocity deficit.

The self-similar Gauss profile was discovered in experiments in wind tunnels, in numerical calculations, and in operating wind farms. An analytical model with a self-similar Gauss profile was obtained in [20]:

$$\frac{\Delta U}{U_\infty} = \left(1 - \sqrt{1 - \frac{c_T}{8(k^* x/d_0 + \varepsilon)^2}}\right) \tag{22}$$
$$\times \exp\left(-\frac{1}{2(k^* x/d_0 + \varepsilon)^2}\left\{\left(\frac{z - z_h}{d_0}\right)^2 + \left(\frac{y}{d_0}\right)^2\right\}\right).$$

The analytical models (19)–(22), which were proposed by different authors to predict the deficit of the streamwise velocity in the of wind turbine, were implemented in open source software code FLORIS. FLORIS code is written on Python and is available on github.com [18].

In this paper, the calculation was carried out for a model wind farm with 3 wind turbines according to the Jimenez model, taking into account the influence of changes in the yaw angle for the nacelle [7].

An example of the calculation for a wind farm in the FLORIS package is shown in Fig. 10. The size of the numerical domain was selected with 500 m in the OY direction and 3200 m in the OX direction. The velocity was defined as 8 m/s at the inlet of the numerical domain. The wind farm consisted of 3 NREL 5 MW wind turbines. They were located on the same straight line, respectively, with coordinates 0, 7D, 14D.

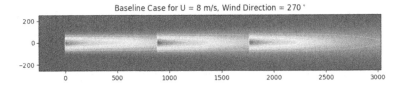

Fig. 10. Initial position of wind turbines in wind farm.

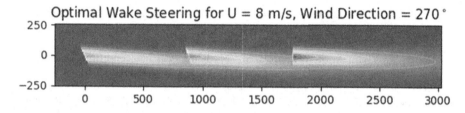

Fig. 11. Velocity value field for 3 wind turbines.

The wind direction was 270°. The initial position of wind turbines is shown in Fig. 10.

The possible range of yaw angle changes for wind turbines is from 0 to 25°. We used the SLSQP (Sequential Least SQuares Programming) optimization algorithm. SLSQP is a nonlinear, gradient-based algorithm that can handle inequality constraints. Iterations are generated by solving quadratic sub-problems. The solution to the optimization problem, the best option is to rotate the nacelle for the first wind turbine is at a 25-degree angle, for the 2nd wind turbine the yaw angle is 19°, for the 3rd wind turbine the yaw angle is 4.3° (Fig. 11). The total power gain amounted to 15.1%.

On of the future directions of work could be dealt with using Globalizer Software, developed at the Lobachevsky National Research University for multi-extreme optimization problems, see the paper [22] to solve the optimization problems for wind farms.

4   Conclusion

This approach, using meteorological data, wind farm data, open-source software WRF-ARW and FLORIS, allows us to formulate the problem of choosing the optimal parameters for the operation of wind turbines in the wind farm. In the future, it is planned to apply the chosen approach to solve the problem of choosing the optimal parameters for the operation of 3 Japanese wind turbines in the wind farm in the Arctic town Tiksi, taking into account the wind rose and complex terrain.

References

1. Skamarock, W.C., et al.: A description of the advanced research WRF model version 4. UCAR/NCAR (2019). https://doi.org/10.5065/1DFH-6P97
2. Jiménez, P.A., Navarro, J., Palomares, A.M., Dudhia, J.: Mesoscale modeling of offshore wind turbine wakes at the wind farm resolving scale: a composite-based analysis with the weather research and forecasting model over horns rev. Wind Energy **18**(3), 559–566 (2014). https://doi.org/10.1002/we.1708

3. Fitch, A.C., et al.: Local and mesoscale impacts of wind farms as parameterized in a mesoscale NWP model. Mon. Weather Rev. **140**(9), 3017–3038 (2012). https://doi.org/10.1175/mwr-d-11-00352.1

4. Ivanov, A., Strijhak, S., Zakharov, M.: Modeling weather conditions in the port area and coastal zone of Tiksi bay. Proc. Inst. Syst. Program. RAS **31**(6), 163–176 (2019). https://doi.org/10.15514/ispras-2019-31(6)-9

5. González, J.S., Rodríguez, A.G., Mora, J.C., Burgos Payán, M., Santos, J.R.: Overall design optimization of wind farms. Renew. Energy **36**(7), 1973–1982 (2011). https://doi.org/10.1016/j.renene.2010.10.034

6. Meyers, J., Meneveau, C.: Optimal turbine spacing in fully developed wind farm boundary layers. Wind Energy **15**(2), 305–317 (2011). https://doi.org/10.1002/we.469

7. Gebraad, P.M.O., et al.: Wind plant power optimization through yaw control using a parametric model for wake effects-a CFD simulation study. Wind Energy **19**(1), 95–114 (2014). https://doi.org/10.1002/we.1822

8. Goit, J.P., Meyers, J.: Optimal control of energy extraction in wind-farm boundary layers. J. Fluid Mech. **768**, 5–50 (2015). https://doi.org/10.1017/jfm.2015.70

9. Bokharaie, V.S., Bauweraerts, P., Meyers, J.: Wind-farm layout optimisation using a hybrid Jensen–LES approach. Wind Energy Sci. **1**(2), 311–325 (2016). https://doi.org/10.5194/wes-1-311-2016

10. Park, J., Law, K.H.: A data-driven, cooperative wind farm control to maximize the total power production. Appl. Energy **165**, 151–165 (2016). https://doi.org/10.1016/j.apenergy.2015.11.064

11. Park, J., Law, K.H.: Bayesian ascent: a data-driven optimization scheme for real-time control with application to wind farm power maximization. IEEE Trans. Control Syst. Technol. **24**(5), 1655–1668 (2016). https://doi.org/10.1109/tcst.2015.2508007

12. Hou, P., Hu, W., Soltani, M., Chen, C., Chen, Z.: Combined optimization for offshore wind turbine micro siting. Appl. Energy **189**, 271–282 (2017). https://doi.org/10.1016/j.apenergy.2016.11.083

13. Hou, P., Hu, W., Soltani, M., Chen, Z.: Optimized placement of wind turbines in large-scale offshore wind farm using particle swarm optimization algorithm. IEEE Trans. Sustain. Energy **6**(4), 1272–1282 (2015). https://doi.org/10.1109/tste.2015.2429912

14. Bastankhah, M., Porté-Agel, F.: Wind farm power optimization via yaw angle control: A wind tunnel study. J. Renew. Sustain. Energy **11**(2) (2019). https://doi.org/10.1063/1.5077038

15. Bastankhah, M., Porté-Agel, F.: Experimental and theoretical study of wind turbine wakes in yawed conditions. J. Fluid Mech. **806**, 506–541 (2016). https://doi.org/10.1017/jfm.2016.595

16. Fleming, P.A., Ning, A., Gebraad, P.M.O., Dykes, K.: Wind plant system engineering through optimization of layout and yaw control. Wind Energy **19**(2), 329–344 (2015). https://doi.org/10.1002/we.1836

17. Astolfi, D., Castellani, F., Natili, F.: Wind turbine yaw control optimization and its impact on performance. Machines **7**(2), 41 (2019). https://doi.org/10.3390/machines7020041

18. NREL: FLORIS. Version 2.2.3 (2020). https://github.com/NREL/floris

19. Jensen, N.: A note on wind generator interaction. No. 2411 in Risø-M, Risø National Laboratory (1983)

20. Bastankhah, M., Porté-Agel, F.: A new analytical model for wind-turbine wakes. Renew. Energy **70**, 116–123 (2014). https://doi.org/10.1016/j.renene.2014.01.002
21. Frandsen, S., et al.: Analytical modelling of wind speed deficit in large offshore wind farms. Wind Energy **9**(1-2), 39–53 (2006). https://doi.org/10.1002/we.189
22. Gergel, V., Barkalov, K., Sysoyev, A.: Globalizer: a novel supercomputer software system for solving time-consuming global optimization problems. Numer. Algebra Control Optim. **8**(1), 47–62 (2018). https://doi.org/10.3934/naco.2018003

Multidimensional Interpolation Methods in Simulation Planning for Modeling

Elena Glazunova⬡, Andrey Deulin⬡, Mikhail Kulikov$^{(\boxtimes)}$⬡,
Nikolay Starostin⬡, and Andrey Filimonov⬡

The Russian Federal Nuclear Center—All-Russian Scientific Research Institute
of Experimental Physics, Sarov, Russia

Abstract. For a variety of applications natural tests can be properly replaced with simulations and simulated object with a model. Even though simulations promise many benefits they also require the most computational power. Simulations are performed with massive parallel computing machines which are also often referred to as supercomputer. However, even they may lack compute resource to solve practical problems, especially if a simulation process is arranged in a non-efficient manner. The paper addresses one of the issues of efficient simulations, namely proper simulation planning for modeling. The method exploits a modification of Shepard method for model interpolation as well as a priori evaluation of data points to be simulated in order to increase model accuracy. The evaluation is performed using a Voronoi diagram based on previous seeds matching simulations. A Voronoi diagram is constructed in a multidimensional space by radial growth from the seeds outward with additional correction. Compared to naive grid-based approaches the method reduces the number of required simulations by six times.

Keywords: Design of experiments · Surrogate modeling ·
Metamodeling · Multidimensional interpolation · Sequential sampling

1   Introduction

High-fidelity simulations in engineering design reduce the need for naturals tests. On the other hand, replacing natural tests with simulations often requires significant computing resource. Just one iteration to calculate the sharp edge and cosine gust simulations lasting more than 8300 CPU hours (4 h of wall clock time on 2016 cores) is a good example [1]. A major way to cut the need for computing power is to apply metamodeling. A metamodel (known as surrogate model) approximates system response from a limited number of selected original model responses directed by the design of experiments (DoE). The accuracy of metamodels correlates to the experimental designs applied. Proper DoE drives towards reducing the number of virtual tests without compromising the metamodel accuracy. While building a metamodel, DoE can embody two response sampling techniques, namely one-stage sampling and sequential sampling [2].

© Springer Nature Switzerland AG 2021
D. Balandin et al. (Eds.): MMST 2020, CCIS 1413, pp. 376–388, 2021.
https://doi.org/10.1007/978-3-030-78759-2_31

One-stage sampling relies on Classical design of experiments and Space-filling methods [3]. The idea of Classical designed experiments is to mitigate the effect of a random error on the approval or rejection of a hypothesis. Classical experimental design methods include such approaches as Central composite, Box-Behnken and Plackett-Burman design [4], alphabetical optimal [5] and others. For supercomputer high-fidelity simulations Space-filling methods that provide uniform filling of the space are preferable. There is a variety of Space-filling methods available, e.g. Latin Hypercube [6,7], Orthogonal Arrays [8], Hammersley sequence [9], Uniform designs [10]. One-stage sampling approaches provide uniform distribution but due to the difficulty of determining adequate and sufficient sampling size they are often redundant. To eliminate this effect Monte Carlo-like processes can be considered (regardless of their inefficiency).

Another way to reduce sampling size is using sequential sampling. Sequential sampling (known as adaptive sampling) prefers so-called significant regions over regions with trivial inner dependencies. Preference is worked out of the prior samples. Sequential sampling usually performs better in terms of sampling set size. Due to these features, sequential sampling has gained popularity in recent years [11–15].

The paper suggests a sequential DoE method. The latter one implements a modification of Shepard method for model interpolation as well as a priori evaluation of the data points to be simulated to increase the model accuracy. The problem statement is introduced in Sect. 2. Various interpolation algorithms for the problem are discussed in the Sect. 3. Pre-sampling based on Voronoi diagram borders points estimation and construction of approximated Voronoi diagram based on seeds matching simulations introduced in the Sect. 4. Yet another pre-sampling algorithm based on points estimation lying on intersection of Voronoi diagram borders and base vectors considered in the Sect. 5. Two algorithms of obtaining the $i + 1$ iteration batch's points discussed in the Sect. 6. Finally, the application of the presented algorithms for solving the model problem and the practical problem of the motion of a body in a liquid are considered in the Sect. 7.

2   Problem Statement

The metamodel creation problem can be stated as an optimization problem which consists in minimizing the error between metamodel prediction $\tilde{F}$ and the real measurement F as well as the number of experiments K. The input data for this task are as follows:

1. list and scope of model variable parameters $x \in X \subset R^n$.
2. list of unknown functions $y = F(x), y \in R^m$.
3. size of sequential sampling batch L. Application of $L > 1$ provides the ability of parallel simulation. Denote batch count by $Kb = \lceil K/L \rceil$.
4. training set of points obtained in previous iterations (base points):

$$P_i = (x_j, y_j), y_j = F(x_j), i = 1, ..., Kb, j = 1, ..., i \times L.$$

5. maximum number of simulation experiments $K_{max}, K \leq K_{max}$, maximum number of batch $Kb_{max} = \lceil K_{max}/L \rceil$.
6. the required accuracy A of metamodel:

$$y_{max_j} = max(F_j(x)), j = 1, ..., m, y_{max} \in R^m;$$

$$y_{min_j} = min(F_j(x)), j = 1, ..., m, y_{min} \in R^m;$$

$$|F_j(x) - \tilde{F}_{Kb_j}(x)| < A(y_{max_j} - y_{min_j}), \forall x \in X, j = 1, ..., m.$$

On each i step of sequential sampling the solution of the following problems is provided:

1. constructing smooth interpolation functions $\tilde{F}_i(x), x \in X$, passing throw base points of i iteration, $\tilde{F}_i(x_i) = y_i = F(x_i), (x_i, y_i) \in P_i$.
2. constructing a generalized error estimation function $e(x)$ for the constructed interpolations in the model parameter space. Pre-sampling with constructed error estimation function:

$$e(x) = \sqrt{\Sigma_{j=1,...,m}(F(x) - \tilde{F}_i(x))^2}, x \in X, e(x) \in R, i = 1, ..., Kb.$$

3. obtaining the $i + 1$ iteration batch's points from the pre-sampled points, $i < Kb$.

3   Multidimensional Interpolation Method Selection

The function value y_j to be interpolated is known at given base points $x_j = (x_{j1}, ..., x_{jn}), (x_j, y_j) \in P_i, i = 1, ..., Kb$ and the interpolation problem consist in yielding values at arbitrary points. Since in general the base points may not correspond to the nodes of some regular grid, the general interpolation methods that do not use a regular grid are of interest. For this class of problems, the following interpolation algorithms are well known:

1. Gaussian process regression (Kriging)—method of interpolation for which the interpolated values are modeled by a Gaussian process governed by prior covariances [16];
2. inverse weighted distance method (IDW)—assigned values to unknown points are calculated with a weighted average of the values available at the known points, where the weights of the known points are inversely proportional to the distance to these points [17];
3. natural neighbor interpolation—method based on the Voronoi diagram for a set of spatial points, the weights of the known points are selected in proportion to a fraction of the Voronoi diagram cell that would fall inside the Voronoi diagram cell corresponding to the desired point [18];
4. Shepard's method—as an estimate of the value at an arbitrary point, it uses the weighted average of the polynomial approximating functions constructed around each of the known point by the least squares method (both linear functions and second and third degree polynomials can be used) [19].

These algorithms, for a relatively small computational cost, provide the calculation of "smooth" interpolations of target characteristics that depend on a significant number of variable parameters of the model of a high-tech product. It is experimentally established that the second-order Shepard method demonstrates the best results for constructing interpolation functions for the problems under consideration, both in terms of accuracy and speed of calculation.

4   Pre-sampling Based on Voronoi Diagram Borders Points Estimation

One of the main issues during sequential sampling is error estimation between metamodel prediction $\tilde{F}_i$ and the real measurement F, as well as obtaining the $i+1$ iteration batch points at which the obtained dependencies have, with high probability, the maximum error.

At each iteration only the data about the base points can be considered reliable since in general case the features of the desired dependencies are a priori unknown. At these points, the difference between the obtained metamodel and real measurement is 0. The problem of calculating the error estimate of the interpolation functions is reformulated in terms of the contribution of each base point to the obtained dependence.

As in [15] space partitioning based on Voronoi diagrams is considered and the evaluation of possible error $\tilde{e}(x)$ considered as the difference between interpolation function $\tilde{F}_i(x)$ and a new interpolation function $\tilde{F}_i^{oj}(x)$ is constructed without $(x_j, y_j) \in P_i$ corresponding to each cell:

$$\tilde{e}_j(x) = \sqrt{\Sigma_{l=1,\dots,m}(\tilde{F}_i(x) - \tilde{F}_i^{oj}(x))^2}, x \in X, e(x) \in R, i = 1, \dots, Kb, (x_j, y_j) \in P_i.$$

In contrast to [15], estimation of the error $\tilde{e}_j(x_v)$ where x_v are points on the cell boundaries of a Voronoi diagram is considered. Thus, estimation is based on the assumption of smoothness of the functions $F(x)$ and $\tilde{F}_i(x)$ so the error evaluation function $e(x)$ near the base points (except for the neighbourhood itself to a base points) should also be smooth, as it is defined as the distance. Therefore, most likely, the interpolation functions in general case have the maximum deviation from the target functions at points that lie at the maximum distance from base points being the boundaries of the cells of the Voronoi diagram for a given set of base points. The boundary points where the "error" exceeds the required accuracy are included in pre-selected points PB_i.

Construction of approximated Voronoi diagram based on seeds matching simulations:

1. the accuracy ε is selected depending on the required metamodel accuracy A (so the step in the space of variable parameters is selected). Thus a regular grid in the space of variable parameters is implicitly set (there is no need to store the nodes of the regular grid).
2. for each base point, the nearest nodes of the generated grid are denoted as "base node".

3. each "base node" becomes the source of the "wave", which in the first cycle includes the nodes located from the reference nodes at a distance of no more than ε and in the second cycle at a distance of no more than 2ε and so on;

4. the meeting of two "waves" from different sources determines the quasi-boundary points.

5. the reverse wave from quasi-boundary points determines the area along which the boundary of the domains of the Voronoi diagram passes.

5   Pre-sampling Based on Points Estimation Lying on Intersection of Voronoi Diagram Borders and Base Vectors

The algorithm described in previous section shows good results in terms of determining the points most likely to contain the maximum error but it requires consideration of a significant number of points. Based on the fact that the "close" points that were selected during pre-sampling stage will still be eliminated, the number of calculations can be significantly reduced. For this purpose as in [12], only the vertices of the Voronoi diagram can be considered. In this paper, we propose a different approach based on determining the intersection points of the rays released collinearly to the base vectors and the boundaries of the Voronoi diagram (which, generally speaking, does not require the construction of the Voronoi diagram itself). The main steps of the algorithm are:

1. search rays $r_{j,k}, k = 1, ..., 3^n - 1$, are defined for each base point $x_j \in P_i$. The search rays are chosen to be collinear to the base vectors, consisting of all possible combinations of $1, -1, 0$ on all coordinate axes (obviously, by adding and excluding the base vectors, the construction of the metamodel can be easily parametrized in order to increase or decrease the accuracy and performance).

2. for each base point x_j, the distances $d_{j,k}$ to the nearest points in the directions k are determined. For each direction, the points that fall in hypercone with cone angle $60°$ are selected.

3. the metamodel is constructed using one of the methods presented in Sect. 3.

4. for each base point $(x_j, y_j) \in P_i$, an interpolation function $\tilde{F}_i^{oj}$ is constructed. (For each direction, at least two points located at the minimum distance are selected).

5. the first stage of the pre-sampling is performed, on which the search rays are selected. The selected set of rays include rays $r_{j,k}$ that satisfy the following condition:

$$\sqrt{\Sigma_{l=1,...,m}(\tilde{F}_i(x_r) - \tilde{F}_i^{oj}(x_r))^2} > A\sqrt{\Sigma(max(x_j) - min(x_j))^2},$$

$$(x_j, y_j) \in P_i, x_r \in r_{j,k}, \sqrt{\Sigma(x_r - x_j)^2} = d_{j,k}/2.$$

6. the corner points of the hypercube constructed around the input data are added to the selected points of the first stage (if the base points don't already contain them).
7. for each of the directions pre-selected at the first stage, the intersection of the base ray with the boundary of the Voronoi diagram is considered x_{rv} (binary search can be used for this purpose, instead of building the Voronoi diagram directly).
8. border points where the "error" $\tilde{e}_j(x_{rv})$ exceeds the specified accuracy are included in the pre-selection of points PB_i.

6   Obtaining the Next Iteration Batch's Points

6.1   Batch Points Sampling

The second stage of sampling consists in selecting the required number of points from pre-sampling points set. The selection of a given number of points from the pre-selection is based on the following requirements:

1. it's advisable to select the points with the highest values of the error estimate;
2. it's impractical to select points located "close" to each other in the parameter space.

Two methods are proposed for the final selection of points:

1. dichotomy method for point selection;
2. iterative method for selecting points based on the convolution of normalized criteria.

6.2   Sampling Based on Dichotomy Method

The main steps of the dichotomy method for selecting points are:

1. select an initial interval of threshold that determines the allowed distance between the selected points: the minimum allowable value (left boundary of the interval) is defined as 0, the maximum value (right border of the interval) is defined as the magnitude distance between a pair of maximally distant points in the multidimensional space of parameters;
2. select the current threshold value—the center of the range of acceptable threshold values;
3. all obtained points are sorted in descending order of the error estimate values;
4. the point with the maximum error estimate is selected—it is placed in the set of "selected points", and points for which the distance to the selected point does not exceed the current threshold value are excluded from the set of points;
5. as long as the ordered set of "boundary points" is not empty, go to step 4, otherwise step 6;

6. if the number of "selected points" exceeds the specified limit, the left boundary of the interval is assumed to be equal to the current threshold value, otherwise the right boundary of the interval is assumed to be equal to the current threshold value;
7. until the left border of the interval is equal to the right border, go to step 2, otherwise exit.

6.3 Iterative Sampling Method

The main steps of the iterative point selection method are:

1. at each iteration of the method, until the required number of points are selected, the value of the criterion function is calculated. The selection criterion is the additive convolution of 3 normalized criteria in the range from 0 to 1:
   (a) the estimation of error $\tilde{e}_j(x)$;
   (b) the distance to the nearest point of base points set or the points selected in the previous iteration;
   (c) the percentage of original data points and points selected at the previous iteration fall within hypersphere centered in the considered point and a radius equal to the average minimum distance (the average distance from the points to their closest points).
2. the point with maximal criterion value falls into the final set.

7   Algorithms Complexity

7.1   Shepard's Method Complexity

Since the Sheppard method is based on the application of the least squares method for each of the p points, $p = |P_i|$, the complexity of constructing the interpolation function will be $O(n^2 p)$. Thus construction of interpolation function $\tilde{F}_i$ and p functions $\tilde{F}_i^{oj}(x)$ (Sect. 4 and 5) required $O(n^2 p^2)$).

7.2   Construction of Approximated Voronoi Diagram Complexity

Construction of approximated Voronoi diagram based on seeds matching simulations as in Sect. 4 in the worst case requires iterating through all the points of quasi-grid. Count of points of quasi-grid is $q_k = (max(x_{j_k}) - min(x_{j_k}))/\varepsilon, k = 1, ..., n, x_j \in P_i$. Thus complexity of constructing is $O(c^n)$ where c is constant. Count of pre-sampled points s by this method in the worst case are nearly count of points of quasi-grid $s = a\Sigma_k(q_k)$ where $a < 1$ is constant.

7.3 Obtaining Points Lying on Intersection of Voronoi Diagram Borders and Base Vectors Complexity

Count of considered points lying on intersection of Voronoi diagram borders and base vectors complexity as in Sect. 5 corresponds to count of rays for each of p points, $p = |P_i|$. Thus complexity of construction is $O(p(3^n))$. Count of pre-sampled points s by this method in the worst case corresponds to the number of rays $s = p(3^n - 1)$.

7.4 Sampling Based on Dichotomy Method Complexity

Complexity of sampling method depends of count of pre-sampled points s (Sect. 7.2, 7.3). Before sampling, it is necessary to calculate the distances between the pre-selected points—$O(s^2)$. At each iteration, it is necessary to select L points and at each selection, exclude points that are too close by performing calculations for each of s points. The estimation of the number of iterations— $O(log(s))$. Thus, the total complexity of procedure is $O(s^2) + O(sLlog(s))$. Since L is less than s by definition, the total complexity is $O(s^2)$.

7.5 Iterative Sampling Method Complexity

Complexity of sampling method depends of count of pre-sampled points s (Sect. 7.2, 7.3). For each of pre-sampled points the criterion—$O(s)$ and the nearest distance between pre-sampled points and base points—$O(sp), p = |P_i|$ have to be calculated. On each of L itterations the update of distance between remained pre-sampled points, base points and sampled points may need $O(L(s - L/2))$. Thus the complexity is $O(sp)$.

8 Experiments

8.1 Model Experiments

As an illustration of the work of the presented algorithms, an example of a multiextremal function is considered:

$$y = (sin(x_1^2) + cos(x_2^2))e^{-(x_1^2 + x_2^2)/8}.$$

The 3D view of the presented function is shown in Fig. 1.

Figure 2 shows a comparison of the results of the pre-sample points algorithm using error estimates on the border of a Voronoi diagram with final sampling based on dichotomy method (triangles) and pre-sampling based on points estimation lying on intersection of Voronoi diagram borders and base vector with iterative sampling method (circles) based on the analysis of the metamodel constructed by base points (rectangles). In the figure, the function values are shown in grayscale, where black is the minimum value and white is the maximum value.

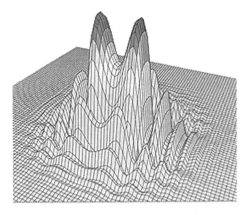

Fig. 1. 3D view of a multiextremal three-dimensional function

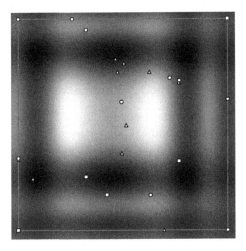

Fig. 2. Comparison of the algorithms

Figure 3 shows the result of points sampling, superimposed on a graph showing the difference between the constructed metamodel on i iteration and analytical model. In the figure, the values of the error function are shown in greyscale, where black is the minimum value and white is the maximum value, black also shows the areas where the interpolation function is within the specified accuracy (10% of the difference between the minimum and maximum value). The figure shows that presented algorithms accurately enough determine the areas of the greatest "error" and offer the points for their correction.

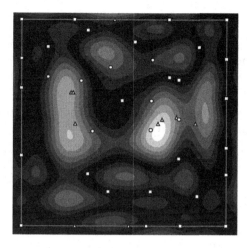

Fig. 3. Difference between the constructed metamodel on i iteration and analytical model

To evaluate the results of the algorithm application, the average errors in the grid nodes of a metamodel constructed by sequential sampling and a metamodel constructed by interpolation of the similar number of points uniformly distributed in space were compared. The results of the tests with following parameters: $n = 2, m = 1, A = 0.1, Kb_{max} = 100, L = 10, Kb = 17$, are shown in Fig. 4.

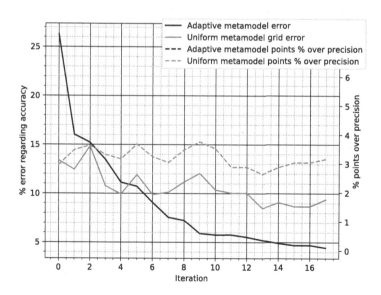

Fig. 4. Comparison of the average errors reduction

The graphs below show the following results:

1. An increase in the average accuracy of the obtained solution by more than 2 times with metamodel build by sequential sampling.
2. The accuracy of function behavior reflection is about 5 times higher.

8.2 Application of Algorithms for the Practical Problem of the Body Motion in a Liquid

The proposed algorithm was tested on the practical problem of determining the hydrodynamic characteristics of a body moving in a liquid. The following parameters are used as variable parameters: speed of movement, angular velocity, angle of attack, and drift angle. The coefficients of forces and moments acting on the model are estimated as hydrodynamic characteristics (objective functions). Thus, the following characteristics were considered:

1. the dimension of the problem parameter space is $n = 4$;
2. number of objective functions is $m = 6$;
3. function characteristics are multi-extreme, smooth functions;
4. accuracy is $A = 0.1$;
5. maximum points $Kb_{max} = 100, L = 10$.

To evaluate the results of the algorithm application, the average errors in the grid nodes of a metamodel created by sequential sampling and a metamodel constructed by interpolation of the similar number of points uniformly distributed in space were compared. The results of the tests are shown in Fig. 5.

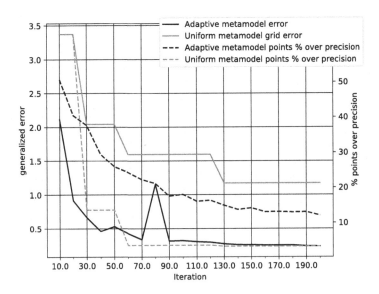

Fig. 5. Comparison of the average errors reduction

The graphs below show the following results:

1. the reduction by a factor of 6 is observed in the scope of virtual testing—the average accuracy of the interpolation function based on a uniform grid with 130 iterations corresponds to the interpolation accuracy based on the points obtained by the Shepard interpolation algorithm on the Voronoi diagrams with 20 iterations;

2. a good level of interpolation accuracy of the desired dependencies is observed—the interpolation function for 200 iterations based on a uniform mesh never reached the accuracy of interpolation functions constructed on the points obtained by the Shepard interpolation algorithm using the Voronoi diagrams with 90 iterations.

References

1. Ghoreyshi, M., Greisz, I., Jirasek, A., Satchell, M.: Simulation and modeling of rigid aircraft aerodynamic responses to arbitrary gust distributions. Aerospace **5**(2), 43 (2018)
2. Fedorov, V.V.: Theory of Optimal Experiments. Academic Press, New York (1972)
3. Wang, G.G., Shan, S.: Review of metamodeling techniques in support of engineering design optimization. J. Mech. Des. **129**(4), 370–80 (2007)
4. Myers, R.H., Montgomery, D.: Response Surface Methodology: Process and Product Optimization Using Designed Experiments. Wiley, Toronto (1995)
5. Mitchell, T.J.: An algorithm for the construction of "D-Optimal" experimental designs. Technometrics **16**(2), 203–210 (1974)
6. McKay, M.D., Bechman, R.J., Conover, W.J.: A comparison of three methods for selecting values of input variables in the analysis of output from a computer code. Technometrics **21**(2), 239–245 (1979)
7. Tang, B.: Orthogonal array-based Latin hypercubes. J. Am. Stat. Assoc. **88**(424), 1392–1397 (1993)
8. Hedayat, A.S., Sloane, N.J.A., Stufken, J.: Orthogonal Arrays: Theory and Applications. Springer, New York (1999). https://doi.org/10.1007/978-1-4612-1478-6
9. Meckesheimer, M., Booker, A.J., Barton, R.R., Simpson, T.W.: Computationally inexpensive metamodel assessment strategies. AIAA J. **40**(10), 2053–2060 (2002)
10. Fang, K.T., Lin, D.K.J., Winker, P., Zhang, Y.: Uniform design: theory and application. Technometrics **39**(3), 237–248 (2000)
11. Burnaev, E., Panov, M.: Adaptive design of experiments based on Gaussian processes. In: Gammerman, A., Vovk, V., Papadopoulos, H. (eds.) SLDS 2015. LNCS (LNAI), vol. 9047, pp. 116–125. Springer, Cham (2015). https://doi.org/10.1007/978-3-319-17091-6_7
12. Duchanoy, C.A., Calvo, H., Moreno-Armendáriz, M.A.: ASAMS: an adaptive sequential sampling and automatic model selection for artificial intelligence surrogate modeling. Sensors **20**, 5332 (2020)
13. Picheny, V., Ginsbourger, D., Roustant, O., Haftka, R.T., Kim, N.: Adaptative designs of experiments for accurate approximation of a target region. J. Mech. Des. **132** (2010)
14. Chen, R.J.W., Sudjianto, A.: On sequential sampling for global metamodeling in engineering design. In: Proceedings of DETC 2002, Montreal, Canada, September 29–October 2 (2002)

15. Xu, S., et al.: A robust error-pursuing sequential sampling approach for global metamodeling based on voronoi diagram and cross validation. J. Mech. Des. **136**(7), 071009 (2014)
16. Cresssie, N.: Spatial prediction and ordinary kriging. Math. Geol. **20**(4), 405–421 (1988)
17. Shepard, D.: A two-dimensional interpolation function for irregularly-spaced data. In: Proceedings of the 23rd ACM National Conference, pp. 517–524. ACM, New York (1968)
18. Park, S., Linsen, L., Kreylos, O., Owens, J.D., Hamann, B.: Discrete Sibson interpolation. IEEE Trans. Visual Comput. Graphics **12**(2), 243–253 (2006)
19. Dell'Accio, F., Di Tommaso, F.: Scattered data interpolation by Shepard's like methods: classical results and recent advances. Proc. Kernel-Based Methods Function Approx. **9**, 32–44 (2016)

The Optimized Finite Element Dynamical Core of the Arctic Ocean Sea Ice Model

Sergey Petrov$^{(\boxtimes)}$ and Nikolay Iakovlev

Marchuk Institute of Numerical Mathematics of the Russian Academy of Sciences
(INM RAS), 119333 Moscow, Russia
director@mail.inm.ras.ru
https://www.inm.ras.ru/en/

Abstract. This paper introduces the dynamic block of a new high spatial resolution sea ice model. Arctic region irregular triangular grid generating technology based on the INMOST package is presented. The mesh refined in areas of high ice concentration, at the coastline, and in narrow passages. The parallel finite element implementation of the model with several time integration schemes for the momentum balance equation is presented. For the ice mass and compactness transport equations the Taylor-Galerkin-Flux-Correction scheme was implemented. An optimization of the time integration scheme for the momentum equation is proposed, which allows to accelerate the standard stationary mEVP method used in modern ice models (CICE, LIM, FESIM). The idea of the proposed nonstationary mEVP-opt method is to approximate the iteration parameter to the locally optimal one obtained from the estimate of the integration step in the approximation of the linearized transition operator. The numerical experiment to reproduce the most computationally complex mode of slow ridging was performed. The result is compared to the Picard method one with 10 pseudo-iterations, which gives high accuracy at high computational costs. It is shown that the new mEVP-opt method provides a significant reduction in the computation time compared to mEVP with a slight increase in the number of operations. Represented dynamic core will be generalized for a multicategory ice thickness case, supplemented by thermodynamics blocks, ice thickness redistribution due to ridging, and data assimilation procedure. It could be used as a high-quality sea ice forecast tool.

Keywords: Sea ice · Viscous-plastic rheology · INMOST · Ani2D · Ani3D · Supercomputer modeling · Ocean modeling

1 Introduction

Ice models play an important role in predictive systems and climate models, and the presence of ice significantly affects heat and mass transfer between the

Supported by Moscow Center for Fundamental and Applied Mathematics at the INM RAS, Moscow, Russia.

ocean and the atmosphere. Ice dynamics is of practical interest for the design of offshore hydrocarbon production facilities and for forecasting the routes of Arctic expeditions.

The presented work consists of two parts. The first part is devoted to the construction of a computational domain triangulation - the area of the Arctic Ocean with all adjacent seas and a detailed description of the coastline. The following is a description of a parallel system for interpolating geodata onto a model grid, which is used to fill the model with data and generate output. Basically, geodata in the world scientific community are usually distributed in the netCDF format on a rectangular grid with depth levels. The proposed model is built on an unstructured triangular mesh, the process of interpolating scalars and vectors from rectangular to triangular mesh and vice versa requires a special explanation.

The second part of the work is devoted to the description of the optimization of the most widespread method of numerical integration of the momentum balance equation. The active development of ice models began after the introduction of equations for the dynamics of sea ice with viscous-plastic rheology [6]. Subsequently, many methods were proposed for the numerical integration of the momentum balance equations [2,7], and their optimizations as well [10]. In this paper, we present a nonstationary method based on the classical mEVP-approach [2] and the idea of a local decrease in the square of the residual norm of a linearized functional, which increases the convergence rate. This method can be applied to optimize computations in sea ice dynamics blocks. The main advantage of the finite element approach for modeling sea ice is the ability to accurately account for coastline heterogeneity, the ability to thicken the mesh in the area of solution features and areas of the user interest. The physical formulation of this problem leads to ill-conditioned discrete operators, which affects the computational complexity of the solution. Therefore, optimizations of methods like mEVP need to be developed for efficient and economical use of ice models in conjunction with ocean models.

2   Building a Model Grid

The Arctic region, located above 45° latitude, was selected as the computational domain. Coastal contours were taken from the open coastline database GSHHG [17] using the GMT package [16]. We used the roughest resolution available as a starting point. However, practice has shown, that there are inconsistencies in the presented coastal contours, which are expressed in the self-intersection of coastline segments. Another disadvantage is the presence of narrow bays and adjacent coastlines forming sharp corners. To eliminate this drawbacks, the following smoothing technique is used, which pursues the main goal: one can delete some vertices in a broken coastline, but can not move them. Firstly, this requirement is necessary for writing an automated coastline smoothing program for an arbitrary "bad" contour. Secondly, it leaves the possibility of adding a more detailed part of a broken coastline from the same database between existing vertices. The following parameterized procedure is executed:

1. If the length of the segment exceeds the specified value, then its end point is deleted;

2. If two adjacent line segments form an angle less than the predetermined value, then the corner vertex is removed;

3. Fix a vertex and a natural number n. If among the n of the following segments there is a segment close to the selected point at a distance less than a fixed value, then this segment and all previous points (up to the fixed point) are deleted;

4. Each island with less than 5 line segments is removed.

At the initial stage of building the grid, the river mouths were removed, and the computational domain was closed through the water passing along 45° latitude. Then modified coastline data submited to the procedure for constructing a regular triangulation by Ani-2D AFT package [3]. The regular triangulation is shown in Fig. 1.

mesh

Fig. 1. Regular triangulation.

Regular triangulation has an obvious drawback - there is practically no ice at low latitudes, an excessive number of nodes in such places leads to unnecessary computations. Thus, a mesh thickening procedure was applied in an area with a potentially high ice concentration. We used data on satellite measurements of ice concentration for the last 10 years [12] and special functionality of the Ani-2D AFT package, which allows one to set the desired local size of the triangle at a specified point.

The boundary between high and low concentration ice is rather sharp. In order to smooth the boundary, satellite data on ice concentration at the nodes of a rectangular grid were transformed using the discrete Laplace operator

$$a_{i,j}^{\text{new}} = -\frac{a_{i-1,j} + a_{i+1,j} + a_{i,j-1} + a_{i,j+1} - 4a_{i,j}}{h^2},$$

where h is spacial resolution of the rectangular grid. The minimum possible concentration is considered equal to $a_{\min} = 0.05$. Maximal size of a triangle was set 5 times larger than the minimal one $d_{\max} = 5d_{\min}$. If $a(x,y)$ is the value of the bilinear interpolant of concentration at the point (x,y), then the desired size of the triangle at this point $d(x,y)$ is calculated by

$$d(x,y) = d_{\max} + \frac{d_{\min} - d_{\max}}{1 - a_{\min}} \cdot (a(x,y) - a_{\min}).$$

The result of triangulation with a mesh refining to an area with high potential ice concentration for $d_{\min} \approx 10\,\text{km}$ is shown in Fig. 2.

mesh

Fig. 2. Thickened triangulation.

The constructed model area consists of 329 670 nodes, 642 387 triangles, and 17 049 boundary edges. It is clear that solutions to problems of similar dimensions need to be designed taking into account the parallel architecture of the supercomputer.

Note that the use of geographic (longitude/latitude) coordinates as model ones is difficult due to the singularity at the North Pole. To work around this problem, model coordinates differ from geographic ones by rotating the North Pole to the geographic equator. This is achieved by choosing the Euler angles $(-30°, -90°, 0°)$.

3   Features of the INMOST Package

The INMOST [4] software package is developed and maintained at the INM RAS. This package is designed for massively parallel modeling on grids of arbitrary structure, both in 2D and 3D. INMOST is well optimized for using finite element and finite volume approximations. The INMOST package is written in the C++ and includes MPI and OpenMP functionality for multiprocessor and multithreaded computations. To construct an efficient parallel division of the computational domain into subdomains corresponding to different processes, minimizing the number of exchanges, this model uses the ParMETIS [9] library integrated into INMOST. The mesh data corresponding to elements (node, edge, triangle) is stored in an ordered format, which provides optimal search algorithms. Figure 3 shows the division of the model grid into 20 processes.

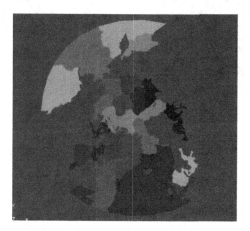

Fig. 3. Decomposition of the model grid by 20 processes.

The library PETSc [1], also integrated into INMOST, is used for the parallel solution of linear systems.

As mentioned earlier, most geodata are distributed in netCDF format on a rectangular grid with multiple levels. To implement the dynamic block of the ice model, one need to know the initial distribution of ice concentration and height, as well as the distribution of the ocean level every few steps of model integration. We decided to fill the model with data from the operational oceanic European forecast system TOPAZ4 [15], which is part of the European Earth operational forecast program Copernicus. This choice is due to the openness of the resource for non-commercial research, regular updates and support. TOPAZ4 data are received online every hour, which allows to validate the developed model according this data in the future.

The netCDF [14] standard library allows parallel reading and writing from a netCDF file. TOPAZ4 data is located on a rectangular grid in stereographic projection onto the tangent plane to the North Pole of the Earth. Each processor reads only the part of the rectangular grid it needs, which optimizes further calculations in terms of time and memory used. The result of parallel bilinear interpolation of scalar fields is shown in Fig. 4 and Fig. 5.

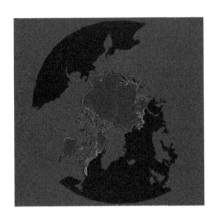

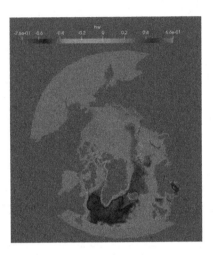

Fig. 4. Distribution of sea ice concentration at 00:00 on April 1, 2020 according to TOPAZ4 data, interpolated on the model grid.

Fig. 5. Ocean level distribution at 00:00 on April 1, 2020 according to TOPAZ4 data, interpolated to the model grid.

In addition to scalar fields, the dynamic core of the model requires regular updating of the oceanic and atmospheric forcing. The data on the zonal velocity components of the boundary layer of the atmosphere and ocean are shown in Fig. 6 and Fig. 7 respectively.

4   Optimized Method for Numerical Integration of the Sea Ice Momentum Balance with Viscous-Plastic Rheology

Various numerical integration methods are used in dynamic blocks of modern global sea ice models with viscous-plastic rheology. The following two are most common: mEVP [2], VP-Picard [8]. This paper makes a comparison of these methods qualitatively and in terms of the residual norm in a square model area of 1000 km size with artificial external forcing, periodic in time and space, which, despite its simplicity, implements a computationally complex mode of slow ridging. Also an alternative accelerated method mEVP-opt is proposed. It is based on the classical mEVP method and the idea of local decay of the squared residual norm of the linearized operator.

4.1   Governing Equations

The 2D system of sea ice dynamics with viscous-plastic rheology consists of momentum balance and two transport equations - ice concentration and mass advection [6]:

$$
\begin{cases}
m(\partial_t + f\mathbf{k}\times)\mathbf{u} = a\tau - C_d a\rho_0(\mathbf{u} - \mathbf{u_0})|\mathbf{u} - \mathbf{u_0}| + \mathbf{F} - mg\nabla H \\
\tau = C_a \rho_a |\mathbf{u_a}|\mathbf{u_a} \\
F_l = \frac{\partial \sigma_{kl}}{\partial x_k},\ l = 1,2 \\
\sigma_{kl}(\mathbf{u}) = \frac{P_0}{2(\Delta + \Delta_{min})}\left[(\dot\varepsilon_d - \Delta)\delta_{kl} + \frac{1}{e^2}(2\dot\varepsilon_{kl} - \dot\varepsilon_d \delta_{kl})\right] \\
\dot\varepsilon_{kl} = \frac{1}{2}(\partial_k u_l + \partial_l u_k);\ \ \dot\varepsilon_d = \dot\varepsilon_{kk} = \dot\varepsilon_{11} + \dot\varepsilon_{22} \\
\dot\varepsilon_s = ((\varepsilon_{11} - \dot\varepsilon_{22})^2 + 4\dot\varepsilon_{12}^2)^{1/2};\ \ \Delta = (\dot\varepsilon_d^2 + \frac{1}{e^2}\dot\varepsilon_s^2)^{1/2} \\
P_0 = p^* h e^{-C(1-a)} \\
\sigma_{1(2)} = \sigma_{11} \pm \sigma_{22},\ \dot\varepsilon_{1(2)} = \dot\varepsilon_{11} \pm \dot\varepsilon_{22} \\
\partial_t a + \nabla \cdot (\mathbf{u}a) = 0,\ a \leq 1 \\
\partial_t m + \nabla \cdot (\mathbf{u}m) = 0 \\
m = ah\rho,
\end{cases}
\tag{1}
$$

where $\mathbf{k}$ is the unit vertical vector, f is the Coriolis parameter, m, a, h mass, concentration and height of the ice, $\mathbf{u}, \mathbf{u_0}, \mathbf{u_a}$ - ice, water and air velocities, ρ, ρ_0, ρ_a - ice, water and air densities, σ_{ij} - stress tensor components, $\dot\varepsilon_{ij}$ are the components of the strain rate tensor, $e = 2$ is the ellipticity parameter, P_0 is the pressure, H is the ocean level.

The transport equations are solved by the conservative Taylor-Galerkin finite element scheme with Flux-Correction technology [11] applied.

4.2   mEVP Method

The essence of the mEVP-approach [2] for solving the momentum balance equation is to apply the explicit Euler scheme in time separately for the main components of the stress tensor and velocities:

$$
\alpha(\sigma_1^{p+1} - \sigma_1^p) = \frac{P_0}{\Delta^p + \Delta_{min}}(\dot\varepsilon_1^p - \Delta^p) - \sigma_1^p
$$

$$
\alpha(\sigma_2^{p+1} - \sigma_2^p) = \frac{P_0}{(\Delta^p + \Delta_{min})\cdot e^2}\dot\varepsilon_2^p - \sigma_2^p
$$

$$
\alpha(\sigma_{12}^{p+1} - \sigma_{12}^p) = \frac{P_0}{(\Delta^p + \Delta_{min})\cdot e^2}\dot\varepsilon_{12}^p - \sigma_{12}^p
\tag{2}
$$

$$
\beta(\mathbf{u}^{p+1} - \mathbf{u}^p) = -\mathbf{u}^{p+1} + \mathbf{u}^n - \Delta t f \times \mathbf{u}^{p+1}
$$

$$
+\frac{\Delta t}{m}\left[\mathbf{F}^{p+1} + a\tau + C_d a\rho_0(\mathbf{u}_0^n - \mathbf{u}^{p+1})|\mathbf{u}_0^n - \mathbf{u}^p| - mg\nabla H^n\right].
$$

Here the index p corresponds to the local iteration, and n is the number of the global time step. Typical parameter values used in models: $\alpha = \beta = 500$

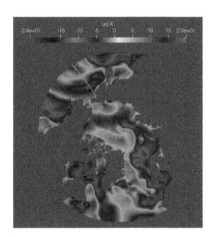

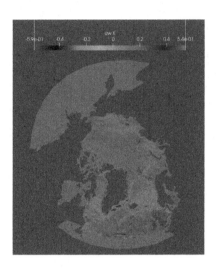

Fig. 6. Distribution of the zonal wind speed component at 00:00 on April 1, 2020 according to TOPAZ4 data, interpolated to the model grid.

Fig. 7. Distribution of the zonal water speed component at 00:00 on April 1, 2020 according to TOPAZ4 data, interpolated to the model grid.

with $N_{\mathrm{mEVP}} = 500$ pseudoiterations. The described process will be referred as mEVP-500. The main advantage of this approach is no need to solve a system of linear equations at each iteration. This property is obtained through the use of a lumped (diagonalized matrix with diagonal entries equal to row sum) mass matrix after applying the Galerkin method for spatial discretization.

4.3   VP-Picard Method

The second most popular approach for solving the momentum balance equation is the Picard iteration method [8]. The values of bulk and shear viscosities are taken from the previous pseudo-iteration and are used to form a new approximation of the stress tensor components. These components are then plugged directly into the momentum balance equation and new values for the velocities are calculated. Thus, an implicit time scheme is organized, which allows avoiding significant restrictions on the integration step. However, the described scheme requires solving a linear system at each pseudo iteration, which significantly enhance computations. The Picard method with 10 pseudo-iterations will be called VP-10. Figure 8 shows a comparison of the relative residual norm of the VP-10 and mEVP-500 methods. One can see that the solution obtained by the former tends to the latter once. Moreover, VP-10 gives a more accurate result. Convergence requires at least 1400 iterations of mEVP-500.

4.4   mEVP-opt Method

Discretized in time and space momentum balance equation from (1) according to the mEVP method (2) can be written in the standard form

$$\alpha(\mathbf{x}^{k+1} - \mathbf{x}^k) + A(\mathbf{x}^k)\mathbf{x}^k + \mathbf{f}(\mathbf{x}^k) = \mathbf{b}, \tag{3}$$

which is a stationary simple iteration method with parameter α. The vector $\mathbf{x} = (\sigma_1^T, \sigma_2^T, \sigma_{12}^T, \mathbf{u}^T, \mathbf{v}^T)^T$ consists of global nodal values of stress and velocity components. The residual of the equation is written as $\mathbf{r}(\mathbf{x}) = A(\mathbf{x})\mathbf{x} + \mathbf{f}(\mathbf{x}) - \mathbf{b}$. The idea of the optimized mEVP-opt method is to organize a non-stationary simple iteration method (3), where the optimal value of the α_{opt} parameter is estimated by minimizing the squared residual norm in the linear approximation [13]:

$$
\begin{aligned}
||\mathbf{r}(\mathbf{x}^{k+1})||^2 &= ||A(\mathbf{x}^{k+1})\mathbf{x}^{k+1} + \mathbf{f}(\mathbf{x}^{k+1}) - \mathbf{b}||^2 \\
&= ||A(\mathbf{x}^{k+1})[\mathbf{x}^k + \frac{1}{\alpha}\left(\mathbf{b} - A(\mathbf{x}^k)\mathbf{x}^k - \mathbf{f}(\mathbf{x}^k)\right)] + \mathbf{f}(\mathbf{x}^{k+1}) - \mathbf{b})||^2 \\
&\approx ||\left[A(\mathbf{x}^k)\mathbf{x}^k + \mathbf{f}(\mathbf{x}^k) - \mathbf{b}\right] - \frac{1}{\alpha}A(\mathbf{x}^k)\left[A(\mathbf{x}^k)\mathbf{x}^k + \mathbf{f}(\mathbf{x}^k) - \mathbf{b}\right]||^2 \\
&= ||\mathbf{r}(\mathbf{x}^k) - \frac{1}{\alpha}A(\mathbf{x}^k)\mathbf{r}(\mathbf{x}^k)||^2 \\
&= ||\mathbf{r}(\mathbf{x}^k)||^2 - \frac{2}{\alpha}\left(\mathbf{r}(\mathbf{x}^k), A(\mathbf{x}^k)\mathbf{r}(\mathbf{x}^k)\right) + \frac{1}{\alpha^2}||A(\mathbf{x}^k)\mathbf{r}(\mathbf{x}^k)||^2 \\
\Rightarrow \alpha_{\mathrm{opt}} &= \left[\frac{\left(\mathbf{r}(\mathbf{x}^k), A(\mathbf{x}^k)\mathbf{r}(\mathbf{x}^k)\right)}{||A(\mathbf{x}^k)\mathbf{r}(\mathbf{x}^k)||^2}\right]^{-1}.
\end{aligned}
\tag{4}
$$

The derived estimate is rough, due to the rigid assumption of the linearity. Direct use of the optimal iteration parameter at each step $\alpha = \alpha_{\mathrm{opt}}$ (4) causes the method to be unstable. Therefore, we suggest to use the following step recalculation procedure, starting with $\alpha^0 = \alpha_{\mathrm{def}}$ ($\alpha_{\mathrm{def}} = 500$ in case of mEVP-500)

$$\alpha^{k+1} = \alpha^k + \frac{(\alpha_{\mathrm{opt}}^k - \alpha^k)}{C_\alpha \alpha_{\mathrm{def}}}.$$

We used value $C_\alpha = 2.0$. Also on the new α^{k+1} following restrictions are imposed

1. Decreasing step: $\alpha^{k+1} < \alpha^k$;
2. The relative decrement should not exceed some predetermined value ε: $\frac{|\alpha^{k+1} - \alpha^k|}{\alpha_{\mathrm{def}}} < \varepsilon$. In code we use $\varepsilon = 0.05$;
3. The stability condition for the mEVP method [5].

If any of the conditions listed above is not satisfied, the next step is assigned equal to the previous one $\alpha^{k+1} = \alpha^k$.

4.5 Numerical Experiments

For the numerical experiment, a square domain was chosen. Size of square is $L = 1000$ km, that approximately corresponds to the real Arctic size. The initial conditions and external forcing are set as $u_a = 5 + (\sin(\frac{2\pi t}{T}) - 3)\sin(\frac{2\pi x}{L})\sin(\frac{\pi y}{L})$, $v_a = 5 + (\sin(\frac{2\pi t}{T}) - 3)\sin(\frac{2\pi y}{L})\sin(\frac{\pi x}{L})$, $T = 4$ days, according to [5]. The water velocity is set as $u_0 = 0.1\frac{2y-L}{L}$, $v_0 = -0.1\frac{2x-L}{L}$. All velocities are measured in m/s. Ocean level is calculated according to geostrophic balance. The initial ice height is 2 m throughout the entire domain. The ice concentration increases linearly from 0 to 1 in the easterly direction.

Figure 9 shows the residual advantage of the mEVP-opt method over the standard mEVP. One can see that the convergence of mEVP-opt is achieved in 800–1000 iterations.

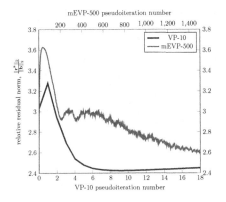

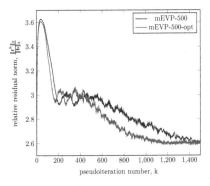

Fig. 8. The relative residual norm depending on the number of the local iteration at the time t = 7 h. Blue line - VP-10 method, red line - mEVP-500. (Color figure online)

Fig. 9. The relative residual norm depending on the number of the local iteration at the time t = 7 h. Blue line - mEVP-500 method, red line - mEVP-500-opt. (Color figure online)

Figure 10 shows a qualitative picture of the better convergence of mEVP-opt to the VP-10 solution compared to the standard mEVP method. It demonstrates that on the top of Fig. 10b the region of the maximum velocity obtained by the mEVP method is not reproduced. One can also notice that the left border of positive velocities is overestimated. Thus, we can conclude that 800 iterations of the mEVP method are not enough to obtain a precise solution. These disadvantages are not present in the velocities, calculated by the mEVP-opt method, that is shown in Fig. 10c.

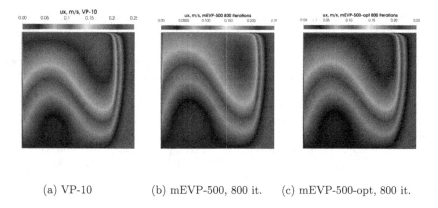

(a) VP-10   (b) mEVP-500, 800 it.   (c) mEVP-500-opt, 800 it.

Fig. 10. Distribution of the zonal ice velocity component after 7 h of modeling.

5 Conclusion

This paper describes the process of constructing a triangulation of the Arctic region with a thickening of the grid in the area of potentially high ice concentration in detail. To construct the triangulation, the Ani2D package and satellite data on ice concentration over the past 10 years were used. The dynamic core of the developed model is based on the INMOST package. The results of model grid decomposition and interpolation of scalar and vector fields are presented. An optimized mEVP-opt method for numerical integration of the sea ice momentum balance equation with visco-plastic rheology is proposed. Qualitatively and quantitatively (by residual), the advantage of this method over the standard mEVP is shown at slightly higher computational costs.

Acknowledgement. The work was performed at the INM RAS with financial support of the Moscow center of fundamental and applied mathematics (agreement with the Ministry of education and science of Russia No. 075-15-2019-1624). Authors express gratitude to the specialists from INM RAS: P. Perezhogin for valuable advice, A. Danilov for help with mesh construction and V. Kramarenko for help with INMOST software package.

References

1. Balay, S., et al.: PETSc Web page (2019). https://www.mcs.anl.gov/petsc
2. Bouillon, S., Fichefet, T., Legat, V., Madec, G.: The elastic-viscous-plastic method revisited. Ocean Model. **71**, 2–12 (2013). https://eprints.soton.ac.uk/359113/
3. Danilov, A.: Unstructured tetrahedral mesh generation technology. Comput. Math. Math. Phys. **50**, 139–156 (2010). https://doi.org/10.1134/S0965542510010124
4. Danilov, A., Terekhov, K., Konshin, I., Vassilevski, Y.: Parallel software platform inmost: a framework for numerical modeling. Supercomput. Front. Innov. **2**(4) (2016). https://superfri.org/superfri/article/view/87

5. Danilov, S., et al.: Finite-element sea ice model (FESIM), version 2. Geosci. Model Dev. **8**(6), 1747–1761 (2015). https://doi.org/10.5194/gmd-8-1747-2015. https://gmd.copernicus.org/articles/8/1747/2015/

6. Hibler III, W.D.: A dynamic thermodynamic sea ice model. J. Phys. Oceanogr. **9**(4), 815–846 (1979). https://doi.org/10.1175/1520-0485(1979)009⟨0815: ADTSIM⟩2.0.CO;2

7. Hunke, E.C., Dukowicz, J.K.: An elastic viscous plastic model for sea ice dynamics. J. Phys. Oceanogr. **27**(9), 1849 (1997). https://doi.org/10.1175/1520-0485(1997)027%3c1849:AEVPMF%3e2.0.CO;2

8. Hutchings, J.K., Jasak, H., Laxon, S.W.: A strength implicit correction scheme for the viscous-plastic sea ice model. Ocean Model. **7**(1/2), 111–133 (2004). https://doi.org/10.1016/S1463-5003(03)00040-4

9. Karypis, G., Kumar, V.: A fast and high quality multilevel scheme for partitioning irregular graphs. SIAM J. Sci. Comput. **20**(1), 359–392 (1999). https://doi.org/10.1137/S1064827595287997

10. Kimmritz, M., Danilov, S., Losch, M.: The adaptive EVP method for solving the sea ice momentum equation. Ocean Modell. **101**, 59–67 (2016). https://doi.org/10.1016/j.ocemod.2016.03.004. http://www.sciencedirect.com/science/article/pii/S1463500316300038

11. Kuzmin, D., Turek, S.: Flux correction tools for finite elements. J. Comput. Phys. **175**(2), 525–558 (2002). https://doi.org/10.1006/jcph.2001.6955. http://www.sciencedirect.com/science/article/pii/S0021999101969554

12. Meier, W.N., Fetterer, F., Windnagel, A.K.: Near-real-time NOAA/NSIDC climate data record of passive microwave sea ice concentration, version 1 (2017). https://doi.org/10.7265/N5FF3QJ6

13. Olshanskij, M.: Lekcii i uprazhneniya po mnogosetochnym metodam. Fizmatlit (2005)

14. Rew, R.K., Davis, G.P., Emmerson, S.: NetCDF User's Guide, An Interface for Data Access, Version 2.3, April 1993. https://www.unidata.ucar.edu/

15. Sakov, P., Counillon, F., Bertino, L., Lisæter, K.A., Oke, P.R., Korablev, A.: TOPAZ4: an ocean-sea ice data assimilation system for the North Atlantic and Arctic. Ocean Sci. **8**(4), 633–656 (2012). https://doi.org/10.5194/os-8-633-2012. https://os.copernicus.org/articles/8/633/2012/

16. Wessel, P., et al.: The generic mapping tools version 6. Geochem. Geophys. Geosyst. **20**(11), 5556–5564 (2019). https://doi.org/10.1029/2019GC008515. https://agupubs.onlinelibrary.wiley.com/doi/abs/10.1029/2019GC008515

17. Wessel, P., Smith, W.: A global, self-consistent, hierarchical, high-resolution shoreline database. J. Geophys. Res. **101**, 8741–8743 (1996). https://doi.org/10.1029/96JB00104

Performance of Supercomputers Based on Angara Interconnect and Novel AMD CPUs/GPUs

Artemiy Shamsutdinov[1,2], Mikhail Khalilov[1,2], Timur Ismagilov[3],
Alexander Piryugin[3], Sergey Biryukov[1,2], Vladimir Stegailov[1,2,4(✉)],
and Alexey Timofeev[1,2,4]

[1] Joint Institute for High Temperatures of RAS, Moscow, Russia
[2] National Research University Higher School of Economics, Moscow, Russia
[3] JSC NICEVT, Moscow, Russia
[4] Moscow Institute of Physics and Technology, Dolgoprudny, Russia

Abstract. A low-latency high bandwidth interconnect that makes a unified system from a collection of nodes is a heart of any modern supercomputer. At the moment, Infiniband is the main commercially available type of interconnect without any other real competition world-wide. Proprietary interconnects are known to stand behind effcient supercomputer systems. Since 2016, the supercomputer centre of JIHT RAS deploys systems based on the Angara interconnect developed in Moscow by JSC NICEVT. In this paper, we present the performance analysis for two recently upgraded supercomputers in JIHT RAS that are based on two types of Angara interconnect and modern AMD Epyc CPUs and AMD Instinct MI50 GPUs. The general properties of Angara interconnects are described and compared with Infiniband FDR. The details of HPL benchmark runs on both systems are analysed.

Keywords: Angara · Infiniband · Epyc · MI50 · HPL · Scalability · Efficiency

1 Introduction

In the past few years, the growth in the computing power of supercomputers is provided not so much by the processor frequency as by the increase in the number of computational nodes and the number of cores. For this reason, the contribution of high-speed interconnect to the maximum computing performance of a supercomputer permanently increases. This trend makes it promising to create high-performance computing systems with interconnects that provide the lowest latency and the highest throughput. Currently, there are several interconnects that provide low latency (about 1 µs) and high throughput (several tens of GBytes per second). Among the former leaders of this industry we can mention Quadrics [1] (1996–2009), Myrinet [2] (since 1995) and Intel Omni-Path [3] (2015–2019). Currently, the market is dominated by Infiniband [4] (since 2000)

© Springer Nature Switzerland AG 2021
D. Balandin et al. (Eds.): MMST 2020, CCIS 1413, pp. 401–416, 2021.
https://doi.org/10.1007/978-3-030-78759-2_33

with a small share of other types of interconnects, e.g. NUMAlink [5] (since 1996) and RapidIO [6] (since 2000). At the moment, we can refer to two types of interconnect under development in Russia: the Angara interconnect is developed by JCS NICEVT [7,8] and the SMPO-10G is developed by the Russian Federal Nuclear Center – All-Russian Research Institute of Experimental Physics [9]. During the last several years, the Angara interconnect has obtained a history of practical usage [10,11]. In 2021–2022 the second generation of the Angara interconnect is expected to be released.

In this article, we present a comparative analysis of the Angara network and the Infiniband FDR network as parts of two supercomputers at the JIHT RAS. The comparison is based not only on the basic characteristics of the network but also on the results of the HPL benchmark. The main computations for the benchmark are performed on modern AMD Epyc CPUs on the Fisher super-computer on AMD Instinct MI50 GPUs on the Desmos supercomputer. For the efficient execution of the benchmark, the optimal parameters of the benchmark configuration are selected, and some methods of speeding up the calculation are performed, including the frequency boosting technology.

Our comparative analysis of the effectiveness of interconnects has two main parts. The first part consists of studying the basic characteristics of two inter-connects on the supercomputers under study. These tests include traditional measurements such as point-to-point latency for several node locations and node-to-node connectivity, which is especially important for the Angara network. In addition to them, we consider some of the most frequently used operations in par-allel calculations, such as Allreduce and AlltoAll. The purpose of this extended set of microtests is to characterize various aspects of interconnect usage. The second part of the performance evaluation consists of application-level tests. We use the HPL benchmark. We not only present the overall performance results but also show the performance of the processors during the benchmark run.

The rest of this paper is organized as follows. In Sect. 2, we give an overview of modern interconnects with an emphasis on the new Angara interconnect. In Sect. 3, we describe in detail the supercomputers on the basis of which the com-parative analysis of interconnects is carried out. Sections 4 presents a comparison of Angara interconnect and Infiniband FDR based on micro-tests. Sections 5 and 6 present interconnects comparison based on the HPL benchmark on two differ-ent supercomputers. Then we draw conclusions in Sect. 7.

2   Related Work

In this paper, we present the results for the Angara network with torus topology and for its variant with the fat-tree topology.

Torus topologies of the interconnect have several attractive aspects in com-parison with fat-tree topologies. In 1990s, the development of supercomputers had its peak during the remarkable success of Cray T3E systems based on the 3D torus interconnect topology [12] that was the first supercomputer that pro-vided 1 TFlops of sustained performance. In June 1998, Cray T3E occupied 4 of top-5 records of the Top500 list. In 2004, after several years of the dominance of Beowulf clusters, a custom-built torus interconnect appeared in the IBM Blue-Gene/L supercomputer [13]. Subsequent supercomputers of Cray and IBM had torus interconnects as well (with the exception of the latest Cray XC series based on the Aries interconnect with the Dragonfly topology). Fujitsu designed K Computer and the current No.1 in Top500 Fugaku that both are based on the Tofu torus interconnects [14,15]. The Aurora Booster and Green Ice Booster supercomputers are based on the Extoll torus interconnect [16].

The development of a new type of supercomputer-oriented interconnect hardware is a complex endeavor requiring simultaneous development of the corresponding software stack. This software stack should enable the correct and efficient operation of all the software packages of end-users. From the economical point of view, the usage of the main type of presently commercially available interconnects (the Mellanox Infiniband) is always cheaper than the development of new technology. Due to these complexity of development and economical considerations, the number of supercomputer interconnect types is quite limited. Among the novel types of interconnects in addition to those listed above, we can mention, for example, the Atos BXI [17], the TaihuLight interconnect [18] and the Cray Slingshot interconnect [19]. These interconnects are available for purchase within the supercomputer installation only that emphasizes the high importance of the development of the corresponding technologies. An interesting review of high-performance computing development trends has recently been published with a focus on the supercomputer interconnects [20].

Table 1. The main characteristics of the Fisher and Desmos supercomputers

Cluster	Fisher	Fisher	Fisher	Desmos	Desmos
Compute Nodes	Host1-17	Angr1-20	Angr21-40	Host1-32	Host17,26-32
Chassis	Supermicro 1023US-TR4	Gigabyte H262-Z62	Gigabyte H262-Z67	Supermicro 1018GR-T	Supermicro 1018GR-T
Processor	2 x Epyc 7301 16c	2 x Epyc 7301 16c	2 x Epyc 7662 64c	Xeon E5-1650v3 6c	Xeon E5-1650v3 6c
GPU	–	–	–	Instinct MI50 32 GB	Instinct MI50 32 GB
Memory	256 GB	128 GB	256 GB	32 GB	32 GB
Interconnect	Infiniband FDR	Angara switch	Angara switch	Angara 4D-Torus	Infiniband FDR
MPI	OpenMPI-3.1.0	MPICH 3.2 for Angara	MPICH 3.2 for Angara	MPICH 3.2 for Angara	OpenMPI-4.0

The hybrid supercomputer Desmos in JIHT RAS was the first supercomputer based on the Angara network with the in-depth analysis of performance for various applications [11,21–23]. The MPI communication over the Angara network was reviewed for the systems with torus topology, Desmos including [24,25].

In September 2018, Desmos (equipped with AMD FirePro S9150 GPUs) was ranked as No. 45 in the Top50 list of supercomputers (the open-source HPL-GPU benchmark based on OpenCL [26] was used for running LINPACK).

3   Hardware

The main details about the supercomputers Desmos and Fisher are given in Table 1 and illustrated in Figs. 1 and 2. Fisher has a segment with Infiniband FDR and air cooling and a segment with Angara and immersion cooling [27].

20 nodes of Fisher (angr21-40) are based on a new type of servers with Epyc 7662 64 core CPUs. The novelty of this hardware is the reason of unstable work of some chassis that is why we do not consider the nodes angr21-24. The unstable work of PSUs in the nodes angr29-32 caused lower CPU frequencies (see below).

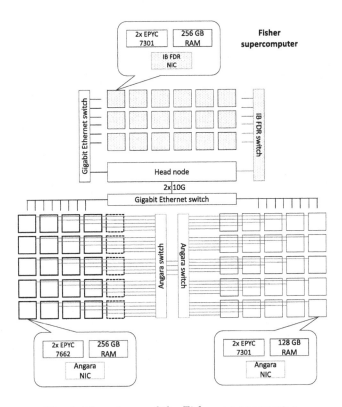

Fig. 1. The scheme of the Fisher supercomputer.

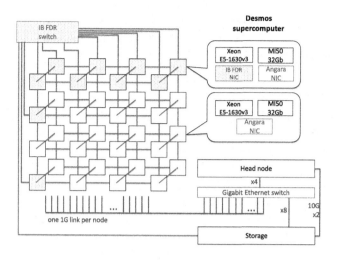

Fig. 2. The scheme of the Desmos supercomputer.

4   Comparison of Angara and Infiniband FDR

Message Passing Interface (MPI) is a de-facto standard for writing large-scale scientific codes. Various HPC applications, including LINPACK, utilize MPI for inter-process communication and synchronization during computations. Hence the performance of MPI library, which is usually highly optimized for an underlying interconnect, plays an important role in the overall application performance.

Ohio State University (OSU) benchmark was selected in order to study initial Angara and InfiniBand FDR performance under typical MPI workloads and compare these interconnect architectures face-to-face. Figure 3 show the results of comparison Angara and IB FDR interconnects installed both in Desmos and Fisher supercomputers on osu_latency and osu_bw tests, correspondingly. The osu_latency benchmark measures the time between the MPI_Send is started at source node and the corresponding MPI_Recv completed on destination node. The osu_bandwidth benchmark is aimed to estimate the amount of data that could be transmitted between MPI ranks for a fixed period of time using MPI Point-to-point operations. Each test included 1 pair of MPI ranks located on 2 different nodes.

P2P latency tests with small-message size with Angara MPI show a 1.2–1.5x speedup in comparison with IB FDR. Higher CPU frequency allows the software layer of the networking stack to generate smaller messages for a fixed period of time. Thus the CPU installed on compute node could also make a significant impact on the performance of network operations. AMD Epyc CPUs used in the Fisher cluster have 1.5x less base frequency than Xeon Haswell CPUs in Desmos. This fact matches with the observation of the similar 1.5–2x P2P latency gap for both interconnects installed on these supercomputers.

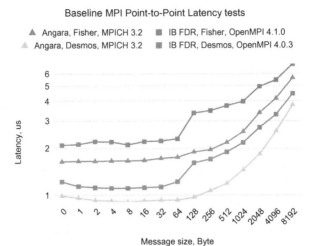

Fig. 3. Comparison of IB FDR and Angara interconnects using MPI point-to-point latency for small message size on Fisher and Desmos supercomputers.

Experiments on the Fisher supercomputer show that MPI rank placement could affect the latency and bandwidth of point-to-point operations because of the Angara switch inner topology (Fig. 4). It is possible to choose a process mapping when data packets have to flow through the multiple crossbar segments for nodes connected to the same switch. Each crossbar segment adds a constant latency (<150 ns). On the Fisher cluster we observe up to 400 ns P2P latency slow down for small (<64 B) MPI messages due to this hardware constraint.

Allreduce and AlltoAll operations are amongst the most important communication patterns in parallel computing. The performance comparison of these MPI operations on the Fisher cluster with IB FDR and Angara interconnects is presented on Figs. 5 and 6. Small-medium (≤16 KB) message Allreduce shows up to 40% less latency with Angara network on Fisher and Desmos supercomputers while scaling up to 16 compute nodes with 1 process per node. For large message sizes the performance results for the both networks are approximately the same.

5   High-Performance Linpack Benchmarks on Fisher

High-Performance Linpack (HPL) is a benchmark used for measuring a system's fp64 computing power. It is the most popular solution for evaluating high-performance computing systems due to its high scalability when running on multiple compute nodes. The TOP-500 rating is based on HPL results, including the most powerful supercomputers in the world. In this rating, not only the obtained result in Gflops is important, but also the ratio of the obtained performance to the peak one, which is called the system efficiency. One of the essential

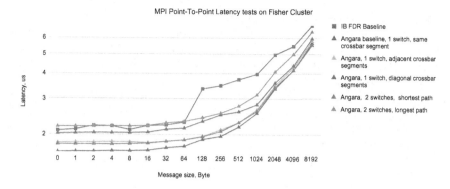

Fig. 4. Comparison of IB FDR and Angara interconnects using MPI point-to-point latency on the Fisher supercomputer.

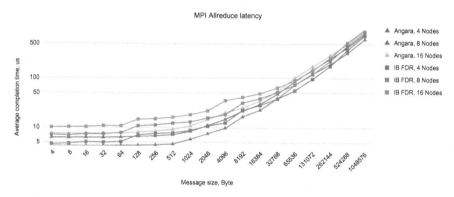

Fig. 5. Comparison of IB FDR and Angara interconnects using MPI Allreduce latency on the Fisher supercomputer.

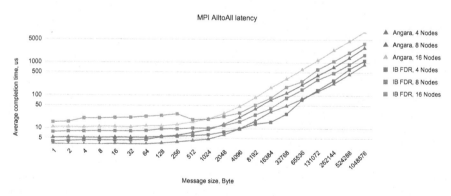

Fig. 6. Comparison of IB FDR and Angara interconnects using MPI AlltoAll latency on the Fisher supercomputer.

components for obtaining high results on a supercomputer is the interconnect, which is responsible for the speed of data exchange between computing nodes. This section discusses how the use of different interconnects Infiniband FDR and Angara Switch affects the resulting performance in the HPL benchmark and the results achieved in the HPL benchmark on new Fisher compute nodes with AMD Epyc 7662 processors. For our tests, HPL is explicitly built for the Fisher supercomputer, using MPICH3.2-angara library with the Angara interconnect support for compute nodes in immersion segment, AMD Optimizing C/C++ Compiler (AOCC) version 2.2, and AMD BLIS 2.2 library with the support of the new Zen 2 architecture. For the Fisher's air segment with Infiniband FDR HPL is built with OpenMPI 3.1.0 library.

5.1 Comparison of Performance on Fisher's Air Segment and Fisher's Immersion Segment

In this subsection, we study how the use of different interconnects affects the resulting performance in HPL. For our tests, we use compute nodes in different segments of Fisher, air segment and immersion segment, with different interconnects. The Fisher air segment (host1-16) and the Fisher immersion segment (angr1-20) have the same CPU (2 x AMD Epyc 7301), so only the interconnect and used cooling system can affect the achieved results in HPL. To verify that all nodes deliver similar performance and are ready for HPL tests on multiple nodes (e.g., see [26]), five runs of HPL are executed on each node with AMD Epyc 7301, and the average value of performance in Gflops is taken. Peak performance of one node with two AMD Epycs 7301 equals 563,2 Gflops, with a base clock frequency of 2.2 GHz. Configuration of HPL benchmark for single node launch: SMT is enabled, 8 MPI processes per node, 8 OpenMP threads per MPI process. With these parameters, the best performance is achieved in both single node and multiple nodes tests. The size of Ns (problem size) is set at 85000, size of Nb (block size) is set at 512 as the optimal value for both small and big problem sizes. Results of single node tests are shown in Fig. 7.

The average performance of nodes angr1-20 is 535,80 Gflops with average efficiency of 95,13%, which is slightly higher, then nodes host1-16 with average performance of 523,91 Gflops and average efficiency of 93,02%. Since interconnect is not used during single node runs, only the immersion cooling system could be the reason for an additional performance boost by providing better temperatures and better average clock speeds.

In order to compare performance of interconnects Angara and Infiniband FDR, HPL is tested on multiple nodes of Fisher (Fig. 8). The numerical values of the HPL performance and efficiency for the range of the computational nodes numbers are presented in the Table 2. Configuration of HPL run on multiple nodes: 8 MPI processes per node, 8 OpenMP threads per mpi process. Size of Ns (problem size) takes approximately 85% of available memory. Size on Nb (block size) is set at 512.

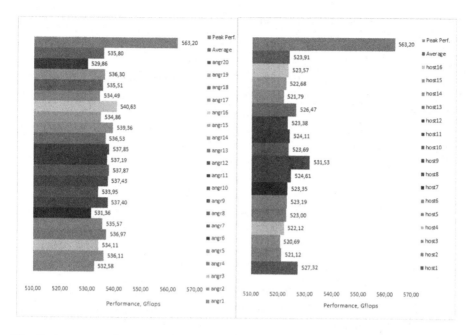

Fig. 7. Average performance of single node tests on angr1-20 (immersion segment) and host1-16 (air segment).

Overall performance in Gflops is slightly higher on the nodes with the Angara interconnect, and the average difference in efficiency is 0,65%. This difference in efficiency can be explained by the fact that the immersion segment has better cooling systems, allowing processors to achieve higher clock speeds. There is no detected effect that could give a preference to one of the two interconnects under consideration.

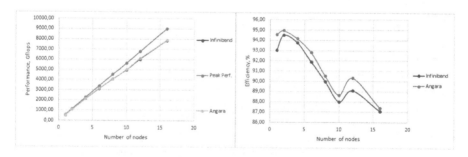

Fig. 8. Comparison of HPL performance and efficiency for IB FDR and Angara interconnects on the supercomputer Fisher.

Table 2. HPL performance and efficiency on Fisher nodes with IB FDR and Angara interconnects.

Interconnect	Infiniband FDR		Angara	
Number of nodes	Performance, Gflops	Efficiency, %	Performance, Gflops	Efficiency, %
1	523	93,02	532	94,56
2	1064	94,49	1069	94,93
4	2112	93,76	2121	94,17
6	3104	91,88	3135	92,78
8	4054	89,98	4077	90,50
10	4956	88,00	4992	88,65
12	6020	89,08	6103	90,31
16	7847	87,08	7875	87,39

5.2   Fisher Immersion Segment's Results with Epyc 7662

In this subsection, we study the HPL results achieved on Fisher compute nodes with new AMD EPYC 7662 processors with the Angara interconnect. AMD EPYC 7662 is a Zen 2 architecture processor released in 2020, with 64 cores and peak performance of 2048 Gflops. The peak performance of a node with two AMD Epyc 7662 is 4096 Gflops, which means that with a large number of nodes, the results obtained in the HPL benchmark for CPUs can compete with the results of the HPL benchmark for GPUs. Configuration of HPL run on a single node: SMT is disabled, 8 MPI processes per node, 16 OpenMP threads per mpi process. Multiple combinations of MPI and OpenMP are tested: 2 MPI + 64 OpenMP, 4 MPI + 32 OpenMP, 8 MPI + 16 OpenMP, 30 MPI + 4 OpenMP. Results on the single node are similar for 2 MPI + 64 OpenMP and 8 MPI + 16 OpenMP, but when running on multiple nodes, 8 MPI + 16 OpenMP performs better. A combination of 16 MPI + 8 OpenMP is not using all cores under load properly. The size of Ns is set at 165184, size of Nb is set at 232. The results of a single node run are shown in Fig. 9.

The average performance of nodes angr21-40 is 3461,7 Gflops with an average efficiency of 84,51%. Low efficiency on Epyc 7662 compared to Epyc 7301 can be explained by low clock speeds on all cores during the HPL run.

Figure 10 shows the average clock speeds of all cores during the HPL run on a single node, where angr1 is the node with Epyc 7301 and angr25 is the node with Epyc 7662. During the run on angr25 average clock speed is 2110 MHz, while the base clock speed of Epyc 7662 is 2000 MHz. The average clock speed on angr1 is 2492 MHz, while the base clock speed of Epyc 7301 is 2200 MHz. Older processors AMD Epyc 7301 achieve higher clock speeds than the new AMD Epyc 7662. This fact can be explained by the fact that while 64-core 7662 CPUs work near the TDP limit for this CPU family, the base frequencies of 16-core 7301 CPUs are further from the TDP limit and the frequency boost is more pronounced.

Results of HPL tests on multiple nodes are shown on Fig. 11 and in the Table 3. Configuration of HPL run on multiple nodes: 8 MPI processes per node, 16 OpenMP threads per MPI process, size of N is set at approximately 85% of available memory (Nb = 232).

For the global HPL run an important role plays a MPICH 3.2-angara backend optimization, in which internal buffers are polled every time only for those processes for which the MPI receive function is called. Also we increase the size of backend internal buffers, and eventually we obtain better global performance.

6 High-Performance Linpack Benchmarks on Desmos

In the previous section, we studied the performance of different interconnects on the supercomputer Fisher, where we used the HPL benchmark for CPUs. However, the highest results in HPL are often obtained by using the HPL benchmark

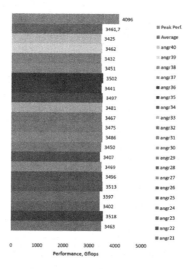

Fig. 9. Average HPL performance on nodes angr21-40 (Fisher segment with liquid immersion cooling).

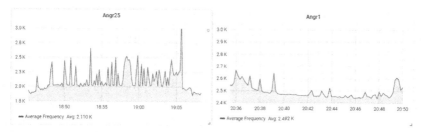

Fig. 10. Average clock speeds of cores during HPL run on nodes angr25 (liquid immersion cooling) and angr1 (air cooling).

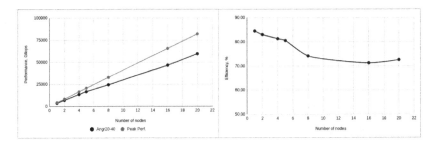

Fig. 11. HPL performance on multiple nodes angr21-40 (the Fisher segment with liquid immersion cooling).

Table 3. HPL performance and efficiency on Fisher nodes with liquid immersion cooling

Number of nodes	Performance, Gflops	Efficiency, %
1	3461	84,51
2	6802	83,03
4	13324	81,32
5	16500	80,57
8	24301	74,16
16	46711	71,28
20	59465	72,59

for GPUs, which can provide higher performance. In this section, we are studying the performance of different interconnects in the HPL benchmark for GPUs on supercomputer Desmos, which compute nodes have a new AMG MI50 GPU with a peak performance of 6.6 Tflops. 8 nodes of Desmos were equipped with an alternative Infiniband FDR interconnect. All tests are done with the experimental HPL binaries compiled with OpenMPI and MPICH for AMD MI50 32 GB provided by AMD. The matrix size in this HPL benchmark is limited by GPU memory. Compute nodes of Desmos have Intel Xeon E5-1650v3 CPU, which peak performance is 336 Gflops, however in HPL benchmark for GPUs, only a small part of the matrix is computed on the CPU, so using an older generation processor shouldn't have a significant impact on the final performance. To verify that all nodes deliver similar performance, five runs of HPL are executed on each node, and the average value of performance in Gflops is taken. Results of single node tests are shown in Fig. 12. Configuration of HPL run on a single node: 1 MPI process per node, 6 OpenMP threads per MPI process. The size of N is set at 63000, size of Nb is set at 384.

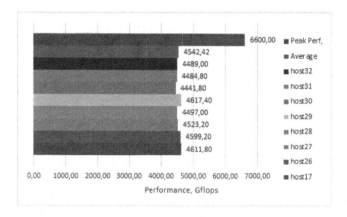

Fig. 12. HPL performance on multiple nodes of Desmos.

The average performance of nodes host17,26-32 is 4542,42 Gflops with average efficiency of 68.82%. Low efficiency can be partially explained by the power limit of 225 W set in the ROCm GPU driver. Higher results on AMD MI50 can probably be obtained by increasing the power limit to 300 W.

To compare performance obtained using Infiniband FDR and Angara interconnects, HPL is tested on multiple nodes of Desmos (Fig. 13). The numerical values of the HPL performance and efficiency for the range of the computational nodes numbers are presented in the Table 4. Configuration of HPL run on multiple nodes: 1 MPI processes per node, 6 OpenMP threads per one MPI process. Ns's size is set at approximately 96% of available memory, the size of Nb is set at 384.

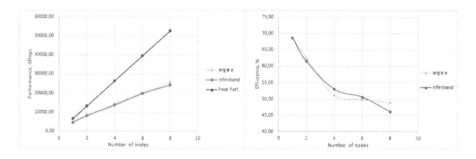

Fig. 13. Comparison of HPL performance and efficiency of IB FDR and Angara interconnects on the Desmos supercomputer.

Table 4. HPL performance and efficiency on Desmos nodes with IB FDR and Angara interconnects.

Interconnect	Infiniband		Angara	
Number of nodes	Performance, Gflops	Efficiency, %	Performance, Gflops	Efficiency, %
1	4542	68,89	4542	68,89
2	8197	61,43	8247	62,48
4	13984	52,97	13478	51,05
6	20036	50,60	19728	49,82
8	24336	46,09	25760	48,79

The difference in performance ranged from 50 to 506 Gflops for 1, 2, 4, 6 nodes, but the difference for eight nodes is 1424 Glops in favor of Angara.

7   Conclusion

MPI benchmarks show that the Angara interconnect in its torus and switch-based variants have latencies competitive with Infiniband FDR. HPL benchmarks of the systems with identical CPUs and GPUs show that the Angara interconnect provides similar performance to Infiniband FDR in HPL benchmark for CPUs and has a slight advantage in HPL benchmark for GPUs on the Desmos supercomputer with torus topology. Angara gives better scaling of the HPL benchmark even for eight nodes. The deployment of the recent 64-core Epyc Rome CPUs and Instinct MI50 GPUs revealed problems in getting HPL performance close to the peak floating-point performance. These problems are expected to be resolved for CPUs by tuning BIOS settings and for GPUs by using the driver with an unlocked power limit.

Acknowledgment. The work is supported by the Russian Science Foundation grant No. 20-71-10127.

References

1. Petrini, F., Feng, W.C., Hoisie, A., Coll, S., Frachtenberg, E.: The Quadrics network: high-performance clustering technology. IEEE Micro **22**(1), 46–57 (2002)
2. Boden, N.J., et al.: Myrinet: a gigabit-per-second local area network. IEEE Micro **15**(1), 29–36 (1995)
3. Birrittella, M.S., et al.: Intel Omni-Path architecture: enabling scalable, high performance fabrics. In: 2015 IEEE 23rd Annual Symposium on High-Performance Interconnects, pp. 1–9. IEEE (2015)
4. Infiniband Trade Association: Infiniband architecture specification. Release **1**, (2000)
5. Laudon, J., Lenoski, D.: The SGI origin: a ccNUMA highly scalable server. ACM SIGARCH Comput. Archit. News **25**(2), 241–251 (1997)

6. RapidIO: An embedded system component network architecture. White Paper (2000)
7. Mukosey, A.V., Semenov, A.S., Simonov, A.S.: Simulation of collective operations hardware support for Angara interconnect. Vestnik Yuzhno-Ural'skogo Gosudarstvennogo Universiteta. Seriya Vychislitelnaya Matematika i Informatika **4**(3), 40–55 (2015)
8. Simonov, A., Brekhov, O.: Architecture and functionality of the collective operations subnet of the Angara interconnect. In: Vishnevskiy, V.M., Samouylov, K.E., Kozyrev, D.V. (eds.) DCCN 2020. LNCS, vol. 12563, pp. 209–219. Springer, Cham (2020). https://doi.org/10.1007/978-3-030-66471-8_17
9. Basalov, V.G., Vyalukhin, V.M.: Adaptive routing system for the domestic interconnect SMPO-10G. VANT. Ser. Mat. Mod. Fiz. Proc. (3), 64–70 (2012)
10. Akimov, V., Silaev, D., Aksenov, A., Zhluktov, S., Savitskiy, D., Simonov, A.: FlowVision scalability on supercomputers with Angara interconnect. Lobachevskii J. Math. **39**(9), 1159–1169 (2018)
11. Stegailov, V., et al.: Angara interconnect makes GPU-based Desmos supercomputer an efficient tool for molecular dynamics calculations. Int. J. High Perform. Comput. Appl. **33**(3), 507–521 (2019)
12. Scott, S.L., et al.: The Cray T3E network: adaptive routing in a high performance 3D torus (1996)
13. Adiga, N.R., et al.: Blue Gene/L torus interconnection network. IBM J. Res. Dev. **49**(2.3), 265–276 (2005)
14. Ajima, Y., Inoue, T., Hiramoto, S., Takagi, Y., Shimizu, T.: The Tofu interconnect. IEEE Micro **32**(1), 21–31 (2012)
15. Ajima, Y., et al.: Tofu Interconnect 2: system-on-chip integration of high-performance interconnect. In: Kunkel, J.M., Ludwig, T., Meuer, H.W. (eds.) ISC 2014. LNCS, vol. 8488, pp. 498–507. Springer, Cham (2014). https://doi.org/10.1007/978-3-319-07518-1_35
16. Neuwirth, S., Frey, D., Nuessle, M., Bruening, U.: Scalable communication architecture for network-attached accelerators. In: 2015 IEEE 21st International Symposium on High Performance Computer Architecture (HPCA), pp. 627–638, February 2015
17. Derradji, S., Palfer-Sollier, T., Panziera, J., Poudes, A., Atos, F.W.: The BXI interconnect architecture. In: 2015 IEEE 23rd Annual Symposium on High-Performance Interconnects, pp. 18–25, August 2015
18. Lin, H., et al.: Scalable graph traversal on Sunway TaihuLight with ten million cores. In: 2017 IEEE International Parallel and Distributed Processing Symposium (IPDPS), pp. 635–645, May 2017
19. Sensi, D., Girolamo, S., McMahon, K., Roweth, D., Hoefler, T.: An in-depth analysis of the Slingshot interconnect. In: 2020 SC20: International Conference for High Performance Computing, Networking, Storage and Analysis (SC), pp. 481–494. IEEE Computer Society, Los Alamitos, November 2020. https://doi.ieeecomputersociety.org/10.1109/SC41405.2020.00039
20. Kozielski, S., Mrozek, D.: Development of high performance computing systems. In: Gaj, P., Gumiński, W., Kwiecień, A. (eds.) CN 2020. CCIS, vol. 1231, pp. 52–63. Springer, Cham (2020). https://doi.org/10.1007/978-3-030-50719-0_5
21. Stegailov, V., et al.: Early performance evaluation of the hybrid cluster with Torus interconnect aimed at molecular-dynamics simulations. In: Wyrzykowski, R., Dongarra, J., Deelman, E., Karczewski, K. (eds.) PPAM 2017. LNCS, vol. 10777, pp. 327–336. Springer, Cham (2018). https://doi.org/10.1007/978-3-319-78024-5_29

22. Kondratyuk, N., Smirnov, G., Dlinnova, E., Biryukov, S., Stegailov, V.: Hybrid supercomputer Desmos with Torus Angara interconnect: efficiency analysis and optimization. In: Sokolinsky, L., Zymbler, M. (eds.) PCT 2018. CCIS, vol. 910, pp. 77–91. Springer, Cham (2018). https://doi.org/10.1007/978-3-319-99673-8_6

23. Kondratyuk, N., Smirnov, G., Stegailov, V.: Hybrid codes for atomistic simulations on the Desmos supercomputer: GPU-acceleration, scalability and parallel I/O. In: Voevodin, V., Sobolev, S. (eds.) RuSCDays 2018. CCIS, vol. 965, pp. 218–229. Springer, Cham (2019). https://doi.org/10.1007/978-3-030-05807-4_19

24. Khalilov, M., Timofeev, A.: Optimization of MPI-process mapping for clusters with Angara interconnect. Lobachevskii J. Math. **39**(9), 1188–1198 (2018)

25. Kondratyuk, N., et al.: Performance and scalability of materials science and machine learning codes on the state-of-art hybrid supercomputer architecture. In: Voevodin, V., Sobolev, S. (eds.) RuSCDays 2019. CCIS, vol. 1129, pp. 597–609. Springer, Cham (2019). https://doi.org/10.1007/978-3-030-36592-9_49

26. Rohr, D., Neskovic, G., Lindenstruth, V.: The L-CSC cluster: optimizing power efficiency to become the greenest supercomputer in the world in the Green500 list of november 2014. Supercomput. Front. Innov. Int. J. **2**(3), 41–48 (2015)

27. Dlinnova, E., Biryukov, S., Stegailov, V.V.: Energy consumption of MD calculations on hybrid and CPU-only supercomputers with air and immersion cooling. In: PARCO, pp. 574–582 (2019)

Author Index

Printed in the United States
by Baker & Taylor Publisher Services